NanoScience and Technology

The series NanoScience and Technology is focused on the fascinating nano-world, mesoscopic physics, analysis with atomic resolution, nano and quantum-effect devices, nanomechanics and atomic-scale processes. All the basic aspects and technology-oriented developments in this emerging discipline are covered by comprehensive and timely books. The series constitutes a survey of the relevant special topics, which are presented by leading experts in the field. These books will appeal to researchers, engineers, and advanced students.

More information about this series at http://www.springer.com/series/3705

Hassan Raza

Nanoelectronics Fundamentals

Materials, Devices and Systems

Springer

Hassan Raza
Centre for Fundamental Research
Islamabad, Pakistan

ISSN 1434-4904 ISSN 2197-7127 (electronic)
NanoScience and Technology
ISBN 978-3-030-32571-8 ISBN 978-3-030-32573-2 (eBook)
https://doi.org/10.1007/978-3-030-32573-2

This Springer imprint is published by the registered company Springer Nature Switzerland AG
The registered company address is: Gewerbestrasse 11, 6330 Cham, Switzerland

Dedicated to Tehseen, Ahmer, and Zuha.

Preface and Acknowledgements

Microelectronics has revolutionized our society and the way of life over the past few decades. The workhorse of this technology has been the metal oxide semiconductor transistor. About 25 years ago, we witnessed a major milestone when this building block crossed the 100 nm limit in commercially available integrated circuit chips. With this landmark, we entered the era of Nanoelectronics!

One should ask why this magic number of 100 nm? The reason is to do with the quantum behavior of electrons. It is well known that the de Broglie wavelength in the semiconductors may be on the order of 100 nm and hence the quantum effects at this length scale become important. Not only the device physics becomes different from that of microelectronics devices, but the novel quantum effects also lead to interesting applications.

The area of nanoelectronics is truly multidisciplinary. On the one hand, one needs to understand the most advanced concepts in quantum physics. Such concepts include understanding quantum mechanics with open boundary conditions, nonequilibrium quantum statistical mechanics, dephasing, dissipation, etc. On the other hand, one needs to develop skills for device analysis, system integration, circuit analysis, and sometimes even architecture design.

Like any other area in physical sciences, nanoelectronics is broadly divided into theory, computation, and laboratory experiments. To elaborate further, theory involves understanding the underlying formalism and its mathematical interpretation. The computation deals with the numerical solution and may be thought of as a *virtual experiment* simulated on a computer. A recent trend in the computational nanoelectronics is the application of high-performance computing to solve large-scale problems. Finally, the laboratory experimentation deals with developing skills and techniques to synthesize, fabricate, and eventually characterize materials, devices, circuits, and systems.

One should note that each of these three broad areas is further divided into subareas. For example in a typical laboratory work, where one has to synthesize materials, the next step is metrology, where one has to use various microscopy and spectroscopy techniques. Once material processing is accomplished, lithography techniques are used to fabricate devices and so on. Hence, a complete design and analysis process may only be achieved by developing an understanding of various areas. This is crucial for the harmony of fundamental concepts underlying the novel nanodevices and their applications.

Usually mastering each one of these areas requires decades of persistence, if not a lifetime. Apart from this, for a successful design process, enough knowledge of subsidiary areas is also essential. This is precisely the motivation behind the book. Our objective is to give the readers a broad yet comprehensive introduction to the theory, computation, and experiments related to the fundamentals of Nanoelectronics.

In this context, this book covers the state-of-the-art discussion about the theoretical understanding, computational modeling, fabrication, characterization, and metrology of nanoelectronics devices and systems. The ten chapters in the book are divided into four parts. Part I discusses Material Properties. Part II deals with the Device Characteristics and Analysis. We discuss Circuits and Systems in Part III, which leads the discussion to various applications. Finally, the discussion about Fabrication, Characterization, and Metrology is included in Part IV.

The book is aimed at senior-level undergraduate students and first-year graduate students in engineering, physics, chemistry, and related disciplines interested in nanoelectronics from materials, devices, and systems perspective. No prerequisites are required. However, knowledge about linear algebra, introductory quantum mechanics, introductory device physics, and computational software like Matlab is helpful. The discussion is kept at a tutorial level for the intended readership.

The author further anticipates that the book will be especially useful for small university programs, where it is not feasible to offer multiple courses on various aspects of nanoelectronics. A course based on this book may provide an avenue for a comprehensive educational curriculum and infrastructure.

I would like to thank my family, Tehseen, Ahmer, and Zuha, to whom this book is dedicated to. Without their understanding, support, and patience over the past 6 years, this book would not have been a reality. Furthermore, I would like to thank Dr. Tehseen Raza for improving the content of the book by extensive proofreading.

It is a pleasure to acknowledge useful discussions with Prof. David R. Andersen, a dear friend and a very valuable colleague. I would also like to acknowledge and appreciate the vision behind the Associateship program at the Abdus Salam ICTP (International Centre for Theoretical Physics), Trieste, Italy. I am grateful to Dr. Sandro Scandolo and Dr. Ralph Gebauer for hosting my Associateship visit at the Abdus Salam ICTP, that has been instrumental in bringing this project to completion. I would like to further thank Ms. Francesca Prelazzi of ICTP for facilitating my Trieste trip. This acknowledgment section would not be complete without remembering Dr. Abdus Salam.

I would like to thank Claus Ascheron, Zachary Evenson, Adelheid Duhm, and Elke Sauer at Springer Verlag for their collaboration and patience to bring this project to fruition. I would also like to thank Ravivarman Selvaraj, Sudhany Karthick and their production team at Springer Verlag for their commitment toward this book.

The initial version of the book was written at the University of Iowa, Iowa City, USA. Significant additions were made in the Abdus Salam ICTP, Trieste, Italy, and the final version was prepared in the Centre for Fundamental Research (CFR), Islamabad, Pakistan.

Islamabad, Pakistan | Hassan Raza

Contents

Part I Material Properties

1 Introduction 3
1.1 Why NANO? 5
1.2 Book Outline 6

2 Atomic Structure 11
2.1 Crystal, Lattice, and Unit Cell 12
2.2 Miller Index 19
2.3 Surface Reconstruction 20
2.4 Reciprocal Space 22

3 Electronic Structure 29
3.1 Heuristic Introduction to Schrödinger Equation 33
3.2 Electrostatics 38
3.3 Real Space Basis Set 42
3.4 Orbital Space Basis Set 49
3.5 Band Structure 58
References 76

Part II Device Characteristics and Analysis

4 Quantum Transport 79
4.1 Wavefunction Approach 80
4.2 Landauer's Approach 86
4.3 Fermi's Function 87
4.4 Quantum Mechanical Transmission 89
4.5 Density of States 91
4.6 Green's Function 99
4.7 Nonequilibrium Green's Function Formalism 102

4.8 Self Energy 105
4.9 Coherent Transport 111
4.10 Incoherent Transport 124
4.11 Selfconsistent Mean Field Transport 133
4.12 Beyond Mean Field Transport 136
References 138

5 Charge Based Devices 139
5.1 pn Junction Diode 140
5.2 Zener Diode 146
5.3 Field Effect Transistor 147
5.4 Resonant Tunneling Diode 156

6 Spin Based Devices 161
6.1 Ferromagnetic Materials 161
6.2 Giant Magnetoresistance Devices 165
6.3 Magnetic Tunnel Junction Devices 168
6.4 Spin Transfer Torque Devices 174

Part III Circuits and Systems

7 Memories 183
7.1 CMOS Inverter 184
7.2 SRAM 185
7.3 DRAM 188
7.4 Flash Memory 190

8 Circuits and Systems 197
8.1 Combinational Circuits 198
8.2 Sequential Circuits 203
8.3 Charge Coupled Devices 205
8.4 VLSI 207
8.5 Power Dissipation 209

Part IV Fabrication, Characterization and Metrology

9 Nanofabrication 215
9.1 Chemical Safety and Environmental Protection 217
9.2 Substrate 218
9.3 Oxidation and Annealing 220
9.4 Photolithography 224
9.5 Electron Beam Lithography 229
9.6 Nanoimprint Lithography 230
9.7 Physical Vapor Deposition 231
9.8 Chemical Vapor Deposition 234

9.9 Molecular Beam Epitaxy 237
9.10 Atomic Layer Deposition 238
9.11 Etching 240

10 Microscopy and Spectroscopy 243
10.1 Scanning Probe Microscopy 243
10.2 Electron Microscopy 249
10.3 Optical Microscopy 253
10.4 Photoemission Spectroscopy 255
10.5 Photon Spectroscopy 258
10.6 Electron Spectroscopy 263
References 265

Index 267

Part I
Material Properties

Chapter 1
Introduction

Nanoelectronics is the study of nanoscale materials, devices, and systems. Modern practice of science and engineering is founded on three pillars, namely theory, computation, and experiments.[1] Nanoelectronics is no exception, where a unified approach towards theory, computation, and experiments is vital for successful and efficient device design as shown in Fig. 1.1. The boundaries between these areas are rather soft and these disciplines indeed work hand-in-hand. For example, prior to initiating device fabrication, a theorist may provide the foundation for the device design. The theory may very well be complemented with a numerical computation of the device characteristics, or a successful amalgam of theory and computation may provide the design rules for device fabrication.

Yet another scenario could be to take the lead with an experiment, the outcome of which may later be understood by using a combination of theoretical models and computational tools, which in turn may be used to design better experiments. In fact, most of the times, the historical progress has been made in this manner. In this context, our objective in this book is to develop a unified approach of theory, computation, and experiments to provide an understanding of the nanoscale materials, devices and systems, covering a broad range of nanoelectronics applications.

Although we discuss various device applications, it is helpful to draw a generic device as shown in Fig. 1.2. Such a device consists of a source, a drain, and a channel. Source and drain regions are the contacts, where one may apply the external bias. With the combination of only source, drain and channel, one may have a two terminal device. Additionally, one could add multiple gates (two shown in Fig. 1.2 making this a four terminal device). If the gates are separated from the channel by a dielectric, one gets a field effect transistor (FET). In the absence of the dielectric, one obtains a Schottky gate transistor. In general, one could add or remove any of the materials selectively to tailor the device structure for a specific application.

Using this generic device structure, we discuss various nanoelectronics applications. Moreover, this discussion may be extended to topics related to chemical and bio

---

[1] Big Data is expected to be the fourth pillar.

H. Raza, *Nanoelectronics Fundamentals*, NanoScience and Technology,
https://doi.org/10.1007/978-3-030-32573-2_1

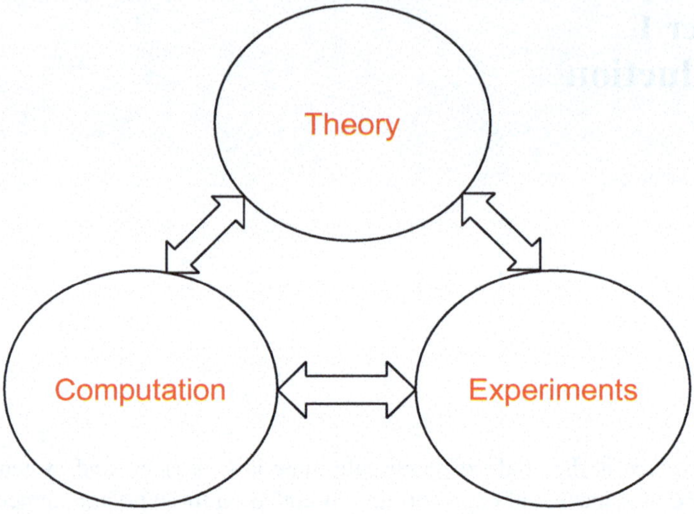

Fig. 1.1 Theory, computation, and experiments—three pillars of the modern practice of science and engineering, as well as nanoelectronics

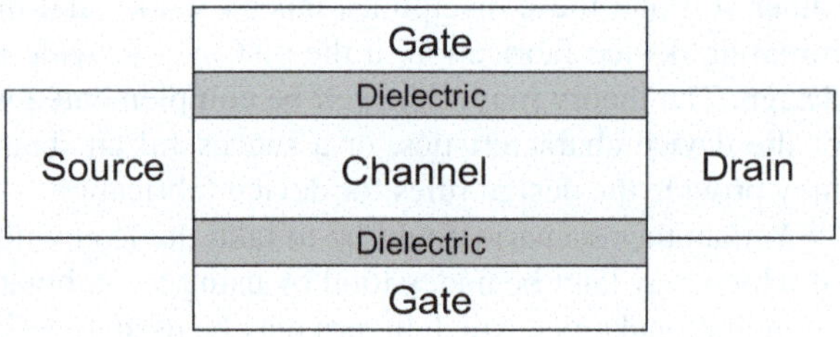

Fig. 1.2 Generic multiterminal device structure

sensors, etc. In such devices, one may incorporate some input/output ports for fluids. Such fluids, for example, may include blood samples for an ISFET (ion sensitive FET) based biosensors, etc.

The materials inside the channel, source, drain, gate and gate dielectric region could be of arbitrary shape and composition. For example, these regions could be made out of silicon nanomaterials, carbon nanomaterials, or heterostructures, etc. In addition to the material composition, one also has to consider the dimensionality of the material. Zero dimensional (0D) nanomaterials include silicon nanodots/nanocrystals, Buckyball (C_{60}), etc. One dimensional (1D) nanomaterials include silicon nanowires, graphene nanoribbons, carbon nanotubes, etc. Two dimensional (2D) nanomaterials include silicon nanomembranes, graphene, etc. Three dimensional (3D) nanomaterials include C_{60} crystals and heterostructures of various materials (e.g. Si/dielectric/metal stack in transistors and flash memories, super lattice LEDs (light emitting diodes), layers of magnetic and nonmagnetic materials in hard disks, etc.). As we discuss later in this book that not only the material of a structure is important but also the dimensionality as well as the arrangement of atoms.

In fact, sometimes the dimensionality governs the electronic properties in a much pronounced manner than the chemical compositions or atomic arrangement itself.

On a broader scale, our objective is to impart skill set in the audience to solve, understand, and design a wide range of devices with various compositions, and functionality. Consider the hypothetical example of a device with C_{60} $0D$ channel, carbon nanotube as 1D contacts, and graphene as the 2D gate material. We further incorporate these materials into devices and apply external bias to study the device operation under nonequilibrium conditions.

The above mentioned generalization is indeed feasible due to the atomic scale details in the nanoscale devices when one uses the bottom up approach. Once one understands how to calculate the atomic scale properties, one may use any type of atoms and their arrangements in the reported theoretical framework. This unified atomic theory of matter helps to treat devices within the bottom up approach, where understanding material properties at the atomic scale is the norm and not the exception. In a way, our objective is to apply this bottom up atomic theory to design and engineer nanoscale materials, devices, and systems.

1.1 Why NANO?

Why nanoscale materials, devices, and systems have become so important, and continue to drive technological innovation? How does the atomic arrangement govern physical, chemical, and electronic properties? How these atoms interact with each other and the surroundings? How one may harness the novel properties to make functional materials, devices, and systems?

To answer these and similar questions, we highlight below some of the important characteristics at the nanoscale,

(1) Size dependent properties. Physical, chemical, and electronic properties vary with the changing size at the nanoscale. As an example, we discuss how the electromagnetic properties of gold change by scaling. If one looks at a 1 μm diameter gold sphere and compares it with say, 1 mm diameter gold sphere; both essentially have similar electromagnetic properties, i.e. these appear gold in the optical range. However, at the nanoscale, the size of the material amongst other factors (e.g. arrangement) determines its properties. In this case, hypothetically speaking, 5 nm diameter gold sphere appears red, whereas 1 nm diameter appears blue. One may imagine using these size dependent effects in optoelectronic devices.

Yet another example is the size dependent magnetic properties of materials. By reducing the size of a ferromagnetic material, one may reach the superparamagnetic limit, where the material losses its spontaneous magnetization. Apart from this, an etchant used for removing a film sometimes fails to etch the nano phases of the same material due to different chemical properties at the nanoscale.

(2) Effect of Dimensionality. Between 0D, 1D, 2D and 3D, the physical and chemical properties are completely different for the same atomic arrangement and composition. For example, 0D benzene is a liquid, 1D carbon nanotube forms a mesh, 2D graphene is a membrane, etc. The dimensionality also affects the electronic properties as we discuss in Chap. 3.

(3) Atomic and Molecular Arrangement. The arrangement of atoms determines the physical, chemical, and electronic properties. It is so fascinating that if one takes few atoms in various configurations, the properties are completely different. Consider the example of the arrangement of carbon atoms in diamond, which is the hardest material known to humans, whereas the same carbon atoms arranged in graphitic arrangement lead to one of the softest materials.

(4) Environment. The proximity effects on nanomaterials are also very important. For example, the C_{60} appears black in powder form. Dissolving the same C_{60} in acetone results in brown solution, whereas dissolving in toluene results in purple solution. One should note that acetone and toluene are both colorless liquids.

(5) Sensitivity/Performance. The key advantage in the use of nanomaterials in devices and sensors is extremely high surface to volume ratio. Since most atoms are on the surface in a nanostructure, the sensitivity is very high. In fact, sensitivities on the order of ppb (parts per billion) are not unheard of in these nanosensors.

On a different note, we would like to clarify a misconception about nanomaterials that these are only used for niche high end applications. While the above scenario is indeed a reality, however macroscale and microscale devices do exist where the building blocks are still at the nanoscale—consider the use of C_{60} crystals in nanosensors. The performance of such macroscale or microscale devices (containing nanomaterials) is still very high compared to the conventional devices.

1.2 Book Outline

This book has been divided into four parts. In part I, we discuss the atomic and the electronic structures of matter at the nanoscale. In part II, we discuss the quantum transport theory and the nanoscale devices, where we solve the Schrödinger equation with open boundary conditions. For the computational part, we focus on computer codes and algorithms by using a commercial computational software Matlab. In part III, we discuss memories, circuits and systems. Finally, part IV deals with the experimental aspects of the nanoscale materials, devices and systems, where we discuss nanofabrication, microscopy, and spectroscopy. The outline of the book is as follows,

Atomic Structure. One may associate physical, chemical, and electronic properties of a material to the atomic arrangement and dimensionality. This serves as the motivation to study the atomic structure of various nanomaterials in Chap. 2.

Electronic Structure. At macroscale, one may understand the material properties by using classical mechanics. However, at atomic, molecular and nanoscale, a new kind of mechanics takes over where the Schrödinger equation is the basis to nonrelativistic Quantum mechanics. We discuss various methods and approximations to the Schrödinger equation in Chap. 3.

Quantum Transport. For electronic structure calculations, the Schrödinger equation is solved with closed boundary conditions; where no electron may be injected in or taken out from the channel. In other words, the channel is isolated, and hence there are no contacts. With contacts, electrons may be injected in and taken out by using an applied bias, thereby making this system open. In this case, one solves the Schrödinger equation with open boundary conditions as discussed in Chap. 4.

Although there are various methods to solve this problem, NEGF (nonequilibrium Green's function) formalism has rightfully become the method of choice due to its rigor and flexibility. NEGF may be coupled with the electronic structure methods for calculating various nonequilibrium properties. Due to atomic and nanoscale details included in the quantum transport of NEGF formalism, this method is the most fundamental. One may approach the limit of semiclassical and classical transport by including the relevant scattering processes in the NEGF formalism.

Within the classical transport regime for microdevices, the drift and diffusion are the dominant mechanisms. For the semi classical transport for submicron devices, Boltzmann's equation is the method of choice. However, these methods are inadequate for nanodevices to say the least.

Combination of NEGF and electronic structure methods may lead to coherent transport, where the phase of an electron is preserved. At room or elevated temperatures, such devices may not be in the coherent regime anymore. For example, interaction of the electron degree of freedom with the phonon (lattice vibrations) degree of freedom leads to incoherent scattering, where the electron wavefunction loses its phase and thereby becomes incoherent. The analogy is that of a coherent laser light. While passing through a medium, it may become diffusive and loose the coherence.

Charge Based Devices. In Chap. 5, we start the discussion with pn junction diodes and develop analytical model for the quantum transport through such devices. Next, we discuss Zener diode. Three terminal devices like FETs and tunnel transistors are presented next with various device parameters. Resonant tunneling diodes based on heterostructures are discussed as well.

Spin Based Devices. The next set of devices are based on the spin of the electron instead of its charge as discussed in Chap. 6—an active area of research called Spintronics. The first one under discussion is GMR (giant magnetoresistance) device, where a paramagnetic metal is sandwiched between two ferromagnetic metallic contacts. The relative magnetic orientation of the two ferromagnetic contacts in parallel and antiparallel configurations leads to a change in resistance. The next device is TMR (tunnel magnetoresistance) device, where the paramagnetic metal is replaced by an insulator (alumina or magnesium oxide), thereby enabling quantum mechanical tunneling. The memory based on GMR and TMR devices is called MRAM (magnetoresistive random access memory), where one has to apply an external magnetic field to switch the magnetization. Next, we discuss a device in which an internal spin polarized current may switch the polarization of a magnet, called Spin Transfer Torque (STT) device. STTRAM is more noise immune and scalable.

Memories. We discuss the nanoelectronics memories in Chap. 7. We start the discussion with the gate stack engineering for nonvolatile flash memory devices.

Furthermore, SRAM (static RAM) and DRAM (dynamic RAM) are discussed as well.

Circuits and Systems. We discuss various circuits and systems in Chap. 8. We start the discussion with logic gates and the associated circuits. Furthermore, CCDs (charge coupled devices) are explored, which have become an important part of everyday life as well as scientific discovery. VLSI (very large scale integration) design and the role of HDL (hardware description languages, e.g. Verilog) is explored. We conclude the chapter with the emerging area of SoC (system on chip), and thermodynamic limits of computing.

Nanofabrication. In Chap. 9, we discuss photolithography, etching, physical vapor deposition (PVD), chemical vapor deposition (CVD), nanofabrication (electron beam, nanoimprint lithography, SPM based techniques), etc. [SPM ≡ scanning probe microscopy]. Next, we study the growth of various nanomaterials by using PVD and CVD techniques. One may use PVD to grow nanodots or nanocrystals. CVD may be used for growing carbon nanotubes, nanowires, graphene, etc. We further discuss the bottom up approaches, like molecular beam epitaxy, atomic layer deposition, etc.

Microscopy and Spectroscopy. We focus on the device and material metrology in Chap. 10. We start the discussion with electron microscopes like scanning electron microscopy (SEM), transmission electron microscopy (TEM) and scanning transmission electron microscopy (STEM), etc. Next, we discuss optical microscopy, in particular confocal microscopy. SPM based techniques (STM ≡ scanning tunneling microscopy, AFM ≡ atomic force microscopy, etc.) are discussed as well. Many of these microscopy techniques may be used to perform spectroscopy as well. Additionally, we discuss FTIR (Fourier Transform Infra Red) spectroscopy, Raman spectroscopy, XPS (X-ray Photoemission Spectroscopy), UPS (UV Photoemission Spectroscopy), ARPES (Angle Resolved PhotoEmission Spectroscopy), etc.

In short, our goal is to understand transport through nanodevices in a unified approach by combining theory, computation, and experiments.

Problems

1.1 What is the number of transistors in the latest Intel chip?

1.2 What is the number of transistors in the latest flash memory chip? You may pick any manufacturer.

1.3 What is the transistor size (critical dimension) in the latest Intel chip?

1.4 What is the transistor size (critical dimension) in the latest flash memory chip? You may pick any manufacturer.

1.5 Name a 0D nanomaterial not mentioned in this chapter.

1.6 Name a 1D nanomaterial not mentioned in this chapter.

1.7 Name a 2D nanomaterial not mentioned in this chapter.

1.8 Provide an example of an emerging nanomaterial not mentioned in this chapter. Describe its application.

1.9 Provide an example of an emerging nanoscale device not mentioned in this chapter. Describe its application.

1.10 Provide an example of an emerging nanoscale system not mentioned in this chapter. Describe its application.

1.11 Provide an example of a nanomaterial, not mentioned in this chapter, of how size affects its physical, chemical, and electronic properties at the nanoscale.

1.12 Provide an example of a nanomaterial, not mentioned in this chapter, of how dimensionality affects its physical, chemical, and electronic properties at the nanoscale.

1.13 Provide an example of a nanomaterial, not mentioned in this chapter, of how atomic and molecular structure affects its physical, chemical, and electronic properties at the nanoscale.

Research Assignment

R1.1 Various applications are discussed in this chapter. Pick a grand problem that humanity faces, and write a one page summary of how nanoscale materials, devices, and/or systems may help solve this problem.

Chapter 2
Atomic Structure

A physical phenomena is always described as a function of time and spatial coordinates $(t, \boldsymbol{r})$.[1] One may Fourier transform the time domain (t) to the frequency domain (ω) without losing any information, where ω (rad/s) is the angular frequency.[2] Since frequency is related to the energy as $E = \hbar\omega$, frequency and energy domains are interchangeable.[3] Similarly, the real space $(\boldsymbol{r})$ may be Fourier transformed to the wavevector space $[\boldsymbol{k} = (k_x, k_y, k_z)]$, which is also called the reciprocal space. The advantage of such a transformation is to compact the information in one domain, provided it is periodic in the other domain.

Thus, if the transformation is carried out without any approximations, any of the following combinations carry the same amount of information,

$$(t, \boldsymbol{r})$$

$$(\omega, \boldsymbol{r})$$

$$(\omega, \boldsymbol{k})$$

$$(E, \boldsymbol{k})$$

$$(E, \boldsymbol{r})$$

It turns out that the combinations $(E, \boldsymbol{k})$ and $(E, \boldsymbol{r})$ are the most useful for device analysis. In fact, $(E, \boldsymbol{k})$ usually written as $E(\boldsymbol{k})$ is referred to as the *Band Structure* and $(E, \boldsymbol{r})$ usually written as $E(\boldsymbol{r})$ is called the *Band Diagram*; two key terms frequently used in this book.

In this chapter, we focus on the periodicity in the arrangement of atoms in the real space $(\boldsymbol{r})$ and the associated reciprocal space $(\boldsymbol{k})$. The discussion about the energy domain (E) follows in the next chapter.

---

[1] Where $\boldsymbol{r} \equiv (x, y, z)$ in Cartesian coordinate system, and $\boldsymbol{r} \equiv (r, \theta, \phi)$ in spherical coordinate system.

[2] $\omega = 2\pi f$, where f (Hz) is the cyclic frequency.

[3] Where $\hbar$ is the reduced Planck's constant given as, $\hbar = h/2\pi$, and h is the Planck's constant given as, $h = 6.62 \times 10^{-34}$ Js.

H. Raza, *Nanoelectronics Fundamentals*, NanoScience and Technology,
https://doi.org/10.1007/978-3-030-32573-2_2

Based on the structural coherence, materials may be classified into the following six categories,

(1) Crystalline. In these materials, there is a perfect order and periodicity of atomic arrangement in 1D, 2D or 3D. The atomic structure also possesses n-fold rotational symmetry where $n = 2, 3, 4, 6$.[4] This rotational symmetry refers to the periodicity in the structure after every $2\pi/n$ rotation. Most of the semiconductors like Si, Ge, GaAs, etc, are single crystalline, although one may also synthesize single crystalline metallic or insulating substrates.

(2) Polycrystalline. These materials have single crystalline structure within a small region called grain, which may have microscale dimensions. However across the grains, crystal orientations differ. Most metals like Au, Al, etc, are polycrystalline. Semiconductor films may also posses polycrystalline phase.

(3) Nanocrystalline. These materials consist of tiny nanocrystals with short range order, otherwise arranged in an amorphous form. Hence, these materials are essentially amorphous with nanocrystalline constituents.

(4) Liquid Crystals. While molecules in liquid phase do not have any specific order, these tend to align along a specific direction in liquid crystals. The niche application for these materials is LCD (liquid crystal display).

(5) Quasi Crystals. These materials possess $n = 5$ rotational symmetry in addition to one or more $n \neq 5$ symmetries. These materials are not periodic; however they do possess structural order and hence are termed as quasi crystals.

(6) Amorphous. These materials do not have any short or long range order at all. Dielectrics are an excellent example of amorphous materials, e.g. SiO_2, etc.

2.1 Crystal, Lattice, and Unit Cell

For device analysis, crystalline materials are the simplest to work with due to the structural periodicity. Consider the example of a 1D crystal as shown in Fig. 2.1a. One should note that a crystal is defined as an arrangement of atoms, whereas a lattice is defined as a collection of points. Here, we may define a one atom unit cell. Since this is the smallest unit cell, it is also known as *primitive unit cell*. Another unit cell with two atoms is shown in Fig. 2.1b. In fact, one could have infinite number of unit cells for an infinite crystal.

Primitive unit cells are also called Wigner Seitz unit cells. In order to understand the Wigner-Seitz method of finding the smallest unit cell possible, consider the examples of the 1D crystals shown in Fig. 2.1a, b. If one connects the nearest neighbors by lines and then draws perpendiculars to the connecting lines, the region bounded by these perpendiculars constitutes the primitive unit cell. In 2D, the intersections of the perpendicular features constitutes a plane; whereas in 3D, this procedure leads to a volumetric shape.

---

[4] $n = 5$ is not allowed since one may not fill a 2D space with pentagons only.

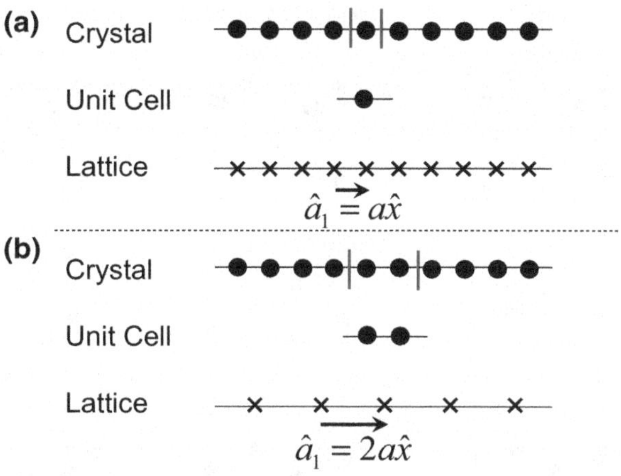

Fig. 2.1 1D crystal. **a** Primitive unit cell, the real space lattice, and the primitive basis vector. **b** Nonprimitive unit cell, the real space lattice, and the basis vector

Hence, for the crystals in 1D, 2D, and 3D, the corresponding unit cells are characterized by length, area, and volume, respectively. For a given material, if the unit cells have the same dimension, the number of atoms, and the arrangement, these are indistinguishable and hence equivalent. In other words, one simply may not perform an experiment that could distinguish between the two unit cells. Moreover, the number of atoms within a unit cell and its physical dimensions are important in determining the electronic structure as discussed in Chap. 3.

In 1D, there is only one way to arrange lattice points in space. Such an arrangement does not have to be rectilinear. It could very well be curvilinear. In 2D, there are five distinct ways to arrange lattice points in space and hence five unique lattices, called Bravais lattices. In 3D, there are fourteen distinct ways of arranging points in space and hence fourteen Bravais lattices as shown in Fig. 2.2.

Once a unit cell is identified, one may define a Bravais lattice. The unique arrangement of points are represented by the following real space translation vectors (***R***) in 1D, 2D and 3D respectively,

$$
\begin{aligned}
\boldsymbol{R}_{1D} &= l_1\hat{a}_1 \\
\boldsymbol{R}_{2D} &= l_1\hat{a}_1 + l_2\hat{a}_2 \\
\boldsymbol{R}_{3D} &= l_1\hat{a}_1 + l_2\hat{a}_2 + l_3\hat{a}_3
\end{aligned}
\tag{2.1}
$$

where $\hat{a}_1$, $\hat{a}_2$, and $\hat{a}_3$ are the *basis vectors* (also called *lattice unit vectors*) in real space and $l_1, l_2, l_3 = 0, \pm 1, \pm 2, \pm 3, \ldots$. Each point in the real space lattice (arrangement of points only) is described by a combination of (l_1, l_2, l_3) in (2.1). If a unit cell (arrangement of atoms) were to be placed at each lattice point, one recovers the crystal. Hence, one may reconstruct a complete crystal based on the information about the unit cell and the lattice.

Since the number of unit cells are infinite for an infinite crystal, the corresponding number of lattices are also infinite. Hence, there is no unique unit cell and lattice for a

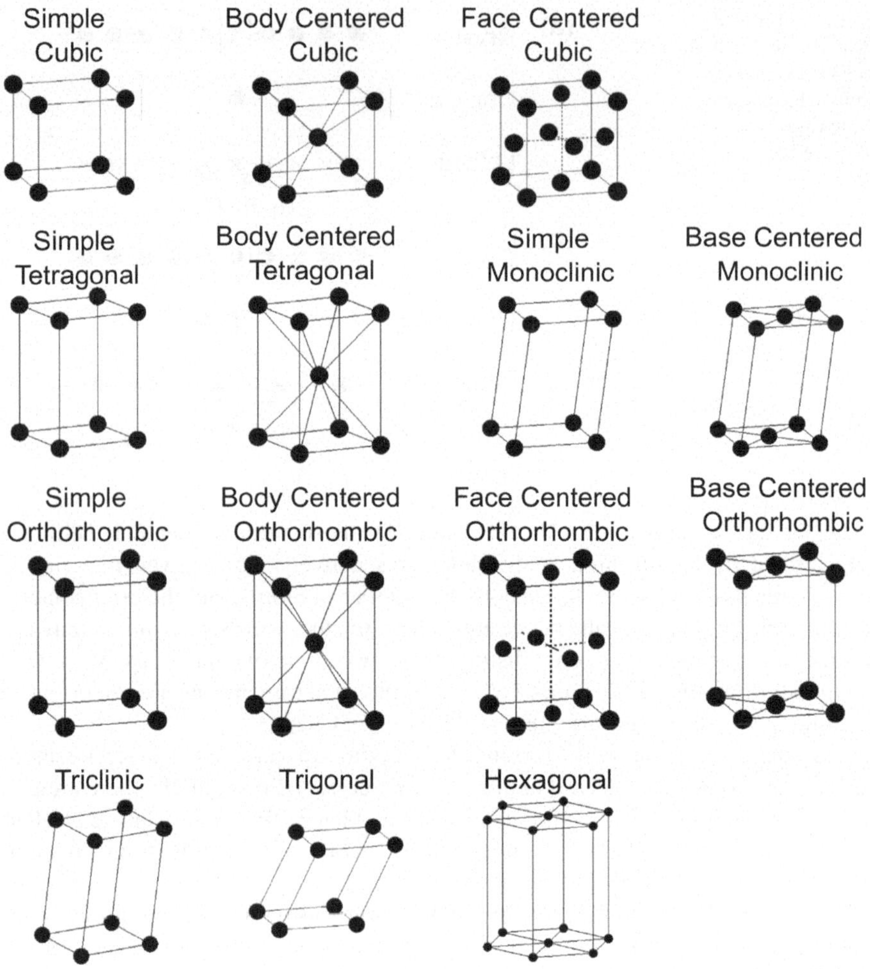

Fig. 2.2 Bravais lattices in 3D

given crystal. However, the primitive unit cell and the associated lattice is still unique. Corresponding to a primitive unit cell, the basis vectors are called primitive basis vectors.[5] The primitive basis vectors for the 1D crystal and corresponding lattice are shown in Fig. 2.1a.

[5]A word of caution that the word crystal is loosely defined and even sometimes used for finite structures without any periodicity. Hence, the reader is encouraged to be careful in the way the word "crystal" is used sometimes.

1D Materials

Let us consider an armchair graphene nanoribbon shown in Fig. 2.3a, which is a 1D structure. The primitive unit cell is highlighted and the corresponding 1D lattice is shown. The nanoribbon edges in the longitudinal (x) direction are of armchair shape, therefore this ribbon is called armchair graphene nanoribbon. If we were to perform a thought experiment of rolling the armchair graphene nanoribbon by joining the two armchair edges with each other, one obtains a nanotube, whose open ends have a zigzag pattern, thereby leading to what is called a zigzag carbon nanotube.

A zigzag graphene nanoribbon is shown in Fig. 2.3b, where the edges in the longitudinal (x) direction have zigzag shape. This nanoribbon leads to an armchair carbon nanotube if conceptually rolled by joining the two zigzag edges with each other. The unit cells for the two kinds of nanoribbons are also shown in Fig. 2.3a, b, for which one obtains the corresponding 1D lattices, shown as well. It is interesting that the unit cell of a carbon nanotube has cylindrical symmetry, and hence is different from that of a nanoribbon but the basis vector and the lattice are the same.

2D Materials

For a 2D crystal with rectangular arrangement of atoms, the primitive basis vectors are given by $\hat{a}_1 = a\hat{x}$ and $\hat{a}_2 = b\hat{y}$ as shown in Fig. 2.4a. With this choice of the unit cell, the corresponding 2D real space lattice is shown in Fig. 2.4b.

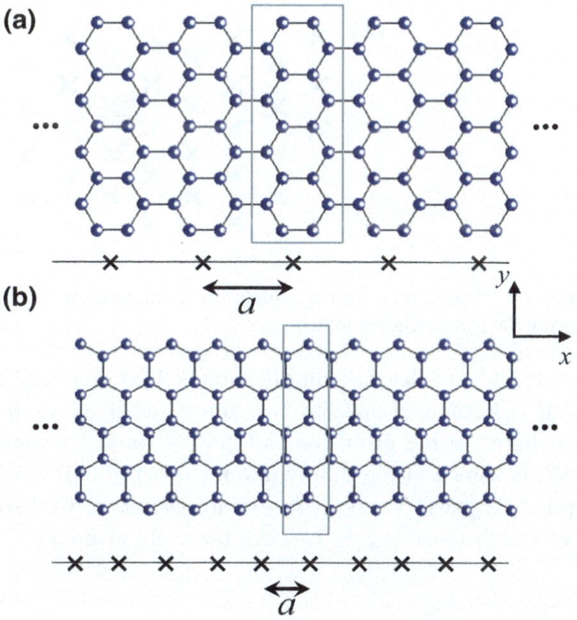

Fig. 2.3 Graphene Nanoribbons **a** Armchair. Crystal, primitive unit cell, and the corresponding lattice. $a = 3a_{cc}$, where $a_{cc} = 1.42$ Åis the carbon-carbon atomic distance, and 1 Å = 0.1 nm. **b** Zigzag. Crystal, primitive unit cell, and the corresponding lattice. $a = \sqrt{3}a_{cc}$

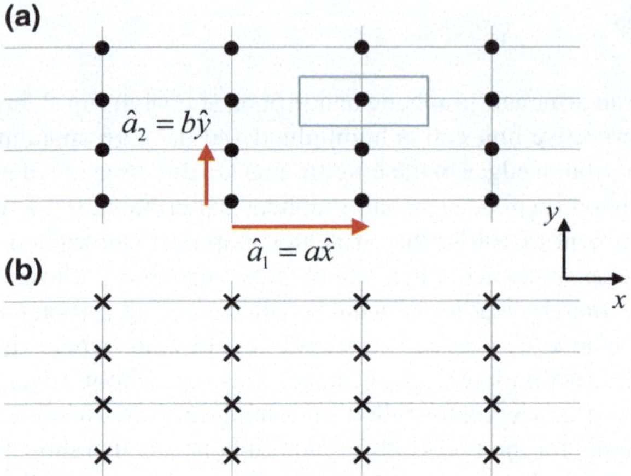

Fig. 2.4 2D crystal. **a** Crystal with primitive unit cell, and the primitive basis vectors. **b** The corresponding lattice

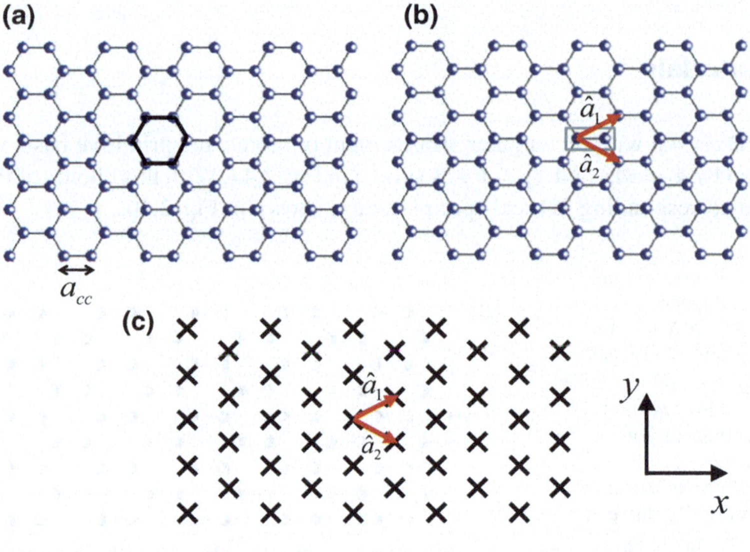

Fig. 2.5 Graphene. **a** Atomic structure with the primitive unit cell, **b** the primitive basis vectors, and **c** the corresponding lattice

Next, consider the primitive unit cell for graphene as shown in Fig. 2.5a. The unit cell is hexagonal and has two atoms, where at each corner one third of the atom is shared in the primitive unit cell. A more convenient way to represent the unit cell is shown in Fig. 2.5b without explicitly outlining the area of the unit cell. The primitive basis vectors ($\hat{a}_1, \hat{a}_2$) are shown as well. In terms of the carbon-carbon atomic distance ($a_{\text{cc}} = 1.42$ Å), these are given as,

$$\hat{a}_1 = \frac{3a_{cc}}{2}\hat{x} + \frac{\sqrt{3}a_{cc}}{2}\hat{y}$$
$$\hat{a}_2 = \frac{3a_{cc}}{2}\hat{x} - \frac{\sqrt{3}a_{cc}}{2}\hat{y} \qquad (2.2)$$

For various combinations of (l_1, l_2), the lattice points are shown in Fig. 2.5c.

3D Materials

Let us consider 3D crystals with cubic symmetry. The simplest case is that of a simple cubic crystal with one atom at each corner as shown in Fig. 2.6a. The primitive unit cell is a cube and the corresponding primitive orthogonal[6] basis vectors are as follows,

$$\begin{aligned}\hat{a}_1 &= a\hat{x} &&= a(1, 0, 0)\\ \hat{a}_2 &= a\hat{y} &&= a(0, 1, 0)\\ \hat{a}_3 &= a\hat{z} &&= a(0, 0, 1)\end{aligned}$$

where a is the length of each side of the cube. Since, each corner atom is shared with eight unit cells, the number of atoms per unit cell is one. Furthermore, there are eight nearest neighbors.

For the body centered cubic (BCC) crystal shown in Fig. 2.6b, there is an additional atom at the center of the cube, i.e. halfway on the body diagonal. It may look like the primitive unit cell is still a cube and such a unit cell has two atoms per unit cell. But, in fact, the primitive unit cell has only one atom and the primitive nonorthogonal[7] basis vectors are given as,

$$\begin{aligned}\hat{a}_1 &= \tfrac{a}{2}(\hat{x} + \hat{y} - \hat{z}) &&= \tfrac{a}{2}(1, 1, -1)\\ \hat{a}_2 &= \tfrac{a}{2}(\hat{x} - \hat{y} + \hat{z}) &&= \tfrac{a}{2}(1, -1, 1)\\ \hat{a}_3 &= \tfrac{a}{2}(-\hat{x} + \hat{y} + \hat{z}) &&= \tfrac{a}{2}(-1, 1, 1)\end{aligned}$$

For a face centered cubic (FCC) crystal shown in Fig. 2.6c, one may have six additional atoms on each face of the cube in addition to the atoms at each corner as in simple cubic crystal. Again, it may seem like the primitive unit cell is still a cube with four atoms within this unit cell. However, the primitive unit cell consists of one atom and the corresponding primitive nonorthogonal basis vectors are given as,

$$\begin{aligned}\hat{a}_1 &= \tfrac{a}{2}(\hat{y} + \hat{z}) = \tfrac{a}{2}(0, 1, 1)\\ \hat{a}_2 &= \tfrac{a}{2}(\hat{x} + \hat{z}) = \tfrac{a}{2}(1, 0, 1)\\ \hat{a}_3 &= \tfrac{a}{2}(\hat{x} + \hat{y}) = \tfrac{a}{2}(1, 1, 0)\end{aligned}$$

[6]For orthogonal basis vectors, $\hat{a}_i \cdot \hat{a}_j = 0$, where $i \neq j$.

[7]For nonorthogonal basis vectors, $\hat{a}_i \cdot \hat{a}_j \neq 0$, where $i \neq j$.

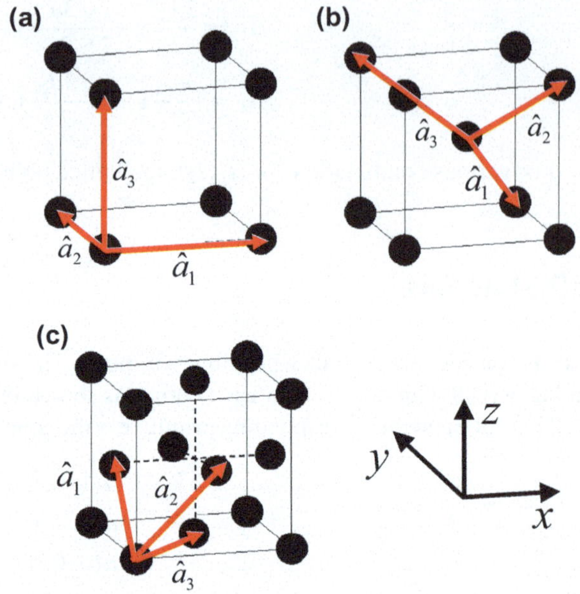

Fig. 2.6 Crystals with cubic symmetry. **a** Simple cubic crystal with primitive basis vectors. **b** Body centered cubic crystal with primitive basis vectors. Examples include Fe, Cr, W, etc. **c** Face centered cubic crystal with primitive basis vectors. Examples include Al, Ni, Co, Ag, Au, Pt, etc

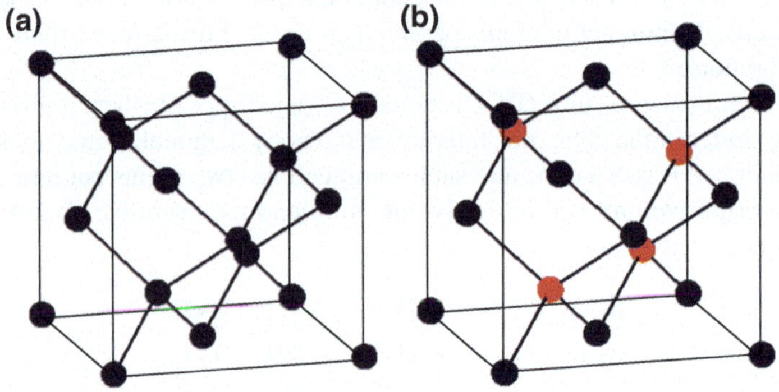

Fig. 2.7 Crystals with cubic symmetry. **a** Diamond crystal. Examples include Si, C, Ge, etc. **b** Zinc Blend crystal. Examples include GaAs, CdTe, etc

The diamond crystal shown in Fig. 2.7a essentially consists of two interpenetrating FCC crystals, where one of the two FCC crystals is displaced along one fourth of the body diagonal. The primitive unit cell now consists of two atoms (due to two FCC crystals) and the lattice remains the same as that of the primitive unit cell of FCC crystal. Indeed, diamond and silicon have the diamond crystal. If the two interpenetrating FCC crystals consist of two different atoms, the resulting crystal is referred to as Zinc Blend. GaAs is an example of such a crystal as shown in Fig. 2.7b, where Ga atoms occupy one FCC crystal and the As atoms occupy the other FCC crystal.

2.2 Miller Index

Let us now consider the concept of Miller index for defining crystal planes and directions in a quantitative manner, where the planes and directions are represented by ( ) and [ ] brackets,[8] respectively. As shown in Fig. 2.8a, let us consider the highlighted plane that has the following intercepts along the three axes,

	x-axis	y-axis	z-axis
Intercept	a	∞	∞
Take inverse	1/a	$1/\infty$	$1/\infty$
Multiply with LCF	1	0	0

Taking inverse and multiplying with the least common factor (LCF) yields Miller indices, which for the above example are (100). The direction perpendicular to this plane is referred to as [100]. Similarly it may be shown that the planes shown in Fig. 2.8b, c have (110) and (111) Miller indices, respectively.

One may further conclude that (100) has five other equivalent planes, six in total, namely (010), (001), $(\bar{1}00)$, $(0\bar{1}0)$, $(00\bar{1})$. Physically one may not distinguish

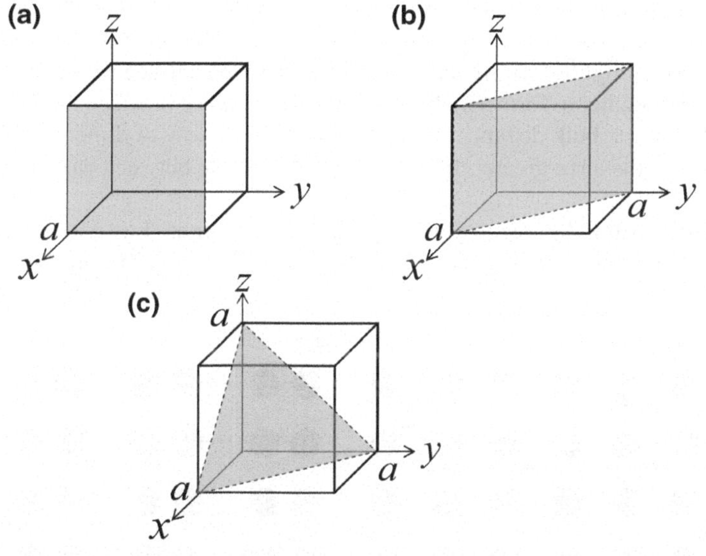

Fig. 2.8 Miller indices for (100), (110) and (111) planes in (**a**), (**b**) and (**c**), respectively

[8]Not to be confused with the use of ( ) and [ ] brackets for defining the range and the domain of a function.

between them since these depend on an arbitrary choice of the three axes. These equivalent planes are referred to as {100}.

Similarly, there are twelve {110} equivalent planes, one of which is highlighted in Fig. 2.8b and {111} has eight equivalent planes, one of which is plotted in Fig. 2.8c. The equivalent directions are represented by <> brackets.

2.3 Surface Reconstruction

The concept of surface reconstruction is relevant to all nanostructures and surfaces. Consider the example of Si(100) surface as shown in Fig. 2.9a. To construct this surface, one may perform a thought experiment of taking bulk Si in 3D and cleaving it along one face of the cubic symmetry. This face is also called cleavage plane. Since periodicity is lost in the cut direction, one may not use Fourier transform in this direction anymore. Once we truncate a structure, the atoms on the surface usually do not follow the bulk lattice structure anymore. These may very well rearrange themselves to reduce the overall surface energy, a phenomena known as surface reconstruction. This phenomenon is applicable to 2D materials like semiconductor surfaces, 1D materials like nanowires and nanotubes, 0D materials like quantum dots, nanocrystals, nanodots, etc.

Consider the Si(100) surface again, where the arrangement of atoms is in a square pattern without reconstruction as shown in Fig. 2.9a. Based on the bulk crystal structure, the distance between two atoms is 3.84 Å.[9] For this surface, the surface unit cell is 2D and is highlighted by the dotted line. With the surface reconstruction, two atoms group together to form a Si dimer, where the distance between the dimer atoms is smaller than the bulk distance, which results in a 1D row of dimers on this plane separated by a distance greater than the average distance between atoms in the bulk form.

The unit cell for this reconstructed surface is highlighted by the dotted line in Fig. 2.9b, that is twice as big as that of the unit cell in Fig. 2.9a in one direction,

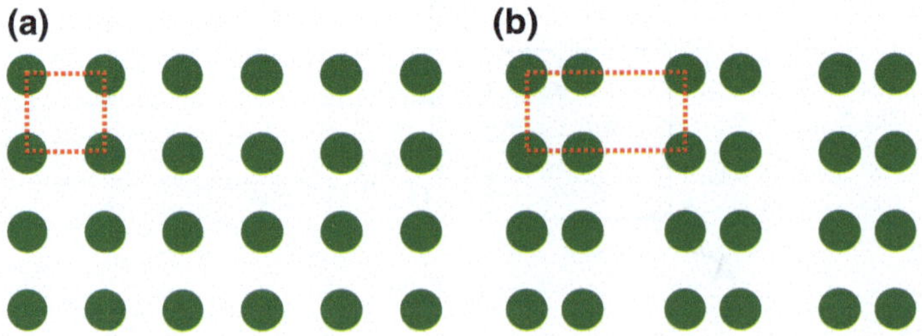

Fig. 2.9 Si(100) surface reconstruction. **a** Unreconstructed surface with bulk lattice constant. **b** (2 × 1) reconstructed surface

[9] 1 Å = 0.1 nm

whereas in the other direction, the length of the unit cell is the same as that of the bulk. For this reason, this specific reconstruction of Si(100) surface is referred to as (2 × 1) reconstruction.

It is interesting to note that Si(111) surface goes through (7 × 7) reconstruction as shown in Fig. 2.10, i.e. the unit cell of the reconstructed surface is seven times the length of the unit cell of the surface without reconstruction. This discovery of Si(111)-(7 × 7) reconstruction gave a compelling evidence of the capability of scanning tunneling microscope (STM) shortly after its invention. It was remarkable that there were several theories about this surface suggesting various other reconstructions and no one expected (7 × 7) reconstruction.

Strictly speaking it is not only the atoms on the surface layer that go through reconstruction, but subsurface atoms also rearrange themselves. In semiconductors, atoms within a few nanometers of the surface may rearrange themselves, which results in significant electronic structure changes.

Similarly, in heterostructures, significant reconstruction could be expected in various materials with the added complexity of matching different lattice constants of various materials. Such reconstruction in heterostructures may result in significant strain, which may result in varying properties. One such example in the use of SiGe alloys in nanoscale transistors for mobility enhancement.

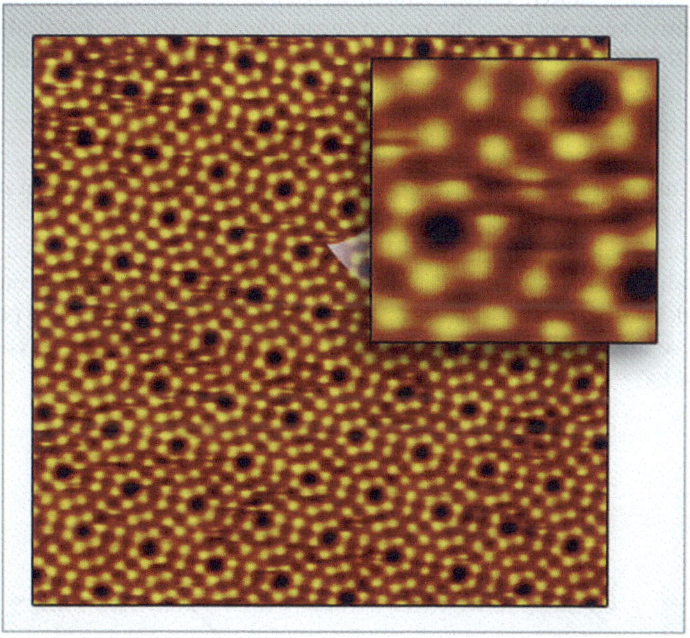

Fig. 2.10 Si(7 × 7) surface reconstruction. (Courtesy of RKH Tech)

2.4 Reciprocal Space

The real space (***r***) information may be further transformed into the reciprocal space (***k***). The advantage here is that broad information in the real space gets transformed into very compact information in the Fourier domain if the atomic structure in the real space is periodic and infinite. In other words, one may transform the real space (x, y, z) information to the reciprocal space (k_x, k_y, k_z) as shown in Fig. 2.11 by using the Fourier transform. The structural symmetry of the unit cell in real space retain the general shape in the reciprocal space. However, the larger dimension in the real space gets squeezed in the reciprocal space and vice versa.

Starting with the Bravais lattice described by the real space translation vector in (2.1), the corresponding translation vectors (***K***) in the reciprocal space are given as,

$$\begin{aligned}\boldsymbol{K_{1D}} &= n_1\hat{b_1}\\ \boldsymbol{K_{2D}} &= n_1\hat{b_1} + n_2\hat{b_2}\\ \boldsymbol{K_{3D}} &= n_1\hat{b_1} + n_2\hat{b_2} + n_3\hat{b_3}\end{aligned} \tag{2.3}$$

where $\hat{b_1}$, $\hat{b_2}$ and $\hat{b_3}$ are the reciprocal space basis vectors. Since $(\hat{a}_1, \hat{a}_2, \hat{a}_3)$ define the real space unit cell, $(\hat{b_1}, \hat{b_2}, \hat{b_3})$ define the reciprocal space unit cell. For a given set of real space primitive basis vectors $(\hat{a}_1, \hat{a}_2, \hat{a}_3)$, $(\hat{b_1}, \hat{b_2}, \hat{b_3})$ give the reciprocal space primitive basis vectors. The region encapsulated by $(\hat{b_1}, \hat{b_2}, \hat{b_3})$ is also known as the *first Brillouin zone*, and may be determined by using the Wigner Seitz method discussed in Sect. 2.1.

With increasing range of the reciprocal unit cells, one may define the second, third, fourth Brillouin zones, etc. However, beyond the first zone, the rest of the zones have redundant information. Therefore, it is instructive to use the first Brillouin zone only. In this book, we simply refer to it as the Brillouin zone.

The dimensionality of the unit cell in real or reciprocal space is tied with the dimensionality of the physical structure. For the 1D, 2D and 3D materials, one obtains 1D, 2D and 3D Brillouin zones, respectively. Since 0D materials have no periodicity (unless these are assembled into a 3D crystal, C_{60} crystals), this discussion is not

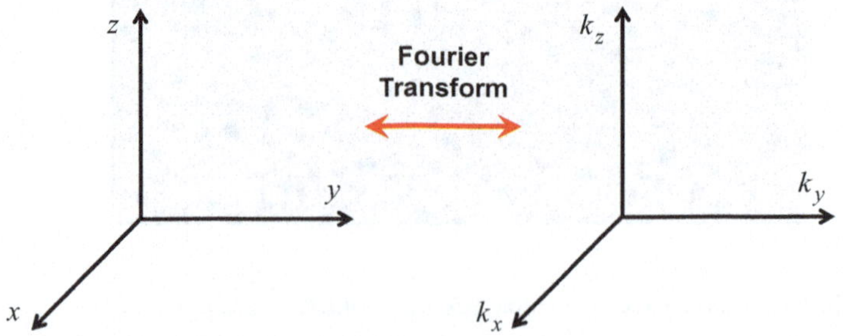

Fig. 2.11 Real space (x, y, z) versus reciprocal space (k_x, k_y, k_z)

relevant to the 0D structures, although one may still calculate quantized wavenumbers for 0D structures, which we discuss later.

The orthogonality condition between the real space and the reciprocal space basis vectors is given as $\hat{a}_i \cdot \hat{b}_j = 2\pi\delta_{ij}$, where, the Kronecker's delta function is defined as,[10]

$$\delta_{ij} = \begin{cases} 1 & i = j \\ 0 & i \neq j \end{cases}$$

With this condition, one may deduce the reciprocal space basis vectors in terms of the real space basis vectors as follows,

$$\begin{aligned} \hat{b}_1 &= \frac{2\pi}{\hat{a}_1 \cdot (\hat{a}_2 \times \hat{a}_3)}(\hat{a}_2 \times \hat{a}_3) = \frac{2\pi}{V_r}(\hat{a}_2 \times \hat{a}_3) \\ \hat{b}_2 &= \frac{2\pi}{\hat{a}_1 \cdot (\hat{a}_2 \times \hat{a}_3)}(\hat{a}_3 \times \hat{a}_1) = \frac{2\pi}{V_r}(\hat{a}_3 \times \hat{a}_1) \\ \hat{b}_3 &= \frac{2\pi}{\hat{a}_1 \cdot (\hat{a}_2 \times \hat{a}_3)}(\hat{a}_1 \times \hat{a}_2) = \frac{2\pi}{V_r}(\hat{a}_1 \times \hat{a}_2) \end{aligned} \tag{2.4}$$

where $\hat{a}_1 \cdot (\hat{a}_2 \times \hat{a}_3) = V_r$ is the volume of the unit cell in real space with [m^3] dimension. Correspondingly, the real space basis vectors in terms of the reciprocal space basis vectors are given as,

$$\begin{aligned} \hat{a}_1 &= \frac{2\pi}{\hat{b}_1 \cdot (\hat{b}_2 \times \hat{b}_3)}(\hat{b}_2 \times \hat{b}_3) = \frac{2\pi}{V_k}(\hat{b}_2 \times \hat{b}_3) \\ \hat{a}_2 &= \frac{2\pi}{\hat{b}_1 \cdot (\hat{b}_2 \times \hat{b}_3)}(\hat{b}_3 \times \hat{b}_1) = \frac{2\pi}{V_k}(\hat{b}_3 \times \hat{b}_1) \\ \hat{a}_3 &= \frac{2\pi}{\hat{b}_1 \cdot (\hat{b}_2 \times \hat{b}_3)}(\hat{b}_1 \times \hat{b}_2) = \frac{2\pi}{V_k}(\hat{b}_1 \times \hat{b}_2) \end{aligned} \tag{2.5}$$

where $\hat{b}_1 \cdot (\hat{b}_2 \times \hat{b}_3) = V_k$ is the volume of the unit cell in the reciprocal space with [m^{-3}] dimension.

In 2D with xy plane, one may assume $\hat{a}_3 = \hat{z}$, which results in the following reciprocal space basis vectors,

$$\begin{aligned} \hat{b}_1 &= \frac{2\pi}{\hat{a}_1 \cdot (\hat{a}_2 \times \hat{z})}(\hat{a}_2 \times \hat{z}) = \frac{2\pi}{A_r}(\hat{a}_2 \times \hat{z}) \\ \hat{b}_2 &= \frac{2\pi}{\hat{a}_1 \cdot (\hat{a}_2 \times \hat{z})}(\hat{z} \times \hat{a}_1) = \frac{2\pi}{A_r}(\hat{z} \times \hat{a}_1) \end{aligned} \tag{2.6}$$

where $\hat{a}_1 \cdot (\hat{a}_2 \times \hat{z}) = A_r$ is the area of the 2D unit cell in real space with [m^2] dimension.

---

[10] Dirac's delta function is a different one as discussed later in this book.

In 1D (along x-direction), $\hat{a_1} = L_r\hat{x}$, $\hat{a_2} = \hat{y}$, and $\hat{a_3} = \hat{z}$, one gets the following reciprocal space basis vector,

$$\hat{b_1} = \frac{2\pi}{\hat{a_1} \cdot (\hat{y} \times \hat{z})}(\hat{y} \times \hat{z}) = \frac{2\pi}{L_r}\hat{x} \tag{2.7}$$

where $L_r = \hat{a_1} \cdot (\hat{y} \times \hat{z}) = \hat{a_1} \cdot \hat{x}$ is the length of the 1D unit cell in real space with [m] dimension.

1D Structures

Consider the real space 1D lattice shown in Fig. 2.12a with the lattice constant a, i.e. $L_r = a$. The real space primitive basis vector is given as $\hat{a_1} = a\hat{x}$. By using (2.7), the reciprocal space primitive basis vector is given as,

$$\hat{b_1} = \tfrac{2\pi}{a}\hat{x}$$

The length of the reciprocal space unit cell is $L_k = 2\pi/a$. The reciprocal space lattice is shown in Fig. 2.12b.

The number of unit cells in the reciprocal space could also be infinite. However, there could only be one primitive unit cell, i.e. Brillouin zone with period $2\pi/a$ and may be defined to have a range of either $(0, 2\pi/a]$ or $(-\pi/a, \pi/a]$; the latter is a more popular choice. One should note that the two end points of the range are redundant and one should retain only one of the points.[11]

2D Structures

Consider the 2D rectangular lattice shown in Fig. 2.13a with the lattice constants a and b in x- and y-directions respectively. The real space primitive basis vectors are given as, $\hat{a_1} = a\hat{x}$ and $\hat{a_2} = b\hat{y}$, and the primitive unit cell is shown as well. By using (2.6), the reciprocal space primitive basis vectors are given as,

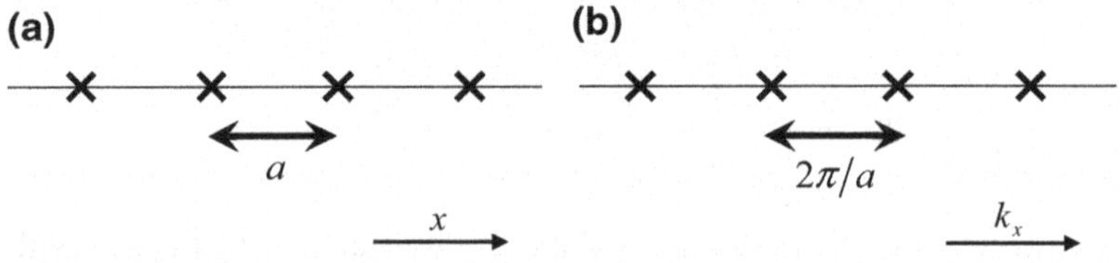

Fig. 2.12 1D. **a** Real space lattice, and **b** reciprocal space lattice

[11] Note the mathematical notation for the range of an interval.] and [brackets mean that the number is included, whereas) and (brackets mean that the number is excluded.

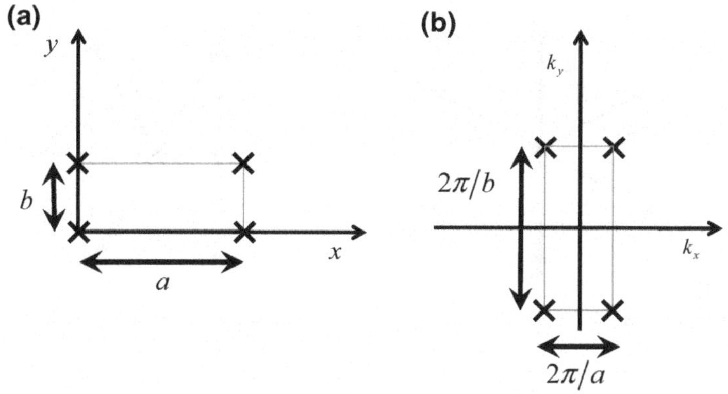

Fig. 2.13 2D. **a** Real space lattice, and **b** reciprocal space lattice

$$\begin{aligned} \hat{b_1} &= \tfrac{2\pi}{a}\hat{x} \\ \hat{b_2} &= \tfrac{2\pi}{b}\hat{y} \end{aligned} \tag{2.8}$$

The corresponding Brillioun zone is given by the following range, $(k_x, k_y) = (-\pi/a, \pi/a], (-\pi/b, \pi/b]$ and shown in Fig. 2.13b.

Consider the example of graphene as shown in Fig. 2.14a, where the primitive basis vectors for the unit cell are given in (2.1). The primitive unit cell is shown as well. By using (2.6), the primitive basis vectors in the reciprocal space are given as,

$$\begin{aligned} \hat{b_1} &= \frac{2\pi}{3a_{cc}}\hat{x} + \frac{2\pi}{\sqrt{3}a_{cc}}\hat{y} \\ \hat{b_2} &= \frac{2\pi}{3a_{cc}}\hat{x} - \frac{2\pi}{\sqrt{3}a_{cc}}\hat{y} \end{aligned} \tag{2.9}$$

The reciprocal lattice is shown in Fig. 2.14b. Using the Wigner-Seitz method, the primitive unit cell in the reciprocal space i.e. the first Brillioun zone, is highlighted by dashed line, which also has a hexagonal symmetry.

3D Structures

Consider the 3D cubic crystal with the following primitive basis vectors in real space, $\hat{a_1} = a\hat{x}$, $\hat{a_2} = b\hat{y}$ and $\hat{a_3} = b\hat{z}$. By using (2.4), the primitive basis vectors in the reciprocal space are given as follows,

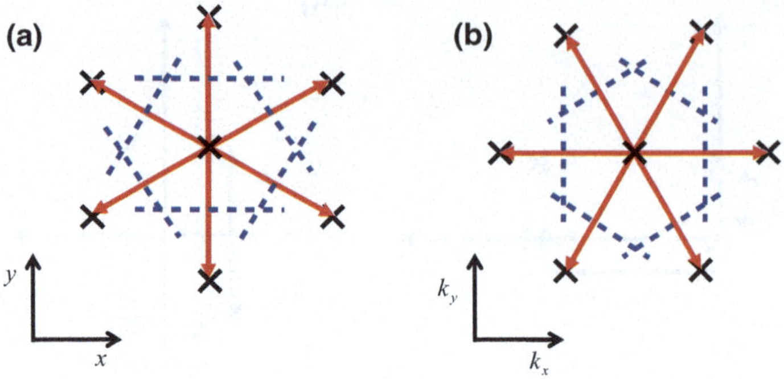

Fig. 2.14 Graphene. Wigner Seitz method is illustrated for both **a** real space lattice, and **b** the reciprocal space lattice

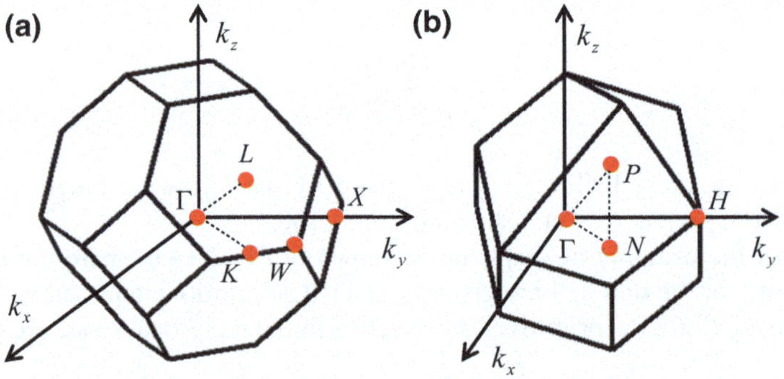

Fig. 2.15 3D. **a** FCC Brillouin zone in the reciprocal space for a real space BCC lattice. **b** BCC Brillouin zone in the reciprocal space for a real space FCC lattice. Various symmetry points and symmetry directions are highlighted

$$
\begin{aligned}
\hat{b_1} &= \tfrac{2\pi}{a}\hat{x} \\
\hat{b_2} &= \tfrac{2\pi}{b}\hat{y} \\
\hat{b_3} &= \tfrac{2\pi}{c}\hat{z}
\end{aligned}
\tag{2.10}
$$

The first Brillouin zone is given by the following range, $k_x, k_y, k_z = (-\pi/a, \pi/a]$, $(-\pi/b, \pi/b]$, $(-\pi/c, \pi/c]$.

For more detailed structures in 3D, the procedure is the similar to the one followed in 1D or 2D. In this context, a real space BCC lattice gives an FCC lattice in the reciprocal space as shown in Fig. 2.15a along with various symmetry points and symmetry directions in the Brillouin zone. Furthermore, a real space FCC lattice gives a BCC lattice in the reciprocal space as shown in Fig. 2.15b highlighting various symmetry points and symmetry directions in the Brillouin zone.

Problems

2.1 Pick an example of a crystalline material not discussed in this chapter. Identify how many folds is the rotational symmetry?

2.2 Pick an example of a polycrystalline material not discussed in this chapter. Identify the size of grains and comment on the grain boundaries.

2.3 Pick an example of a nanocrystalline material not discussed in this chapter. Identify the nanocrystalline constituent.

2.4 What are the constituent materials in liquid crystals?

2.5 Pick an example of a quasicrystalline material not discussed in this chapter. Identify how many folds the rotational symmetry is in addition to the five fold symmetry.

2.6 Draw the Bravais lattices in 2D.

2.7 Draw planes for the following Miller indices (100), (110), (111), (211), (416), and $(\bar{3}12)$.

2.8 List the equivalent planes for (111).

2.9 Derive the real space primitive surface basis vectors and real space surface reciprocal basis vectors for (2×1)-Si(100).

2.10 Derive the real space primitive surface basis vectors and real space surface reciprocal basis vectors for (7×7)-Si(111).

2.11 For $\hat{a}_1 = a\hat{x}$, $\hat{a}_2 = 2a\hat{y}$ and $\hat{a}_3 = 3a\hat{z}$, calculate $\hat{b}_1$, $\hat{b}_2$ and $\hat{b}_3$.

2.12 For an FCC lattice in real space (given $\hat{a}_1$, $\hat{a}_2$, $\hat{a}_3$), show that the reciprocal lattice is a BCC [derive $\hat{b}_1$, $\hat{b}_2$, $\hat{b}_3$ and compare with ($\hat{a}_1$, $\hat{a}_2$, $\hat{a}_3$) of a BCC lattice].

2.13 For a BCC lattice in real space (given $\hat{a}_1$, $\hat{a}_2$ and $\hat{a}_3$), show that the reciprocal lattice is an FCC [derive $\hat{b}_1$, $\hat{b}_2$ and $\hat{b}_3$ and compare with ($\hat{a}_1$, $\hat{a}_2$, $\hat{a}_3$) of an FCC lattice].

2.14 For graphene, given the real space basis vectors ($\hat{a}_1$, $\hat{a}_2$), derive reciprocal space basis vectors ($\hat{b}_1$, $\hat{b}_2$).

2.15 For graphite, given the real space basis vectors ($\hat{a}_1$, $\hat{a}_2$, $\hat{a}_3$), derive reciprocal space basis vectors ($\hat{b}_1$, $\hat{b}_2$, $\hat{b}_3$).

Research Assignments

R2.1 Write a one page summary of the bulk and the surface atomic structure of a single crystalline material, e.g. Au(100). Derive the real space primitive bulk basis vectors and the bulk reciprocal vectors. Also, derive the real space primitive surface basis vectors and surface reciprocal vectors.

R2.2 Write a one page summary of Moiré pattern at the nanoscale in a material of your choice.

R2.3 Write a one page summary of Penrose crystals.

Chapter 3
Electronic Structure

Electronic structure describes how energy levels or spectra are distributed as a function of either real space ($\boldsymbol{r}$) or reciprocal space ($\boldsymbol{k}$). In this chapter, we focus on the *Band Diagram* or *Energy Diagram* $E(\boldsymbol{r})$, and the *Band Structure* $E(\boldsymbol{k})$. The method of choice for calculating the electronic structure of nanomaterials is the Quantum mechanics (also called wave mechanics), where the non-relativistic Schrödinger equation is the norm and not an exception. Before going into the details of this new kind of mechanics of the nanostructures and nanomaterials, let us first review some history.

In early part of the twentieth century, the phenomenon of black body radiation was unexplained by classical or Newtonian mechanics. This phenomenon is really a fancy name for the skewed bell shaped emission spectrum of a hot object, for which the emission intensity depends on the emission wavelength and temperature as shown in Fig. 3.1. By using the classical mechanics, one was only able to fit the spectrum in the long wavelength regime, whereas the short wavelength ultra violet (UV) region was unexplained. This scenario was termed as *UV catastrophe*. In 1905,

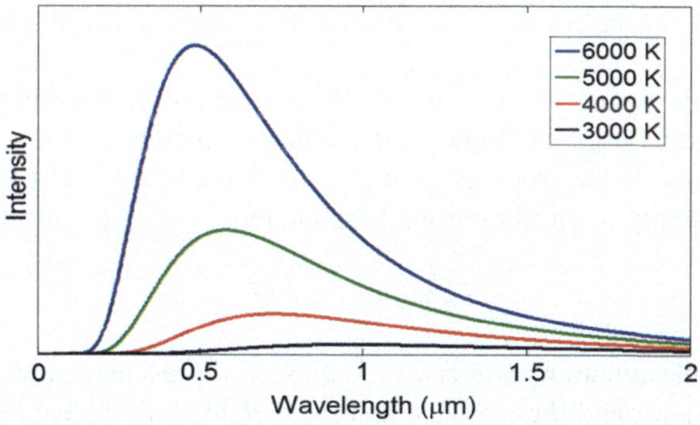

Fig. 3.1 Black body radiation spectrum at various temperatures. Sun's spectrum corresponds to about 5778 K

H. Raza, *Nanoelectronics Fundamentals*, NanoScience and Technology,
https://doi.org/10.1007/978-3-030-32573-2_3

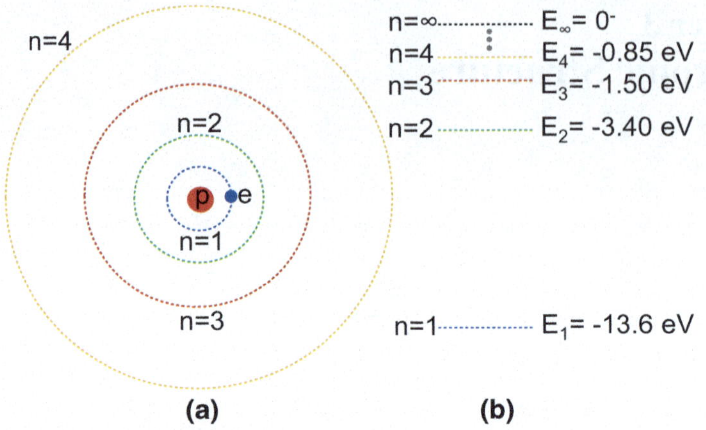

Fig. 3.2 Bohr's model for hydrogen atom. **a** Orbital picture. **b** Energy level picture

in an effort to explain why color of light changes depending on the temperature of the filament of the light bulb, then recently invented by Edison, Planck proposed that for electromagnetic radiation, the energy may be exchanged only in finite quanta, which depends on the frequency as,

$$E = \hbar\omega = hf$$

where ω (rad/s) is the angular frequency, f (Hz) is the cyclic frequency and $\hbar = h/2\pi$ is the reduced Planck's constant, where $h = 6.6262 \times 10^{-34}$ Js is the Planck's constant. The single packet of energy is called photon, which is an elementary particle of light. With this discovery, quantum mechanics was born, where the energy may only be exchanged in a finite manner, in contrast to the continuous energy exchange in the classical mechanics. By using this hypothesis, Planck was able to explain the black body radiation and in this honor, h is named as the Planck's constant. This is such a powerful concept that this constant has become a signature of Quantum mechanics.

In 1913, while trying to explain the discrete emission/absorption lines in the hydrogen spectrum based on Rutherford's planetary picture of atom, schematically shown in Fig. 3.2a, Bohr proposed that the angular momentum (L) of an electron with a linear velocity v, circling around the nucleus is also quantized, and is given as,

$$L_n = m_e v r_n = n\hbar$$

where the principle quantum number $n = 1, 2, 3, \ldots$ represents an individual electron orbit of radius r_n around the nucleus, and $m_e = 9.11 \times 10^{-31}$ Kg is the electron's rest mass. By using this hypothesis, Bohr successfully calculated the energy values (E_n) of the hydrogen atom as follows (also shown in Fig. 3.2b),

$$E_n = -\frac{13.6}{n^2}\ \text{eV}$$

where eV is a unit of energy.[1] There were two intellectual contributions of this model. The first was that the energy levels are quantized and the transitions between these quantized levels determine the peculiar peaks in the hydrogen atom spectrum, something that the classical mechanics does not embody within its framework and thereby sometimes is referred to as continuum mechanics. The second contribution was the notion of quantum numbers (n in this case) and how one may relate the physical observables to these quantum numbers. Indeed now, there are various quantum phenomena specifically related by such behaviors, like *integer quantum Hall effect*, etc. Additionally, this model contributed to the concepts like *well behaved* and *forbidden* quantum numbers. In this specific example, $n = 0$ is a forbidden quantum number, since it results in zero angular momentum. This results in either zero velocity or zero radius, both of which are unphysical.

The next significant development came with the conceptual development of *matter waves*. Louis de Broglie proposed that for a single particle of mass m and velocity v, the linear momentum (p) of such a matter wave is given in terms of the wavelength (λ) as follows,

$$p = \frac{h}{\lambda} = \hbar k$$

where $k = 2\pi/\lambda$ is the magnitude of the wavevector, and has the dimensions of (rad/m).

Combining the above equation with $p = mv$, one may deduce the wavelength associated with a matter wave, as follows $\lambda = h/mv$ and a wave with a wavelength has a mass given as, $m = h/\lambda v$. This gave rise to the notion of *wave particle duality*, i.e. an entity may behave as a particle or a wave under different circumstances. The concept of *matter wave* is also described by the wave packet with a standing wave pattern as shown in Fig. 3.3.

The next contribution came from Heisenberg, who developed the uncertainty principle based on the wave nature of matter after carefully studying the emission and absorption spectra of gases. According to this important principle, for a finite length Δx and finite time Δt, one may not deduce information about the wave with absolute certainty. Since time (t) and space $\boldsymbol{r} = (x, y, z)$ are related to the energy (E) and wavevector $\boldsymbol{k} = (k_x, k_y, k_z)$, respectively by Fourier transform, uncertainty in one results in uncertainty in the Fourier space. Heisenberg proposed that these uncertainties in the two sets of conjugate variables (E, t) and $(\boldsymbol{r}, \boldsymbol{p} = \hbar\boldsymbol{k})$ are related by the following equations,

[1] 1 eV = 1.6×10^{-19} J. The reason behind the wide spread use of the eV unit of energy is its convenience. For example, a voltage applied of a certain value, say 1 V, results in -1 eV shift in the potential energy for an electron, since potential energy of an electron is given as, $-qV$, where V is the potential and $q = 1.6 \times 10^{-19}$ C is the magnitude of the electronic charge.

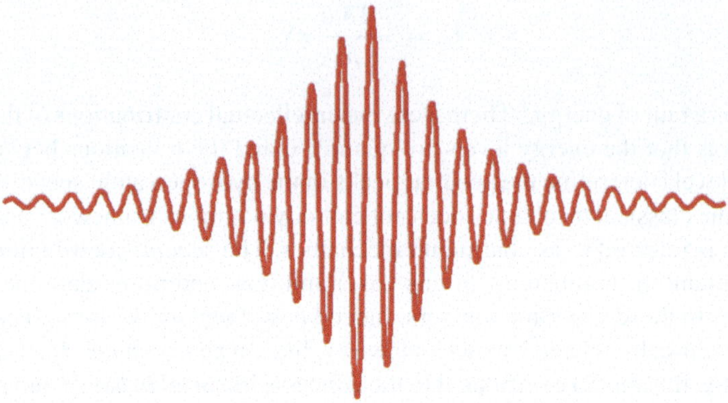

Fig. 3.3 Matter wave packet

$$\Delta t \Delta E \geq \hbar/2$$
$$\Delta x \Delta p_x \geq \hbar/2$$
$$\Delta y \Delta p_y \geq \hbar/2$$
$$\Delta z \Delta p_z \geq \hbar/2$$

This brings interesting situations about the conservation of energy and momentum, which we may otherwise be calculated quite accurately in classical mechanics. However, in quantum mechanics, with an uncertainty in the measurement of time and space, one may not deduce the behavior of energy and momentum deterministically, rather we have to specify it in terms of probabilities. This philosophical concept later on became the soul of quantum mechanics.

The next significant contribution came from Pauli, who proposed that the electrons have inherent magnetism, in addition to the magnetism associated with the orbital motion. This inherent magnetism is associated with the spin quantum number (s). Usually, electrons have two spin orientations, which are 180° out of phase, hence called up-spin (↑-spin) and down-spin (↓-spin). During transition from ↑-spin to ↓-spin, the path associated with the spin dynamics in real space ($\boldsymbol{r}$) could be quite detailed as shown in Fig. 3.4. This viewgraph also elucidates a misconception about spins that they only point in two distinct directions. The spins may point in arbitrary direction within the *sphere* of the spin dynamics and may follow different trajectories, depending on the magnetic field stimuli.

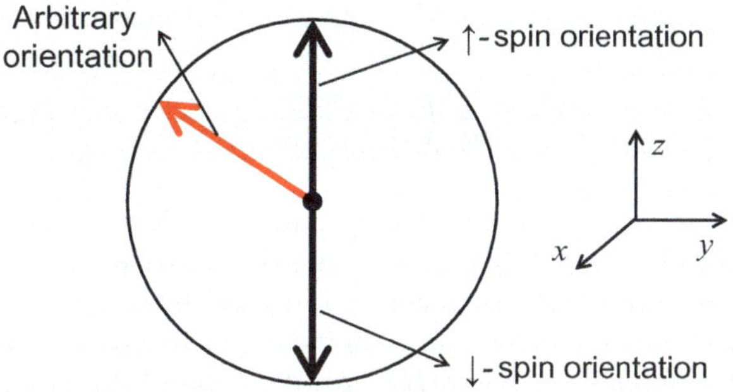

Fig. 3.4 Spin dynamics

3.1 Heuristic Introduction to Schrödinger Equation

Four equivalent descriptions of quantum mechanics were developed by Schrödinger (wave equation), Heisenberg (matrix approach), Dirac-Feynman (path integral formalism), and de Broglie-Bohm (pilot wave theory). While contributions by Heisenberg, Dirac-Feynman, and de Broglie-Bohm are equally noteworthy, we use Schrödinger's approach in this book due to its intuitive appeal.

In classical mechanics, for a single particle with mass m, the kinetic energy is given as, $E_k = p^2/2m = \frac{1}{2}mv^2$, where in 3D, $p^2 = \boldsymbol{p} \cdot \boldsymbol{p} = p_x^2 + p_y^2 + p_z^2$. The potential energy for the electrostatic interaction between two particles with charges q_1 and q_2 separated by a distance r is defined as, $U(r) = Kq_1q_2/r$, where $K = 1/4\pi\epsilon$, ϵ is the permittivity given as $\epsilon = \epsilon_r\epsilon_o$. ϵ_r is the relative permittivity and ϵ_o is the permittivity of free space given as, 8.854×10^{-12} F/m. The total energy, i.e. the sum of the kinetic energy and the potential energy, is termed as a single particle Hamiltonian[2] and is given as,

$$H = \frac{\boldsymbol{p} \cdot \boldsymbol{p}}{2m} + U(\boldsymbol{r}) \tag{3.1}$$

In quantum mechanics, the momenta p_x, p_y and p_z become operators, which are given as,

$$\tilde{p_x} = \frac{\hbar}{i}\frac{\partial}{\partial x}, \quad \tilde{p_y} = \frac{\hbar}{i}\frac{\partial}{\partial y}, \quad \tilde{p_z} = \frac{\hbar}{i}\frac{\partial}{\partial z}$$

making the momentum an operator as follows,

$$\tilde{\boldsymbol{p}} = \hat{x}\tilde{p_x} + \hat{y}\tilde{p_y} + \hat{z}\tilde{p_z} = \frac{\hbar}{i}\left(\hat{x}\frac{\partial}{\partial x} + \hat{y}\frac{\partial}{\partial y} + \hat{z}\frac{\partial}{\partial z}\right).$$

[2]Hamiltonian operator and Hamiltonian matrix (discussed later) should be Hermitian, i.e. $H = H^\dagger$, in order to obtain real energy values for equilibrium conditions. Conceptually, an imaginary energy value or frequency gives rise to wavefunction decay in time domain, and hence the probability of finding a particle becomes time dependent that is not consistent with the equilibrium quantum mechanics.

Hence the Hamiltonian operator, by using (3.1), is given as,

$$\tilde{H} = \frac{-\hbar^2}{2m}\left(\frac{\partial^2}{\partial x^2} + \frac{\partial^2}{\partial y^2} + \frac{\partial^2}{\partial z^2}\right) + U(\boldsymbol{r}) = \frac{-\hbar^2}{2m}\nabla^2 + U(\boldsymbol{r}). \tag{3.2}$$

where ∇^2 is the Laplacian operator. The electrostatic boundary conditions are specified by $U(\boldsymbol{r})$. Although we write Hamiltonian operator in Cartesian coordinate system here, the potential energy $U(\boldsymbol{r})$ and the Laplacian operator ∇^2 may be written in cylindrical or spherical coordinate systems depending on the symmetry. Given the Hamiltonian operator ($\tilde{H}$), time dependent Schrödinger equation is given as,

$$\tilde{H}\Psi = \frac{-\hbar}{i}\frac{\partial \Psi}{\partial t} \tag{3.3}$$

where $\Psi(\boldsymbol{r}, t)$ is the complex wavefunction. Born proposed that $|\Psi(\boldsymbol{r}, t)|^2$ defines the probability density of finding a particle. Therefore, according to the normalization condition, an integral over this quantity equals unity per spin per state, i.e.,

$$\int_{-\infty}^{+\infty} \mathrm{d}\boldsymbol{r}\, |\Psi(\boldsymbol{r}, t)|^2 = \int_{-\infty}^{+\infty} \mathrm{d}\boldsymbol{r}\ \Psi^*\Psi = 1 \tag{3.4}$$

where $\mathrm{d}\boldsymbol{r} = \mathrm{d}x\mathrm{d}y\mathrm{d}z$ is the differential volume. The dimensions of wavefunction are $m^{-3/2}$ in 3D, m^{-1} in 2D and $m^{-1/2}$ in 1D. $\Psi(\boldsymbol{r}, t)$ may be further written as $\Psi(\boldsymbol{r}, t) = \Psi(\boldsymbol{r})e^{-i\omega t} = \Psi(\boldsymbol{r})e^{-iEt/\hbar}$, where $\Psi(\boldsymbol{r})$ is the time independent part of the wavefunction. Substituting in (3.3), the time independent Schrödinger equation is given as,

$$\tilde{H}\Psi = E\Psi \tag{3.5}$$

This equation leads to solutions that are highly non-intuitive. Consider the phenomenon of quantum mechanical tunneling, where an electron or more precisely its wavefunction may penetrate a barrier, i.e. there is a finite probability of finding a particle on the other side of the barrier as shown in Fig. 3.5—a phenomena inconceivable in classical mechanics. Furthermore, the quantum mechanical reflections

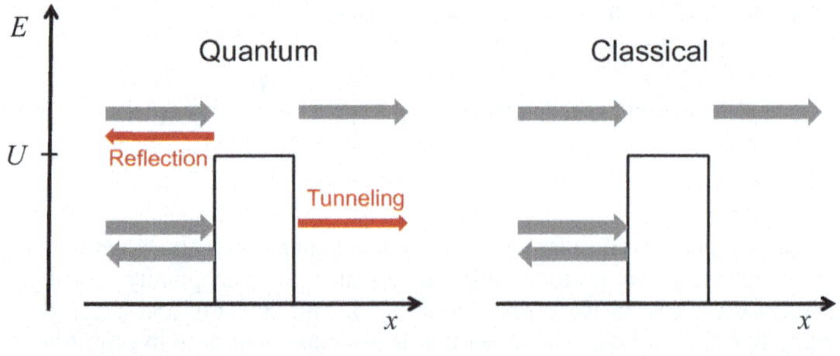

Fig. 3.5 Quantum mechanical tunneling and reflection

are also quite distinct compared to classical reflections. Consider a barrier, where the transmission probability of a classical object through this barrier is zero if the energy of the object is less than the top of the barrier. However, with energy greater than the barrier height, the transmission probability is unity. A quantum particle, however, has a finite probability of transmission even for energies less than the barrier height, which is a signature feature of quantum mechanical tunneling. According to quantum mechanics, this tunneling probability may never be zero, although it could be exponentially small. For energy greater than the barrier height, the probability may still be less than unity, indicative of the quantum mechanical reflections.

The following properties related to wavefunctions and Schrödinger equation are noteworthy,

(1) Given an operator $\tilde{A}$, one may only calculate the expectation value of this operator due to the probabilistic nature of quantum mechanics, which is given as,

$$\langle A(t)\rangle = \int_{-\infty}^{+\infty} \mathrm{d}\boldsymbol{r}\ \Psi^{*}\tilde{A}(\boldsymbol{r}, t)\Psi$$

For a time independent operator, the time independent expectation value is given as,

$$\langle A\rangle = \int_{-\infty}^{+\infty} \mathrm{d}\boldsymbol{r}\ \Psi^{*}\tilde{A}(\boldsymbol{r})\Psi$$

(2) The probability density is always time independent for a closed system, i.e.,

$$|\Psi(\boldsymbol{r}, t)|^2 = |\Psi(\boldsymbol{r})|^2$$

(3) $\Psi(\boldsymbol{r})$ and $\nabla\Psi(\boldsymbol{r})$ should be finite across an interface to eliminate any discontinuity in the second derivative to be used in Schrödinger equation.

(4) Wavefunctions are complex in general, i.e. $\Psi(\boldsymbol{r}) = |\Psi(\boldsymbol{r})|\, e^{i\delta}$. Although the phase does not affect the probability density of an isolated particle, it does give rise to the phenomenon of quantum mechanical coherence when particles interact with each other.

(5) Due to complex nature of the wavefunction, $|\Psi_1(\boldsymbol{r}, t) + \Psi_2(\boldsymbol{r}, t)|^2 \neq |\Psi_1(\boldsymbol{r}, t)|^2 + |\Psi_2(\boldsymbol{r}, t)|^2$, which is one of the key manifestations of quantum mechanics.

(6) Since Schrödinger equation is a linear equation, weighted sum (or linear combinations) of various solutions is also a solution, i.e. if $\Psi_1, \Psi_2, \Psi_3, \ldots, \Psi_M$ are solutions, $\sum_{j=1}^{M} b_j\Psi_j$ is also a solution.

Next we introduce the concept of a basis set, which consists of a finite collection of functions $\phi_j(\boldsymbol{r})$. These functions constituting the basis set may or may not be orthogonal to each other and hence are called orthogonal basis set or nonorthogonal basis set, respectively. The basis set may or may not be normalized, however the final wavefunction should always be normalized. The wavefunction for a certain quantum number n may be written as a linear weighted sum of the basis set as follows,

$$\Psi_n(\boldsymbol{r}) = \sum_{j=1}^{N} c_j^n \phi_j(\boldsymbol{r}) \tag{3.6}$$

where c_j^n are the weighting factors, which are the projections of wavefunction $\Psi_n(\boldsymbol{r})$ on the basis functions $\phi_j(\boldsymbol{r})$. Using a certain N-dimensional basis set (i.e. N basis functions), the operators become square matrices of rank N, where the matrix elements of the Hamiltonian (only the kinetic energy part shown here), potential energy and overlap matrices in 3D are given as,

$$\begin{aligned}
[H]_{ij} &= \frac{-\hbar^2}{2m} \int_{-\infty}^{+\infty} \mathrm{d}\boldsymbol{r}\ \phi_i^*(\boldsymbol{r}) \nabla^2 \phi_j(\boldsymbol{r}) \\
[U]_{ij} &= \int_{-\infty}^{+\infty} \mathrm{d}\boldsymbol{r}\ \phi_i^*(\boldsymbol{r}) \tilde{U}(\boldsymbol{r}) \phi_j(\boldsymbol{r}) \\
[S]_{ij} &= \int_{-\infty}^{+\infty} \mathrm{d}\boldsymbol{r}\ \phi_i^*(\boldsymbol{r}) \phi_j(\boldsymbol{r})
\end{aligned}$$

In 1D, the above equations are simplified to,

$$\begin{aligned}
[H]_{ij} &= \frac{-\hbar^2}{2m} \int_{-\infty}^{+\infty} \mathrm{d}x\ \phi_i^*(x) \frac{\mathrm{d}^2\phi_j(x)}{\mathrm{d}x^2} \\
[U]_{ij} &= \int_{-\infty}^{+\infty} \mathrm{d}x\ \phi_i^*(x) \tilde{U}(x) \phi_j(x) \\
[S]_{ij} &= \int_{-\infty}^{+\infty} \mathrm{d}x\ \phi_i^*(x) \phi_j(x)
\end{aligned} \tag{3.7}$$

For orthogonal basis set, the off diagonal elements of the overlap matrix are zero. Additionally, if the basis set is normalized, the diagonal elements are unity. A basis set that satisfies the conditions of orthogonality and normalization is called orthonormal basis set. For an orthogonal basis set, $[S] = I$, where I is an identity matrix of rank N. Furthermore, substituting (3.6) in the time independent Schrödinger (3.5), one obtains,

$$\sum_{j=1}^{N} c_j^n \tilde{H} \phi_j(\boldsymbol{r}) = E_n \sum_{j=1}^{N} c_j^n \phi_j(\boldsymbol{r})$$

where E_n is the Eigen value for the quantum number $n = 1, 2, 3, \ldots, N$. Multiplying with $\phi_i^*(\boldsymbol{r})$ and integrating,

$$\sum_{j=1}^{N} \int_{-\infty}^{+\infty} \mathrm{d}\boldsymbol{r}\ \phi_i^*(\boldsymbol{r}) \tilde{H} \phi_j(\boldsymbol{r})\ c_j^n = E_n \sum_{j=1}^{N} \int_{-\infty}^{+\infty} \mathrm{d}\boldsymbol{r}\ \phi_i^*(\boldsymbol{r}) \phi_j(\boldsymbol{r})\ c_j^n$$

one obtains,

$$\sum_{j=1}^{N} [H]_{ij}\, c_j^n = E_n \sum_{j=1}^{N} [S]_{ij}\, c_j^n$$

Writing in matrix notation,

$$[H]\left\{c^n\right\} = E_n\,[S]\left\{c^n\right\}$$
$$[S]^{-1}\,[H]\left\{c^n\right\} = E_n\left\{c^n\right\}$$

where $[H]$ and $[S]$ are $N \times N$ matrices and $\{c^n\}$ is an $N \times 1$ column matrix. The above equations give rise to the following Secular equations,

$$[[H] - E_n\,[S]]\,\left\{c^n\right\} = 0$$
$$\left[[S]^{-1}\,[H] - E_n I\right]\,\left\{c^n\right\} = 0$$

where I is an $N \times N$ identity matrix. One may find N Eigen values and N Eigen functions by equating the Secular determinant to zero as follows,

$$det\ [[H] - E_n\,[S]] = 0$$
$$det\ \left[[S]^{-1}\,[H] - E_n I\right] = 0$$

One should note that for orthogonal basis set, since $[S] = I$, the Secular equation and determinant simplify to,

$$[H]\left\{c^n\right\} = E_n\left\{c^n\right\}$$
$$det\ [[H] - E_n I] = 0$$

respectively. In addition to an analytical solution, one may find numerical solution by using a computational software like Matlab. In which case, the following Matlab functions are used for the nonorthogonal and the orthogonal basis sets respectively,[3]

$$[V, D] = eig(inv(S) * H)$$
$$[V, D] = eig(H) \tag{3.8}$$

where the diagonal elements of the matrix $[D]$ give Eigen values (E_n) and the corresponding columns in $[V]$ give Eigen functions (c_j^n) for a certain quantum number n.

One may visualize the basis set to be an N-dimensional Hilbert space, where each basis function is an independent axes as shown in Fig. 3.6 in analogy with the real space. We discuss two types of basis sets in this chapter, namely real space basis sets in Sect. 3.3, and orbital space basis set in Sect. 3.4.

---

[3] *eig* and *inv* are Matlab functions for calculating Eigen values and inverse of a matrix, respectively.

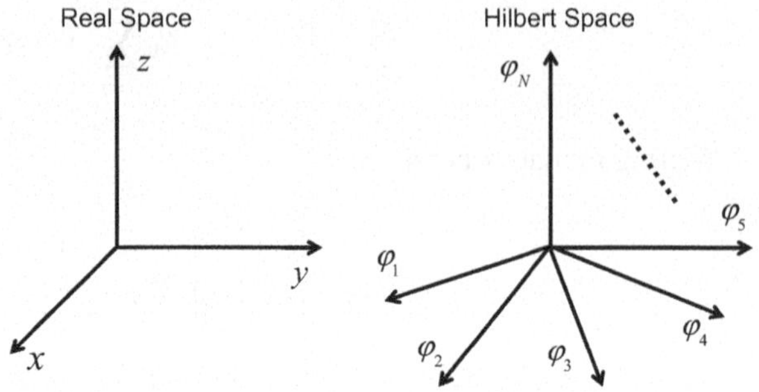

Fig. 3.6 Basis set as Hilbert space in analogy with the real space

3.2 Electrostatics

The electrostatic boundary conditions and the effect of potential is included by using a potential energy matrix $[U]$. Although there is a formal procedure to calculate this potential energy matrix to reflect the external bias, it is useful to note that the potential energy matrix is usually diagonal for orthogonal basis set, whereas for nonorthogonal basis set, discussion is more detailed. Nonetheless for both orthogonal and nonorthogonal basis sets, the potential energy matrix elements $[U]_{ij}$ may be calculated by using (3.7). For orthogonal basis set, one may approximate $U(\boldsymbol{r})$ to be constant at $\boldsymbol{r}$ corresponding to a certain point of the basis set $\phi(\boldsymbol{r})$,

$$[U]_{ij} = \begin{cases} U(\boldsymbol{r}_j) & i = j \\ 0 & i \neq j \end{cases}$$

This approximation results in tremendous convenience and in fact is one of the reasons of the popularity of orthogonal basis set, despite distinct advantages of nonorthogonal basis sets.

In addition, Schrödinger equation may be solved analytically or computationally for a single particle subjected to a certain potential energy (U). In this section, we focus on the analytical solutions in various dimensions (1D, 2D, 3D) for a single particle subjected to various boundary conditions.

Free Particle in 1D

For a free particle with $U = 0$, starting from (3.5), the 1D single particle time independent Schrödinger equation are given as,

$$\frac{\mathrm{d}^2\Psi}{\mathrm{d}x^2} + k_x^2\Psi = 0$$

where $k_x = \sqrt{2mE}/\hbar$ is the 1D wavevector. The general solution for such an equation consists of traveling waves due to free propagation unbounded by any potential and is given as, $\Psi(x) = Ae^{ik_x x} + Be^{-ik_x x}$. The time dependent wavefunction is given as, $\Psi(x,t) = Ae^{i(k_x x - \omega t)} + Be^{-i(k_x x + \omega t)}$. By using a constant phase, one may calculate the group velocity[4] as follows $v_{gx} = \mathrm{d}\omega/\mathrm{d}k_x$. For the first term $e^{i(k_x x - \omega t)}$, the constant phase ($k_x x - \omega t$ = *constant*) gives rise to a positive group velocity. Similarly, it may be shown that the second term $e^{-i(k_x x + \omega t)}$ gives rise to a negative group velocity. Since this particle is free and not subjected to any potential, it does not have any preferential direction of propagation. However, for the sake of discussion, let us assume that the particle is traveling in the positive direction by putting $B = 0$, which results in,[5]

$$\Psi(x) = Ae^{ik_x x} = A\cos(k_x x) + iA\sin(k_x x)$$

By comparing with (3.6), if $\phi_1(x) = \cos(k_x x)$ and $\phi_2(x) = \sin(k_x x)$ are substituted as the basis set, $c_1^n = A$ and $c_2^n = iA$. Furthermore, $\sin(k_x x)$ and $\cos(k_x x)$ functions are orthogonal to each other. Therefore, $[\cos(k_x x), \sin(k_x x)]$ constitutes an orthogonal basis set and a two dimensional Hilbert space.

However, the wavefunction is not normalizable, since A is undefined according to the normalization condition as follows,

$$\int_{-\infty}^{+\infty} \mathrm{d}x \, |\Psi|^2 = 1$$

Since the particle is free to go anywhere in space, it is difficult to locate it. The expectation values for p_x and E may be calculated as follows,

$$\begin{aligned} \langle p_x \rangle &= \tfrac{\hbar}{i} \textstyle\int_{-\infty}^{+\infty} \mathrm{d}x \, \Psi^* \frac{\mathrm{d}\Psi}{\mathrm{d}x} &&= \hbar k_x \\ \langle E \rangle &= \frac{p_x^2}{2m} &&= \frac{\hbar^2 k_x^2}{2m} \end{aligned}$$

Each energy point is two fold degenerate due to $\pm k_x$. Including the spin degree of freedom, the degeneracy is four fold. Usually, one drops the notation of the expectation values and simply refer to the energy and momentum, or $E(k_x)$ diagrams, if the real space and the time domain information is not kept track of. Such a plot for a free particle is shown in Fig. 3.7, which has a parabolic trend. Equating the parabolic $E(k_x)$ to $E = \hbar\omega$, one obtains $\omega = \hbar k_x^2/2m$. The corresponding phase and group velocities are respectively given as, $v_{p_x} = \omega/k_x = \hbar k_x/2m$ and $v_{gx} = \mathrm{d}\omega/\mathrm{d}k_x = \hbar k_x/m$, which makes $v_{gx} = 2v_{px}$.

For a free particle, there is no quantization. However, this particle is still a quantum particle due to the wave behavior with a certain phase. One of the key differences between quantum and classical mechanics is precisely this phase information. These quantum particles may interact and interfere with each other in constructive and destructive manner. This phase coherence plays an important role in addition to the quantization of various physical observables, which makes the quantum devices so

[4]One should note that the phase velocity is simply given as, $v_{px} = \omega/k_x$.

[5]Euler's identity gives, $e^{i\theta} = \cos\theta + i\sin\theta$.

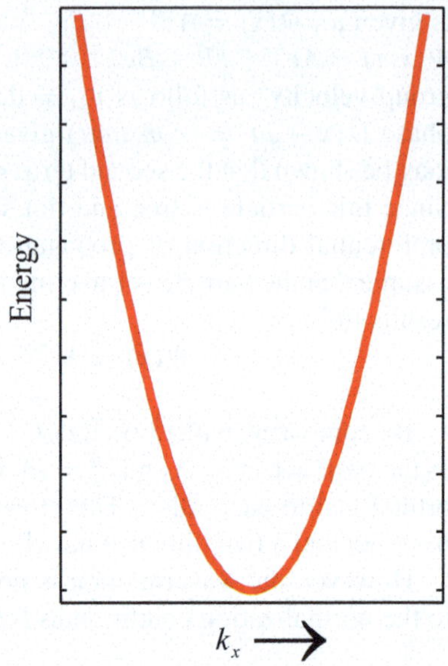

Fig. 3.7 Band structure for a free particle

different from the classical ones. Rightfully, quantum mechanics is also called wave mechanics.

Particle in a Box

Let us now put this quantum particle in a 1D box of length L as shown in Fig. 3.8a with potential energy given as,

$$U = \begin{cases} \infty & -\infty \geq x \geq 0 \\ 0 & 0 < x < L \\ \infty & L \geq x \geq \infty \end{cases}$$

For the wavefunction $\Psi(x)$, the boundary conditions at $x = 0$ and $x = L$ make $\Psi(0) = 0$ and $\Psi(L) = 0$ due to the ∞ potential energies at $x \leq 0$ and $x \geq L$, respectively. For $0 < x < L$, the Schrödinger equation is no different from that of the free particle with $U = 0$ and is given as,

$$\frac{d^2\Psi}{dx^2} + k_x^2\Psi = 0$$

Due to the rigid boundary conditions, one expects the solutions to be standing waves given as follows,[6]

---

[6]A standing wave is a superposition of two waves traveling in opposite directions.

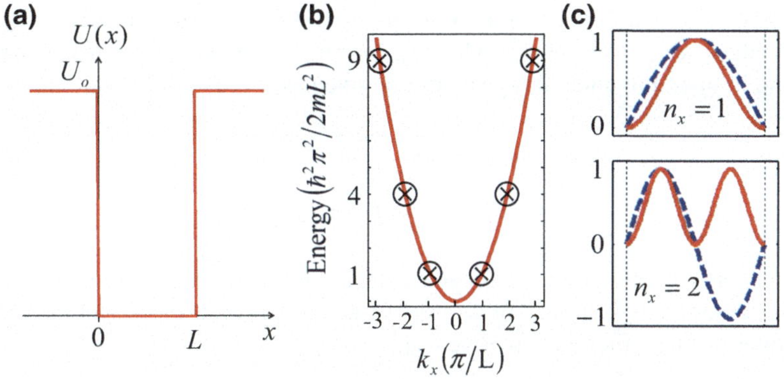

Fig. 3.8 Particle in a 1D box. **a** Potential energy profile. **b** Energy and wavevector quantization shown with ⊗. The dispersion for the free particle is shown as a red line to guide the eye. **c** Wavefunctions (dashed line) and probability densities (solid line) for $n_x = 1$ and $n_x = 2$ are reported. For the wavefunction, the peak value is $\sqrt{2/L}$, whereas for the probability density, the peak value is $2/L$

$$\Psi(x) = A\sin(k_x x) + B\cos(k_x x)$$

The $\sin(k_x x)$ and $\cos(k_x x)$ standing wave solutions are a linear combination of the traveling wave solutions $e^{ik_x x}$ and $e^{-ik_x x}$ within the linear nature of the Schrödinger equation.[7] Applying the boundary condition $\Psi(0) = 0$ gives $B = 0$, whereas $\Psi(L) = 0$ leads to $\sin(k_x L) = 0$ since A may not be zero, otherwise the wavefunction collapses. This leads to $k_{n_x} L = n_x\pi$, where $n_x = \pm 1, \pm 2, \pm 3, \ldots$ is a quantum number. $n_x = 0$ is a forbidden quantum number. $k_{n_x} = n_x\pi/L$ results in discrete momenta and energy values as follows,

$$p_n = \hbar k_{n_x} = \frac{\hbar\pi}{L} n_x$$
$$E_n = \frac{\hbar^2 k_{n_x}^2}{2m} = \frac{\hbar^2\pi^2}{2mL^2} n_x^2$$

$E(k_x)$ diagram is shown in Fig. 3.8b with only a finite combinations of E_n and k_{n_x}, given by the quantum number n_x. Each energy point is two fold degenerate due to $\pm k_x$. Including the spin degree of freedom, the degeneracy is four fold. Finally, the normalized wavefunction for the quantum number n_x by using the normalization condition is given as,

$$\Psi_{n_x}(x) = \sqrt{\frac{2}{L}} \sin\left(\frac{\pi x}{L} n_x\right)$$

and is shown in Fig. 3.8c for $n_x = 1$ and $n_x = 2$ with dashed line. $\left|\Psi_{n_x}\right|^2$ is shown with solid line. For various values of n_x, the wavefunctions are orthogonal to each other.

[7] $\cos(k_x x) = (e^{ik_x x} + e^{-ik_x x})/2$, and $\sin(k_x x) = (e^{ik_x x} - e^{-ik_x x})/j2$.

For a particle in a 2D box of dimensions L_x and L_y with ∞ potential energies at the box edges, $k^2_{n_x,n_y} = k^2_{n_x} + k^2_{n_y}$, where $k_{n_x} = \pi n_x/L_x$ and $k_{n_y} = \pi n_y/L_y$. The corresponding momentum and energy values are given as,

$$p = \hbar k_{n_x,n_y} = \hbar\sqrt{\frac{\pi^2}{L_x^2}n_x^2 + \frac{\pi^2}{L_y^2}n_y^2}$$
$$E = \frac{\hbar^2 k^2_{n_x,n_y}}{2m} = \frac{\hbar^2}{2m}\left(\frac{\pi^2}{L_x^2}n_x^2 + \frac{\pi^2}{L_y^2}n_y^2\right)$$

Each energy point is four fold degenerate. Including the spin degree of freedom, the degeneracy is eight fold. For various values of (n_x, n_y), the wavefunctions are orthogonal to each other and are given as,[8]

$$\Psi_{n_x,n_y}(x, y) = \sqrt{\frac{4}{L_x L_y}} \sin\left(\frac{\pi x}{L_x}n_x\right) \sin\left(\frac{\pi y}{L_y}n_y\right)$$

For a 3D box of dimensions L_x, L_y and L_z with ∞ potential energies at the box edges, $k^2_{n_x,n_y,n_z} = k^2_{n_x} + k^2_{n_y} + k^2_{n_z}$, where $k_{n_x} = \frac{\pi}{L_x}n_x$, $k_{n_y} = \frac{\pi}{L_y}n_y$, and $k_{n_z} = \frac{\pi}{L_z}n_z$. The corresponding momentum and energy values are given as,

$$p = \hbar k_{n_x,n_y,n_z} = \hbar\sqrt{\frac{\pi^2}{L_x^2}n_x^2 + \frac{\pi^2}{L_y^2}n_y^2 + \frac{\pi^2}{L_z^2}n_z^2}$$

$$E = \frac{\hbar^2 k^2_{n_x,n_y,n_z}}{2m} = \frac{\hbar^2}{2m}\left(\frac{\pi^2}{L_x^2}n_x^2 + \frac{\pi^2}{L_y^2}n_y^2 + \frac{\pi^2}{L_z^2}n_z^2\right)$$

Each energy point is eight fold degenerate. Including the spin degree of freedom, the degeneracy is sixteen fold. For various values of (n_x, n_y, n_z), the wavefunctions are orthogonal to each other and are given as,

$$\Psi_{n_x,n_y,n_z}(x, y) = \sqrt{\frac{8}{L_x L_y L_z}} \sin\left(\frac{\pi x}{L_x}n_x\right) \sin\left(\frac{\pi y}{L_y}n_y\right) \sin\left(\frac{\pi z}{L_z}n_z\right)$$

Finally, for a particle trapped in a cylindrical or spherical box, the wavefunction would have the cylindrical or spherical symmetry, respectively, and is given by cylindrical or spherical harmonics.

3.3 Real Space Basis Set

There are various computational methods available for solving Schrödinger equation. All of them involve the use of some kind of a basis set. Using the basis set, the operators and the wavefunctions may be converted into square and column matrices, respectively, and various quantities may be solved by using techniques of Linear Algebra. In real space, two widely used basis sets are finite difference and finite

[8]These wavefunctions correspond to a product state.

element. The finite difference method employs orthogonal brick like orthogonal basis set, whereas the finite element method uses pyramid shaped nonorthogonal basis set.

Free Particle with Finite Difference Basis Set

The basis set and the associated derivatives for the finite difference method is shown in Fig. 3.9. The real space is discretized by using the lattice points with lattice spacing a. The brick shaped function has the width of a and the height of $1/\sqrt{a}$ making it an orthonormal basis set, and is given as,

$$\varphi_i(x) = \frac{1}{\sqrt{a}}[H(x) - H(x-a)]$$
$$\varphi_j(x) = \frac{1}{\sqrt{a}}[H(x-a) - H(x-2a)]$$

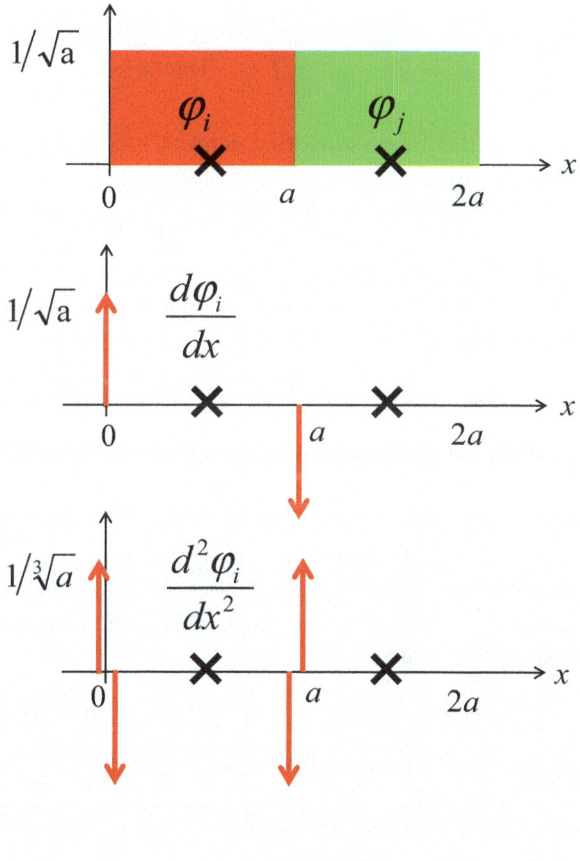

Fig. 3.9 Finite difference basis set for a 1D lattice

where $H(x - x_o)$ is Heaviside function and is defined as,

$$H(x - x_o) = \begin{cases} 1 & x > x_o \\ 0 & x < x_o \end{cases}$$

Taking the first derivative of the basis set gives,

$$\frac{d\varphi_i(x)}{dx} = \frac{1}{\sqrt{a}}[H'(x) - H'(x - a)] = \frac{1}{\sqrt{a}}[\delta(x) - \delta(x - a)]$$

where $\delta(x)$ is Dirac's delta functions defined as follows,

$$\delta(x - x_o) = \frac{dH(x - x_o)}{dx}$$

Some properties of the delta function include,

$$\int_{-\infty}^{+\infty} \delta(x - x_o) = 1$$

$$\int_{-\infty}^{+\infty} f(x)\delta(x - x_o) = f(x_o)$$

We further discuss this function in Chap. 4. Taking the second derivative of the basis set gives,

$$\frac{d^2\varphi_i(x)}{dx^2} = \frac{1}{\sqrt{a}}[\delta'(x) - \delta'(x - a)]$$

where $\delta'(x)$ is Dirac's double delta functions, and is defined as,

$$\delta'(x) = \frac{1}{a}[\delta(x + 0^+) - \delta(x - 0^+)]$$

Using above relations in (3.7), one obtains the following Hamiltonian elements,

$$\begin{aligned} [H]_{ii} &= \tfrac{-\hbar^2}{2ma^2} \textstyle\int_0^a dx \left[-\delta(x - 0^+) - \delta(x - a + 0^+)\right] = 2t_o \\ [H]_{ij} = [H]_{ji} &= \tfrac{-\hbar^2}{2ma^2} \textstyle\int_0^{2a} dx \left[\delta(x - a - 0^+)\right] \qquad = -t_o \end{aligned}$$

where $t_o = \hbar^2/2ma^2$. The tridiagonal Hamiltonian matrix for the 1D finite difference basis set is given as,

$$[H] = \begin{bmatrix} 2t_o & -t_o & 0 & 0 & \cdots \\ -t_o & 2t_o & -t_o & 0 & \cdots \\ 0 & -t_o & 2t_o & -t_o & \cdots \\ \cdots & \cdots & \cdots & \cdots & \cdots \end{bmatrix} \tag{3.9}$$

where the diagonal element is $2t_o$ (also known as onsite energy) and the nearest neighbor hopping element is given as $-t_o$.

In 2D, the basis set remains brick shaped, which gives the onsite energy of $4t_o$ (due to two dimensions) and the nearest neighbor hopping parameter to be $-t_o$. In 3D, the onsite energy is $6t_o$ (due to three dimensions) with the nearest neighbor hopping parameter of $-t_o$. It should be noted that in 2D and 3D, there may be examples where the hopping parameter may not be the same for all the dimensions, consider having different lattice constants and/or effective masses in various dimensions. In this case, the onsite energy is still the sum of twice the $\hbar^2/2ma^2$ parameter in multiple dimensions and the off diagonal elements are negative of this parameter in the respective direction.

Particle in a 1D Box with Finite Difference Basis Set

Consider the particle in a 1D box problem again, which we solve analytically in Sect. 3.4. For the length L of the 1D box, we take a lattice of $N = 25$ points with a lattice spacing a. The 25×25 Hamiltonian is then given as,

$$[H] = \begin{bmatrix} 2t_o & -t_o & 0 & 0 & 0 & \cdots & \cdots & 0 & 0 & 0 \\ -t_o & 2t_o & -t_o & 0 & 0 & \cdots & \cdots & 0 & 0 & 0 \\ 0 & -t_o & 2t_o & -t_o & 0 & \cdots & \cdots & 0 & 0 & 0 \\ 0 & 0 & -t_o & 2t_o & -t_o & \cdots & \cdots & 0 & 0 & 0 \\ 0 & 0 & 0 & -t_o & 2t_o & \ddots & \ddots & 0 & 0 & 0 \\ 0 & 0 & 0 & 0 & 0 & \cdots & \cdots & 2t_o & -t_o & 0 \\ 0 & 0 & 0 & 0 & 0 & \cdots & \cdots & -t_o & 2t_o & -t_o \\ 0 & 0 & 0 & 0 & 0 & \cdots & \cdots & 0 & -t_o & 2t_o \end{bmatrix}$$

With the boundary conditions starting at $x = 0$ (5th point) and $x = L$ (21st point), the 25×25 potential energy matrix element is given as,

$$[U] = \begin{bmatrix} U_o & 0 & 0 & 0 & 0 & \cdots & \cdots & 0 & 0 & 0 \\ 0 & U_o & 0 & 0 & 0 & \cdots & \cdots & 0 & 0 & 0 \\ 0 & 0 & U_o & 0 & 0 & \cdots & \cdots & 0 & 0 & 0 \\ 0 & 0 & 0 & U_o & 0 & \cdots & \cdots & 0 & 0 & 0 \\ 0 & 0 & 0 & 0 & U_o & \ddots & \ddots & 0 & 0 & 0 \\ 0 & 0 & 0 & 0 & 0 & 0 & \ddots & \ddots & 0 & 0 \\ 0 & 0 & 0 & 0 & 0 & \ddots & \ddots & 0 & 0 & 0 \\ 0 & 0 & 0 & \cdots & \cdots & U_o & 0 & 0 & 0 & 0 \\ 0 & 0 & 0 & 0 & \cdots & \cdots & U_o & 0 & 0 & 0 \\ 0 & 0 & 0 & 0 & 0 & \cdots & \cdots & U_o & 0 & 0 \\ 0 & 0 & 0 & 0 & 0 & \cdots & \cdots & 0 & U_o & 0 \\ 0 & 0 & 0 & 0 & 0 & \cdots & \cdots & 0 & 0 & U_o \end{bmatrix}$$

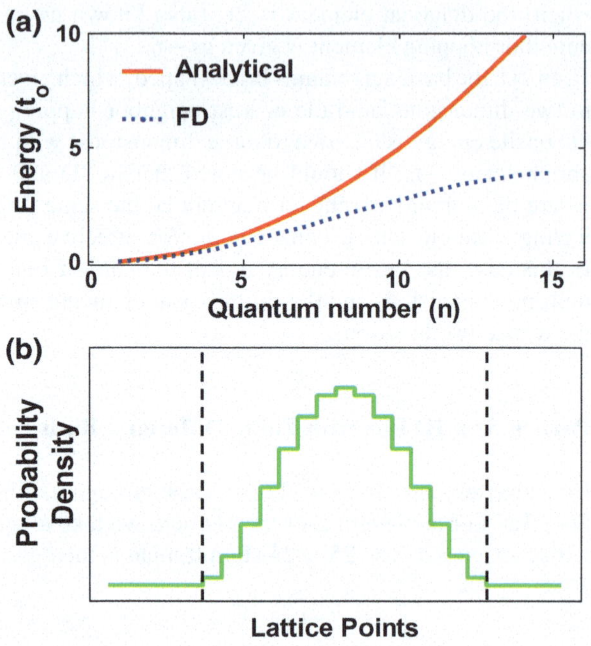

Fig. 3.10 Finite difference method for the particle in a 1D box with potential energy $U_o = 100$ eV. **a** Eigen values (lowest fifteen only, which are associated with the box). **b** Probability density for $n_x = 1$ state

By adding $[H]$ and $[U]$, one may solve for the Eigen values and Eigen vectors in Matlab by using, $[V, D] = eig\,([H] + [U])$. By using the values of the coefficients c_j^n from the Eigen vector (from the columns in matrix $[V]$ for the corresponding diagonal matrix elements in [D]) for a certain Eigen state n_x, the wavefunction is given by (3.6). For $U_o =$ 100 eV, the Eigen values are plotted with respect to the quantum number in Fig. 3.10a. The discretization error is observed compared to the analytical solution, where the maximum value is $4t_o$ for the finite difference method. One should note that the lowest energies for only first fifteen quantum numbers are plotted, since there are fifteen points inside the box. The rest of Eigen values belong to the barrier region. The probability density $|\Psi_1(\boldsymbol{r})|^2$ for the quantum number $n_x = 1$ is plotted in Fig. 3.10b, which reflects that the probability of finding a particle is high within the box region and is minimal in the barrier region, as expected.

We further show the energy spectrum with a barrier height of $U_o = 0.1$ eV in Fig. 3.11a. Due to the discretization error, the energy values deviate from the analytical solution and are smaller in magnitude compared to the higher barrier case shown in 3.10a. The wavefunction also tunnels into the barrier region as shown in Fig. 3.11b. In other words, there is a finite probability of finding the particle in the barrier region, which is a purely quantum mechanical phenomenon called quantum mechanical tunneling.

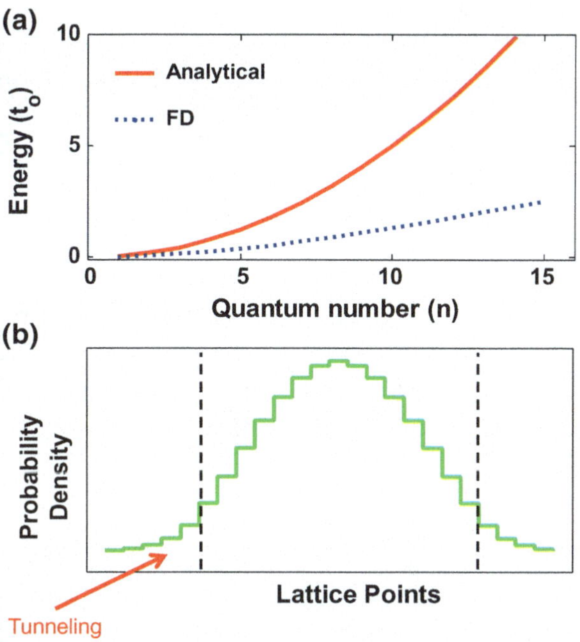

Fig. 3.11 Finite difference method for the particle in a 1D box with potential energy $U_o = 0.1$ eV. **a** Eigen values (lowest fifteen only, which are associated with the box). **b** Probability amplitude for $n_x = 1$ state showing tunneling into the barrier region

Finite Element

The finite element basis set consists of pyramid shaped functions as shown in Fig. 3.12. For the lattice spacing a, the height of basis function is $1/\sqrt{a}$. Since the basis functions on the two nearest neighbor points are overlapping, the overlap and the Hamiltonian matrix elements are given by using (3.7) as follows,

$$[S]_{ii} = \int_{-a}^{a} \mathrm{d}x\ \phi_i^*(x)\phi_i(x) = \frac{2}{3}$$

$$[S]_{ij} = [S]_{ji} = \int_{-a}^{2a} \mathrm{d}x\ \phi_i^*(x)\phi_j(x) = \frac{1}{6}$$

$$[H]_{ii} = \frac{-\hbar^2}{2ma^2}\int_{-a}^{a} \mathrm{d}x\ [-2\delta(x)] = 2t_o$$

$$[H]_{ij} = [H]_{ji} = \frac{-\hbar^2}{2ma^2}\int_{-a}^{2a} \mathrm{d}x\ [+\delta(x-a)] = -t_o$$

Hence, the basis set is not normalized in its current form. The Hamiltonian and the overlap matrices are then given as,

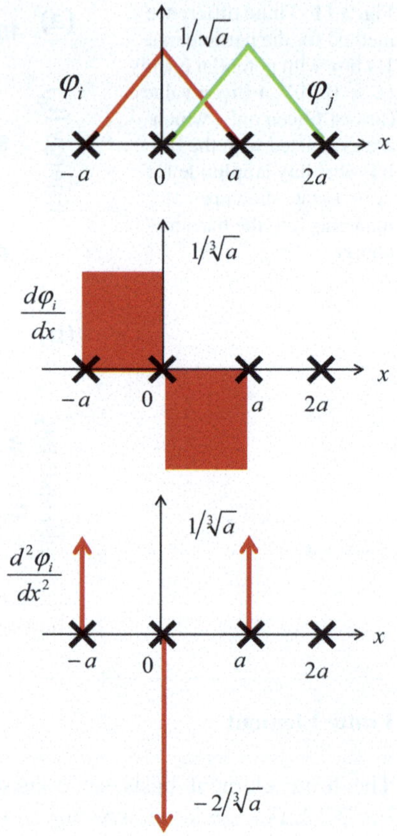

Fig. 3.12 Finite element basis set for a 1D lattice

$$[H] = \begin{bmatrix} 2t_o & -t_o & 0 & 0 & \cdots \\ -t_o & 2t_o & -t_o & 0 & \cdots \\ 0 & -t_o & 2t_o & -t_o & \cdots \\ \cdots & \cdots & \cdots & \cdots & \cdots \end{bmatrix} \tag{3.10}$$

$$[S] = \frac{1}{6}\begin{bmatrix} 4 & 1 & 0 & 0 & \cdots \\ 1 & 4 & 1 & 0 & \cdots \\ 0 & 1 & 4 & 1 & \cdots \\ \cdots & \cdots & \cdots & \cdots & \cdots \end{bmatrix} \tag{3.11}$$

which may be used to calculate the Eigen values and the Eigen functions by using the following Matlab function, $[V, D] = eig\left(S^{-1}H\right)$. Further analysis is left as an exercise for the interested readers.

3.4 Orbital Space Basis Set

Atomic and/or molecular orbitals may also be used as the basis set instead of the real space functions. The key advantage here is the use of orbital symmetries in the choice of the Hamiltonian and the overlap matrices. Consider a hydrogen atom shown in Fig. 3.13 in the Bohr's planetary model, where an electron is orbiting around the nucleus, which consists of a proton. The potential energy for this system is given as, $U(r) = -Kq^2/r$, where q is the magnitude of the electronic charge and r is the distance between electron and proton. The Hamiltonian operator is then given as,

$$\tilde{H} = \frac{-\hbar^2}{2m}\nabla^2 - K\frac{q^2}{r}$$

where $K = 1/4\pi\epsilon_o$ is a prefactor that depends on the permittivity of free space $\epsilon_o = 8.854 \times 10^{-12}$ F/m.

Given the spherical symmetry of the problem, the Laplacian ∇^2 operator is described in the spherical coordinates (r, θ, ϕ). Furthermore, due to the three dimensions, one has to keep track of three quantum numbers, principle quantum number (n), azimuthal quantum number (l) and magnetic quantum number (m) in addition to the intrinsic spin quantum number $s = (\uparrow -spin, \downarrow -spin)$ with the following allowed values,

$$\begin{aligned} n &= 1, 2, 3, \ldots, \infty \\ l &= 0, 1, 2, \ldots, (n-1) \\ m &= -l, -(l-1), \ldots, 0, \ldots, (l-1), l \end{aligned}$$

One should note that $l = 0, 1, 2, 3$ give s, p, d and f orbitals respectively. The value of n is usually specified as the first entry in the notation to describe orbitals. Hence, $1s$-orbital is written as $(n, l, m) = (1, 0, 0)$. For $n = 2$, the four orbitals are given as $(2, 0, 0)$, $(2, 1, -1)$, $(2, 1, 0)$ and $(2, 1, 1)$. In addition, the wavefunction may be

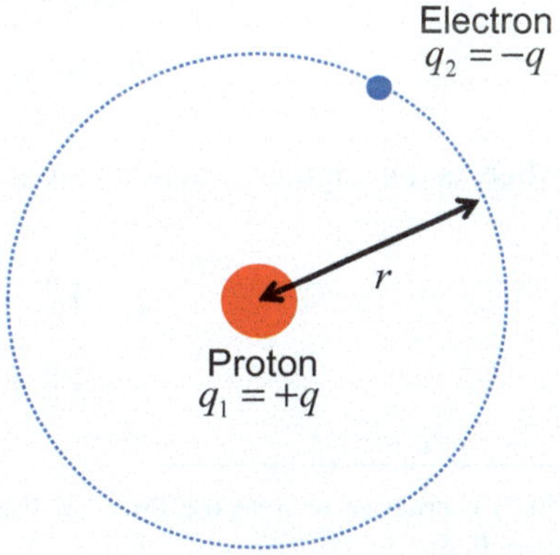

Fig. 3.13 Hydrogen atom

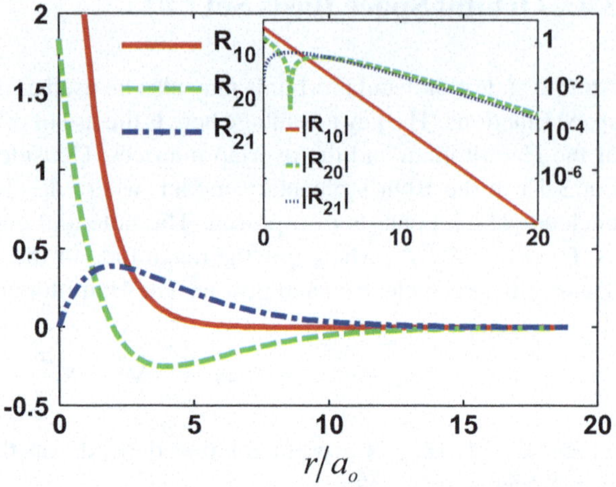

Fig. 3.14 Radial functions for hydrogen wavefunctions on linear scale, and logarithmic scale in the inset

divided into radial $R(r)$ and spherical $Y(\theta, \phi)$ functions.[9] The radial function $R(r)$ of the wavefunction in general depends on (n, l) and the angular function $Y(\theta, \phi)$ depends on (l, m), written as,

$$\Psi_{n,l,m}(r, \theta, \phi) = R_n^l(r) Y_l^m(\theta, \phi)$$

while R_n^l gives the decay of orbital, Y_l^m gives rise to the specific shape. While the angular function Y_l^m is the same for all atoms, the radial function $R(r)$ changes due to a different spherical potential with varying nuclear charge in different atoms. Therefore, for the same quantum numbers (n, l, m), orbitals from two different atoms would have similar shape, although the extent (governed by R_n^l) may be very different.

For hydrogen atom ($a_o = 0.529$Å), the following radial functions are shown in Fig. 3.14,

$$\begin{aligned} R_1^0 &= \frac{2}{a_o^{3/2}} e^{-r/a_o} \\ R_2^0 &= \frac{1}{\sqrt{2} a_o^{3/2}} \left(1 - \frac{1}{2}\frac{r}{a_o}\right) e^{-r/2a_o} \\ R_2^1 &= \frac{1}{\sqrt{24} a_o^{3/2}} \frac{r}{a_o} e^{-r/2a_o} \end{aligned}$$

whereas the angular functions are given as,

$$\begin{aligned} Y_0^0 &= \frac{1}{\sqrt{4\pi}} \\ Y_1^0 &= \sqrt{\frac{3}{4\pi}} \cos(\theta) \end{aligned}$$

[9]It is instructive to note the limits of these variables are given as, $r = [0, \infty]$, $\theta = [0, \pi)$, and $\phi = [0, 2\pi)$.

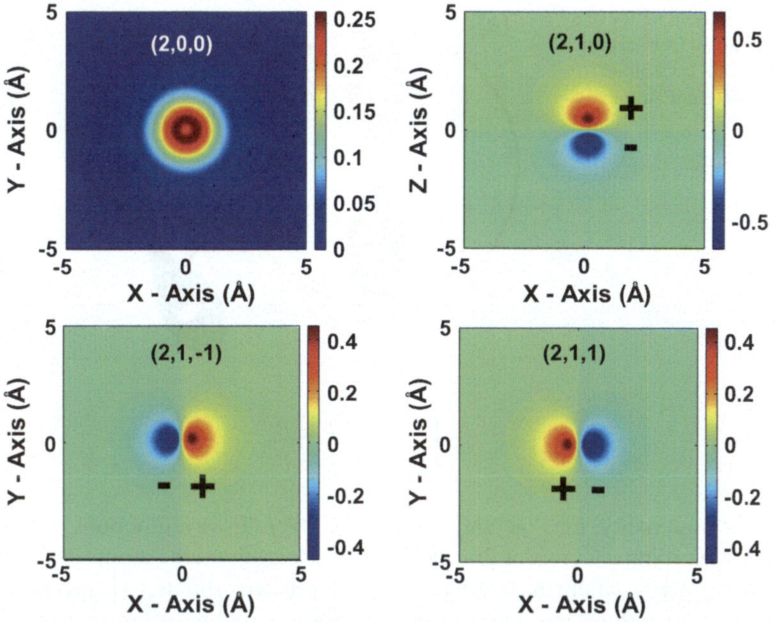

Fig. 3.15 Wavefunction plots of carbon atom for various quantum numbers. Wavefunction and atomic visualization in this text is done by using Hückel-NV [1]

$$Y_1^{\pm 1} = \mp\sqrt{\tfrac{3}{8\pi}}\sin(\theta)e^{\pm j\phi}$$

$(n, l, m) = (2, 0, 0)$ has a spherical symmetry and is termed as a $2s-$orbital. $(2, 1, 0)$ is aligned along z-axis and is referred to as $2p_z$-orbital. $(2, 1, \pm 1)$ are aligned along y-axis. It is a misconception that these are $2p_x$ and $2p_y$ orbitals. It is actually a weighted sum of the angular functions of these two orbitals which gives rise to $2p_x$ and $2p_y$ orbitals as follows,

$$Y(2p_x) = \tfrac{1}{\sqrt{2}}\left(Y_1^{-1} - Y_1^{-1}\right) = \sqrt{\tfrac{3}{4\pi}}\sin(\theta)\cos(\phi)$$
$$Y(2p_y) = \tfrac{i}{\sqrt{2}}\left(Y_1^{-1} + Y_1^{-1}\right) = \sqrt{\tfrac{3}{4\pi}}\sin(\theta)\sin(\phi)$$

This unique wavefunction dependence, in fact, leads to very interesting allotropic forms of carbon like diamond, graphite, and Bucky ball. For carbon atom, these wavefunction plots are shown in Fig. 3.15 for various quantum numbers showing angular as well as radial dependence.

Moreover, the radial functions are usually nonorthogonal. However, the angular functions are orthogonal, i.e. $Y_l^m Y_{l'}^{m'} = \delta_{ll'}\delta_{mm'}$, where $\delta_{ll'}$ is the Kronecker delta function and is defined as,

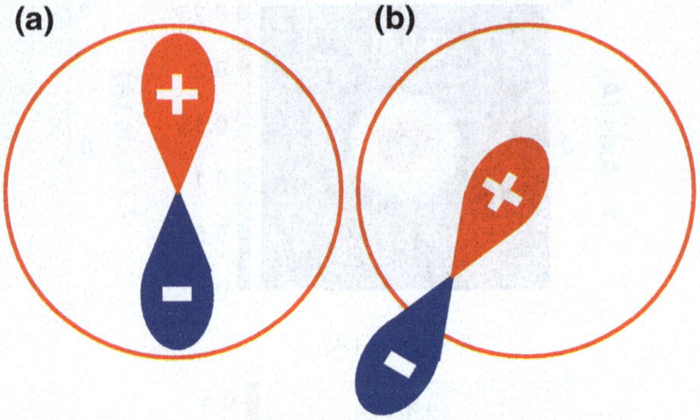

Fig. 3.16 Orthogonality of *s* and *p* orbitals. **a** Orbitals on the same atom. **b** Orbitals from two different atoms

$$\delta_{ll'} = \begin{cases} 1 & l = l' \\ 0 & l \neq l' \end{cases}$$

This orthogonality may actually make the overall wavefunctions orthogonal between two orbitals. Consider 2*s*- and $2p_z$-orbitals of the same atom schematically shown in Fig. 3.16a, where the overlap between these two orbitals, $[S]_{ij}$ given by (3.7), is zero since $2p_z$-orbital has equal amount of positive and negative contributions to the wavefunction product, whereas 2*s*-orbital wavefunction is always positive. However, this overlap may be non zero if either of the two orbitals is displaced, e.g. when these orbitals are from two atoms, schematically shown in Fig. 3.16b.

Such *wavefunction symmetry effects* or *orbital symmetry effects* predominantly determine the device characteristics at nanoscale and hence are quite important to incorporate in any theoretical model.

Atomic Orbitals as Basis Set

Consider the two atom system in Fig. 3.17a with one atomic orbital per atom. By using the atomic orbital wavefunctions (ϕ_a, ϕ_b) as the orthogonal basis set, the Hamiltonian for the complete system in matrix form is given as,

$$[H] = \begin{bmatrix} H_{AA} & H_{AB} \\ H_{BA} & H_{BB} \end{bmatrix}$$

where the Hamiltonian matrix elements of the two atom system are as follows,

$$\begin{aligned}
H_{AA} &= \tfrac{-\hbar^2}{2m} \int_{-\infty}^{+\infty} d\boldsymbol{r}\; \phi_a^*(\boldsymbol{r}) \nabla^2 \phi_a(\boldsymbol{r}) = \epsilon_a \\
H_{AB} &= \tfrac{-\hbar^2}{2m} \int_{-\infty}^{+\infty} d\boldsymbol{r}\; \phi_a^*(\boldsymbol{r}) \nabla^2 \phi_b(\boldsymbol{r}) = -t \\
H_{BA} &= \tfrac{-\hbar^2}{2m} \int_{-\infty}^{+\infty} d\boldsymbol{r}\; \phi_b^*(\boldsymbol{r}) \nabla^2 \phi_a(\boldsymbol{r}) = -t^* \\
H_{BB} &= \tfrac{-\hbar^2}{2m} \int_{-\infty}^{+\infty} d\boldsymbol{r}\; \phi_b^*(\boldsymbol{r}) \nabla^2 \phi_b(\boldsymbol{r}) = \epsilon_b
\end{aligned}$$

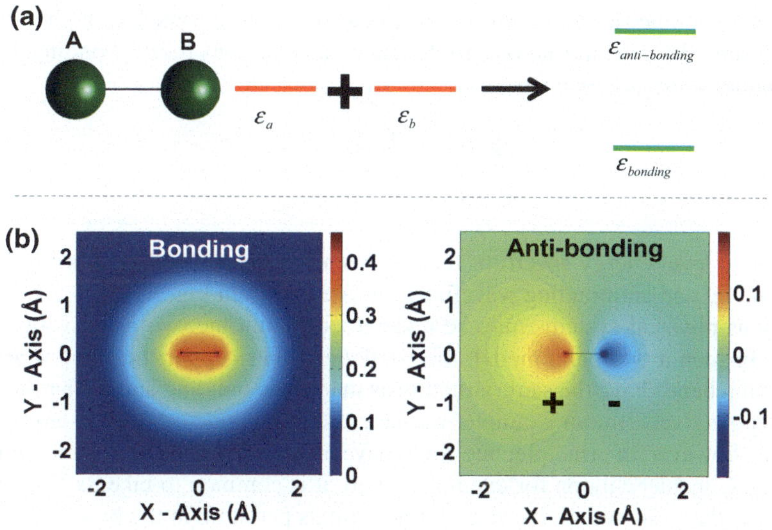

Fig. 3.17 Two atom system. **a** Energy spectrum for a two level model system. **b** Molecular wavefunctions for the bonding and the antibonding states with atomic orbitals as the basis set

which results in,

$$[H] = \begin{bmatrix} \epsilon_a & -t \\ -t^* & \epsilon_b \end{bmatrix}$$

One should note that $H_{AB} = H_{BA}^\dagger$ for Hermitian Hamiltonian matrix. For simplicity, let us assume that both the atoms have the same atomic number, which makes $\epsilon_a = \epsilon_b = \epsilon_o$, giving,

$$[H] = \begin{bmatrix} \epsilon_o & -t \\ -t^* & \epsilon_o \end{bmatrix}$$

This matrix represents a two level model, where ϵ_o is the onsite energy and t is the nearest neightbor hopping parameter. For an orthogonal basis set, $[S] = I$, and the Eigen values are given as $\epsilon_o \pm |t|$.

For the $\epsilon_o - |t|$ Eigen value, the coefficients for the Eigen function are given as,

$$\begin{Bmatrix} c_a \\ c_b \end{Bmatrix} = \begin{Bmatrix} 1 \\ 1 \end{Bmatrix} \frac{1}{\sqrt{2}}$$

where $\sqrt{2}$ is the normalization factor. This symmetric state corresponds to the bonding state (the lower energy state). For the $\epsilon_o + |t|$ Eigen value, the coefficients for the Eigen function are given as,

$$\begin{Bmatrix} c_a \\ c_b \end{Bmatrix} = \begin{Bmatrix} 1 \\ -1 \end{Bmatrix} \frac{1}{\sqrt{2}}$$

This antisymmetric state corresponds to the antibonding state (the higher energy state). Using (3.6), the normalized molecular wavefunctions for the bonding and the antibonding states are given as,

$$\Psi_B(\boldsymbol{r}) = \frac{1}{\sqrt{2}}\left(\phi_a + \phi_b\right)$$
$$\Psi_{AB}(\boldsymbol{r}) = \frac{1}{\sqrt{2}}\left(\phi_a - \phi_b\right)$$

The two level energy spectrum for the two atom system is shown in Fig. 3.17a. The bonding and antibonding wavefunctions are shown in Fig. 3.17b.

Various physical systems may be understood within the two level or two band model. In such a two band model, the coefficient matrices for the valence and the conduction bands have the same structure as that of the bonding and the antibonding states, respectively. In our example, we have taken only one orbital per atom as the basis set, however, in principle, one could have more than one orbital per atom in the basis set. Consider Silicon for example, where it is common to take $3s$, $3p$, and $4d$ orbitals as the basis set, giving rise to nine orbitals per atom in the basis set.

Tight Binding Theory

With a certain choice of the orbital basis set, the Hamiltonian matrix may be written in terms of the onsite energy and the hopping parameters between various orbitals. This description leads to the *tight binding Hamiltonian*, and hence the tight binding theory of electronic structure. Although it is common to include only the nearest neighbor interactions in the tight binding models, in general one could also include second, third, fourth nearest neighbors and beyond.

Consider the 1D chain of six atoms in Fig. 3.18 with the onsite energy of ϵ_o and hopping parameter of $-t$ between the nearest neighbors. The tight binding Hamiltonian is given as,

$$[H] = \begin{bmatrix} \epsilon_o & -t & 0 & 0 & 0 & 0 \\ -t & \epsilon_o & -t & 0 & 0 & 0 \\ 0 & -t & \epsilon_o & -t & 0 & 0 \\ 0 & 0 & -t & \epsilon_o & -t & 0 \\ 0 & 0 & 0 & -t & \epsilon_o & -t \\ 0 & 0 & 0 & 0 & -t & \epsilon_o \end{bmatrix} \quad (3.12)$$

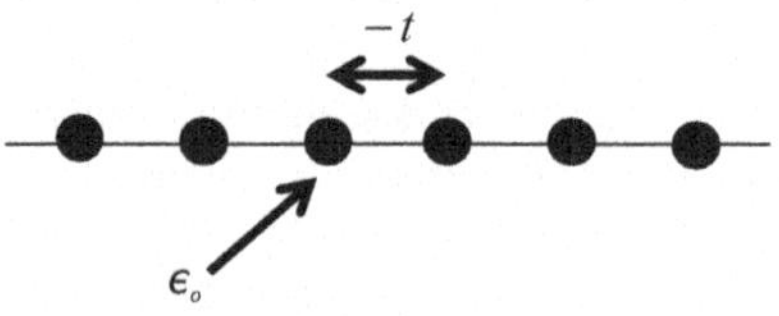

Fig. 3.18 1D atomic chain of six atoms

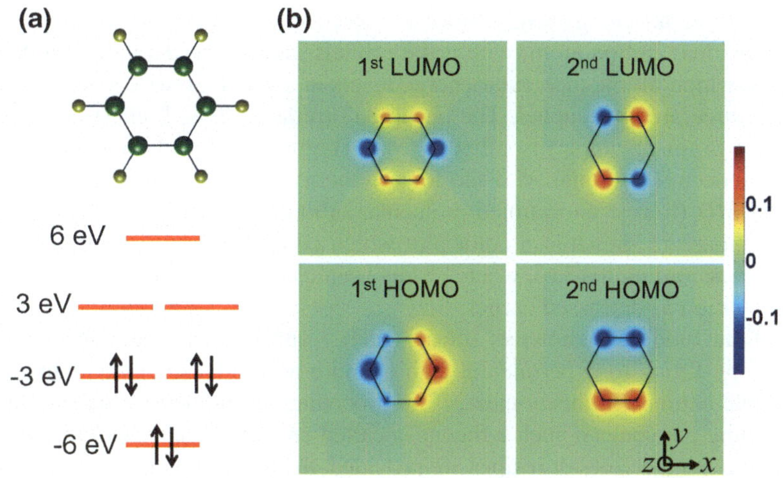

Fig. 3.19 Benzene molecule. **a** Energy levels calculated by using p_z-orbital tight binding Hamiltonian showing doubly degenerate HOMO and LUMO levels at −3 eV and 3 eV, respectively. **b** Molecular wavefunction for HOMO and LUMO levels dominantly have p_z symmetry

The onsite energy ϵ_o depends on the vacuum reference, which is an arbitrary quantity. For convenience, ϵ_o is usually taken as zero. For an inhomogeneous material with a varying vacuum reference, the varying onsite energy may also be included.

This simple *orthogonal tight binding model* is also called *Hückel* model. The models based on nonorthogonal basis sets are called *nonorthogonal tight binding models*. These tight binding models may easily be expanded to 2D and 3D. In the nearest neighbor approximation with one orbital per atom, irrespective of the dimensionality, for nearest neighbors i and j, $[H]_{ij} = -t$ and for the non nearest neighbors i and j, $[H]_{ij} = 0$.

Consider another example of a benzene molecule in Fig. 3.19a. Within the nearest neighbor tight binding model using the p_z-orbital basis set. One may take the onsite energy $\epsilon_o = 0$ and hopping parameter $t = 3$ eV. The Hamiltonian matrix $[H]$ is then given as follows,

$$[H] = \begin{bmatrix} 0 & -t & 0 & 0 & 0 & -t \\ -t & 0 & -t & 0 & 0 & 0 \\ 0 & -t & 0 & -t & 0 & 0 \\ 0 & 0 & -t & 0 & -t & 0 \\ 0 & 0 & 0 & -t & 0 & -t \\ -t & 0 & 0 & 0 & -t & 0 \end{bmatrix}$$

Note that the matrix elements $H(1, 6)$ and $H(6, 1)$ are nonzero in this case, compared to the example in Fig. 3.18, since atoms 1 and 6 are bonded with each other in the benzene molecule. The energy levels, calculated by diagonalizing this Hamiltonian matrix, are shown in Fig. 3.19a. One finds two degenerate energy levels, and two non degenerate energy levels. Since one electron per spin resides in the p_z orbital of an individual carbon atom, within the orbitals under consideration, one has six electrons, which fill the molecular energy levels in a way that three levels are occupied and three are unoccupied. The occupied orbital with the highest energy level (at -3 eV in this example) is called *HOMO* (highest occupied molecular orbital) and the unoccupied orbital with the lowest energy (at $+3$ eV in this example) is called *LUMO* (lowest unoccupied molecular orbital). In Fig. 3.19b, the molecular wavefunctions are shown as a color plot which consists of a weighted sum of the p_z-orbitals according to (3.6), where c^n has been calculated by using the process of diagonalization as discussed earlier.

The tight binding parameters (onsite energy and hopping parameter) may be extracted by fitting to experimental data or more accurate calculations, instead of formal calculations. In this context, if no procedure is outlined for calculating the tight binding parameters, such a theory is called *empirical tight binding theory*. If some intuitive guidelines or recipes are provided, the theory is called *semi empirical tight binding theory*.

The choice of orbital basis set usually depends on the symmetry of the problem at hand. For example, in a particular situation, the energy spectra in the range of interest may belong to only p_z-orbitals, which in fact is the case in graphene, nanoribbons, carbon nanotubes, organic molecules, etc. In this case, it makes sense to use p_z-orbitals as the basis set. In other cases, the situation may be very complex, like Si, where one has to take s, p and even d orbitals in the basis set. Apart from this, the explicit form of the radial function of the wavefunctions may make the calculation computationally intractable, where one could approximate the radial function with an exponential function or Gaussian functions.

To summarize the basis set discussion, we outline the applicability of the widely used basis set as follows,

(1) p_z-orbital basis set. The most widely used basis set due to its suitability for organic molecules (like benzene, Bucky ball, etc.), carbon nanotubes, nanoribbons, graphene, etc.

(2) sp^3 orbitals basis set. Widely used for various semiconductors, like Si, Ge, GaAs, due to multiple orbital contribution to the valence and the conduction bands.

(3) sp^3d orbitals basis set. For some semiconductors, including a d-orbital in the basis set may be required to get more accurate results.

(4) Slater type orbital (STO) basis set. The radial function in the basis set has exponential decay. This type of basis set is appropriate for situations where the radial function in the basis set has predominantly exponential decay, which may be captured by using the following STO basis set,

$$\phi_{n,l,m}(r,\theta,\phi) = (2\zeta)^n \sqrt{\frac{2\zeta}{(2n)!}}\, r^{n-1} e^{-\zeta r} Y_l^m(\theta,\phi)$$

where Y_l^m is the angular function as discussed earlier. Sometimes multiple STOs are used for approximating a single atomic orbital. The advantage of using STOs is the computational simplicity, since most integrals involving exponential functions have analytical solutions.

(5) Gaussian basis set. Instead of STOs, the radial function of the atomic orbitals is approximated by a Gaussian in the following form,

$$\phi_{n,l,m}(r,\theta,\phi) = Ar^{n-1}e^{-\alpha r^2}Y_l^m(\theta,\phi)$$

where α controls the width of the Gaussian function. The advantage here is that the integrals of such functions also have analytical solutions.

(6) STO-nG. The STO basis set (which is an approximation of the atomic orbital) may be further approximated by multiple Gaussian functions. Usually, three Gaussian functions are used, which is termed as STO-3G basis set.

(7) Slater Determinant. Most of the above mentioned basis sets are for a single electron, which are applicable to single electron Hamiltonian and the associated wavefunctions. While most of our discussion is within this *single electron picture*, it is instructive to understand some of the concepts related to *many electron systems*. For these systems, one of the primary requirements is Pauli's exclusion principle, where no two electrons of the same spin may occupy the same state. To incorporate this exchange effect, one may define Slater determinant for N electron system as follows that serves as a wavefunction for the N electron system,

$$\phi_T(\boldsymbol{r}_1,\boldsymbol{r}_2,\cdots,\boldsymbol{r}_N) = \frac{1}{N!}\begin{vmatrix} \phi_1(\boldsymbol{r}_1) & \phi_1(\boldsymbol{r}_2) & \cdots & \cdots & \cdots & \phi_1(\boldsymbol{r}_N) \\ \phi_2(\boldsymbol{r}_1) & \phi_2(\boldsymbol{r}_2) & \cdots & \cdots & \cdots & \phi_2(\boldsymbol{r}_N) \\ \cdots & \cdots & \cdots & \cdots & \cdots & \cdots \\ \cdots & \cdots & \cdots & \cdots & \cdots & \cdots \\ \cdots & \cdots & \cdots & \cdots & \cdots & \cdots \\ \phi_N(\boldsymbol{r}_1) & \phi_N(\boldsymbol{r}_2) & \cdots & \cdots & \cdots & \phi_N(\boldsymbol{r}_N) \end{vmatrix}$$

This simple matrix captures the essence of Pauli's exclusion principle. If any of the two wavefunctions were to be the same, i.e. two rows being equal, the determinant becomes zero, which is an indication of the violation of the exclusion principle.

While the discussion of many body theories is beyond the scope of this book, interested readers are directed to read further about the Hartree Fock Theory (HFT) and Density Functional Theory (DFT). In DFT, energy is a *functional* of electron density. Since electron density is already a function of $\boldsymbol{r}$, one uses the term *functional* and not simply function. A widely used approximation used in DFT is LDA (local density approximation). Detailed discussion of these methods is beyond the scope of this book.

It is important to note that once one chooses a basis set for these methods, the operators become matrices. At which point, all the techniques, learned in this book are applicable.

3.5 Band Structure

In this chapter, we have so far discussed Energy Diagrams $E(\boldsymbol{r})$, where the Hamiltonian is in real space and consequently energy values are plotted as a function of real space ($\boldsymbol{r}$) or real space lattice points. Our objective in this section is to transfer the real space Hamiltonian $H(\boldsymbol{r})$ to the reciprocal space Hamiltonian $H(\boldsymbol{k})$, and subsequently diagonalize the reciprocal space Hamiltonian to calculate the Band Structure $E(\boldsymbol{k})$. In contrast to the prior focus, here we discuss periodic structures in 1D, 2D, and 3D with infinite real space dimensions, which makes $dk \approx 0$ and hence the $\boldsymbol{k}$ values and the energy Eigen values become continuous. Furthermore, the energy Eigen values $E(\boldsymbol{k})$ get grouped together in the form of bands that we plot within the first Brillouin zone.

For a periodic structure, the real space Hamiltonian $H(\boldsymbol{r})$ and the overlap $S(\boldsymbol{r})$ matrices (in real space or orbital space basis sets) may be transformed to the reciprocal space $\boldsymbol{k} = (k_x, k_y, k_z)$ as follows,

$$[H(\boldsymbol{k})] = \sum_{m=1}^{N} [H(\boldsymbol{r})]_{mn} \, e^{i\boldsymbol{k}\cdot(\boldsymbol{d}_m - \boldsymbol{d}_n)}$$

$$[S(\boldsymbol{k})] = \sum_{m=1}^{N} [S(\boldsymbol{r})]_{mn} \, e^{i\boldsymbol{k}\cdot(\boldsymbol{d}_m - \boldsymbol{d}_n)} \tag{3.13}$$

where n is the index for the center unit cell and m is the index for the neighboring unit cells. $(\boldsymbol{d}_m - \boldsymbol{d}_n)$ is the displacement vector and $i\boldsymbol{k} \cdot (\boldsymbol{d}_m - \boldsymbol{d}_n)$ is the phase factor between the center unit cell n and the neighboring unit cells m. One should include all the neighboring unit cells in this summation, where $[H(\boldsymbol{r})]_{mn}$ and $[S(\boldsymbol{r})]_{mn}$ matrices are non zero. For each value of $\boldsymbol{k} = (k_x, k_y, k_z)$, one may analytically diagonalize the $[S(\boldsymbol{k})]^{-1}\,[H(\boldsymbol{k})]$ matrix to find the energy Eigen values for all bands at a certain value of $\boldsymbol{k}$. Alternatively, one may use computational software like Matlab, by using the following function, $[V, D] = eig(inv(S) * H)$, where the diagonal matrix elements of $[D(\boldsymbol{k})]$ contain the energy values for the certain value of $\boldsymbol{k}$ and the columns in $[V(\boldsymbol{k})]$ contain the corresponding Eigen functions $c_j^n(\boldsymbol{k})$. One should note that for an orthogonal basis set, $[S(\boldsymbol{r})] = I$, the overlap matrix in the reciprocal space is also an identity matrix, i.e. $[S(\boldsymbol{k})] = I$, for which, one may either analytically diagonalize to obtain $E(\boldsymbol{k})$ or use, $[V, D] = eig(H)$ in Matlab.

The number of bands depends on the size of $[H(\boldsymbol{k})]$ and $[S(\boldsymbol{k})]$ matrices. For the real space basis set, e.g. finite difference and finite element, the size of $[H(\boldsymbol{k})]$ and $[S(\boldsymbol{k})]$ matrices depends on the number of lattice points within a unit cell. However, for the orbital space basis set, the size of $[H(\boldsymbol{k})]$ and $[S(\boldsymbol{k})]$ matrices not only depends on the number of atoms in a unit cell, but also the number of orbitals per atom. The total number of bands are thus given as, *number of orbitals per atom* $\times$ *number of atoms per unit cell*. For example, if there are two atoms in a unit cell and one uses p_z-orbitals as the basis set, the number of bands are two. However, for the same two atoms per unit cell, if one uses sp^3 basis set (one s-orbital and three p-orbitals), the number of bands would be eight. For sp^3d^5 basis set (one s-orbital, three p-orbitals, and five d-orbitals), one obtains eighteen bands for two atoms per unit cell.

One Dimensional Materials

Consider the 1D lattice with the lattice constant a shown in Fig. 3.20a. For the finite difference basis set, the onsite energy is $2t_o$, whereas the hopping parameter is $-t_o$. $[H(\boldsymbol{r})]$ is calculated next for the center unit cell where $m = n$, for the right unit cell where $m = n + 1$, and for the left unit cell where $m = n - 1$.

$$[H(\boldsymbol{r})] = \left[\begin{array}{c|c|c} 2t_o & -t_o & 0 \\ \hline -t_o & 2t_o & -t_o \\ \hline 0 & -t_o & 2t_o \end{array}\right] = \left[\begin{array}{c|c|c} \cdots & H_{n-1,n} & \cdots \\ \hline \cdots & H_{nn} & \cdots \\ \hline \cdots & H_{n+1,n} & \cdots \end{array}\right]$$

The real space Hamiltonian matrix for the unit cell n is given as $(m = n)$, $[H(\boldsymbol{r})]_{nn} = 2t_o$. The corresponding displacement and phase factor respectively are given as, $(\boldsymbol{d}_n - \boldsymbol{d}_n) = \boldsymbol{0}$ and $i\boldsymbol{k} \cdot (\boldsymbol{d}_n - \boldsymbol{d}_n) = \boldsymbol{0}$, where $\boldsymbol{k} = k_x\hat{x}$.

For $m = n + 1$, $[H(\boldsymbol{r})]_{n+1,n} = -t_o$. The corresponding displacement and phase factor respectively are given as, $(\boldsymbol{d}_{n+1} - \boldsymbol{d}_n) = a\hat{x}$ and $i\boldsymbol{k} \cdot (\boldsymbol{d}_{n+1} - \boldsymbol{d}_n) = ik_xa$.

For $m = n - 1$, $[H(\boldsymbol{r})]_{n-1,n} = -t_o$. The corresponding displacement and phase factor respectively are given as, $(\boldsymbol{d}_{n-1} - \boldsymbol{d}_n) = -a\hat{x}$ and $i\boldsymbol{k} \cdot (\boldsymbol{d}_{n-1} - \boldsymbol{d}_n) = -ik_xa$. Note that $[H(\boldsymbol{r})]_{n\pm2,n}$ and onwards are zero, therefore one does not need to include the contributions of second nearest neighbors and beyond in this calculation. By using (3.13), the Hamiltonian is given as,

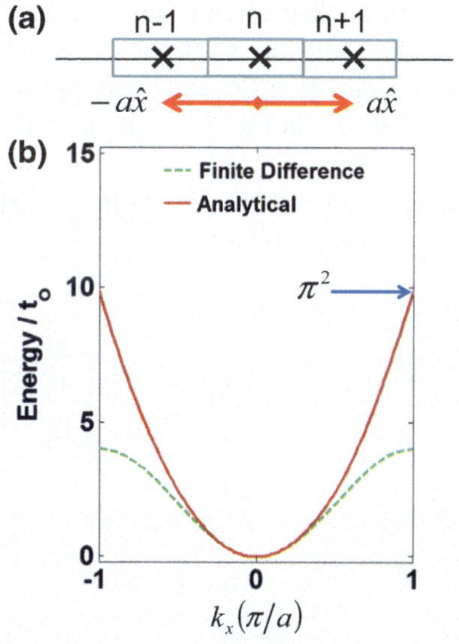

Fig. 3.20 Finite difference band structure for 1D lattice. **a** Unit cells. **b** $E(k_x)$ band structure

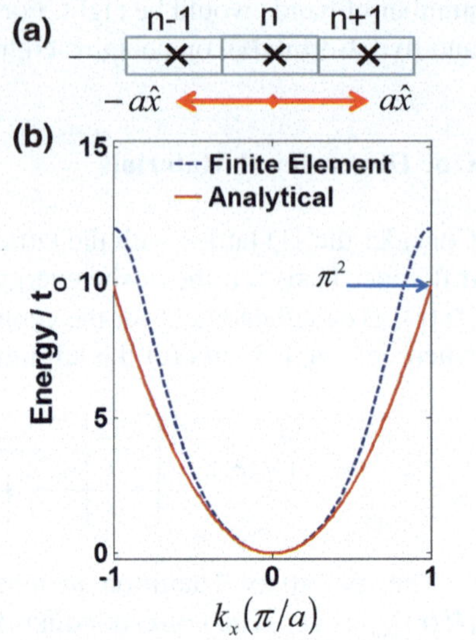

Fig. 3.21 Finite element band structure for 1D lattice. **a** Unit cells. **b** $E(k_x)$ band structure

$$[H(k_x)] = 2t_o - t_o e^{ik_x a} - t_o e^{-ik_x a}$$

and the Eigen values may be summarized as,

$$E(k_x) = 2t_o\,[1 - \cos(k_x a)]$$

which are plotted in Fig. 3.20b. The dispersion for the 1D lattice, by using finite difference method, has a bandwidth of $4t_o$, whereas the bandwidth for the analytical parabolic dispersion is $\pi^2 t_o$, as shown.

Next consider the finite element basis set with a lattice constant a, for which the onsite energy is $2t_o$ and the hopping parameter is $-t_o$. $[H(r)]$ and $[S(r)]$ for the unit cells shown in Fig. 3.21a are given as,

$$[H(\mathbf{r})] = \begin{bmatrix} 2t_o & -t_o & 0 \\ -t_o & 2t_o & -t_o \\ 0 & -t_o & 2t_o \end{bmatrix} = \begin{bmatrix} \cdots & H_{n-1,n} & \cdots \\ \cdots & H_{nn} & \cdots \\ \cdots & H_{n+1,n} & \cdots \end{bmatrix}$$

$$[S(\mathbf{r})] = \frac{1}{6}\begin{bmatrix} 4 & 1 & 0 \\ 1 & 4 & 1 \\ 0 & 1 & 4 \end{bmatrix} = \begin{bmatrix} \cdots & S_{n-1,n} & \cdots \\ \cdots & S_{nn} & \cdots \\ \cdots & S_{n+1,n} & \cdots \end{bmatrix}$$

For which the real space Hamiltonian matrix and overlap matrix elements for $(m = n)$ are given as, $[H(\mathbf{r})]_{nn} = 2t_o$ and $[S(\mathbf{r})]_{nn} = 2/3$.

For $m = n + 1$, $[H(\mathbf{r})]_{n+1,n} = -t_o$ and $[S(\mathbf{r})]_{n+1,n} = 1/6$. For $m = n - 1$, $[H(\mathbf{r})]_{n-1,n} = -t_o$ and $[S(\mathbf{r})]_{n-1,n} = 1/6$. The displacement vectors and phase factors for $m = n - 1, n, n + 1$ are the same as that for the finite difference as discussed earlier. By using (3.13),

$$[H(k_x)] = 2t_o - t_o e^{ik_x a} - t_o e^{-ik_x a}$$
$$[S(k_x)] = \tfrac{2}{3} + \tfrac{1}{6} e^{ik_x a} + \tfrac{1}{6} e^{-ik_x a}$$

Thus the Eigen values calculated by $[S]^{-1}[H]$ are given as,

$$E(k_x) = 6t_o \frac{1 - \cos(k_x a)}{2 + \cos(k_x a)}$$

which are plotted in Fig. 3.21b. The dispersion for the 1D lattice by using finite element method has a bandwidth of $12t_o$, whereas the bandwidth for the analytical solutions is $\pi^2 t_o$. As shown in Figs. 3.20b and 3.21b, the band edge for the $E(k_x)$ of finite difference and finite element basis set is at zero energy. Since onsite energy depends on the vacuum reference, one may symbolically take it as $E_c + 2t_o$, which shifts the band edge to E_c as shown in Fig. 3.22. This band dispersion in fact looks very similar to the band structure around the conduction band edge of most semiconductors.

Consider another example of a 1D crystal of atoms within the nearest neighbor orthogonal tight binding model,[10] which has one atom per unit cell with the atomic

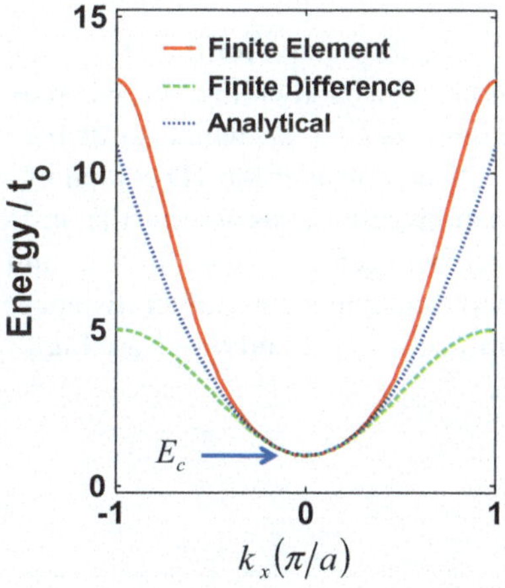

Fig. 3.22 Band edge shift for finite difference and finite element methods to reflect a band edge at $E_c = 1$ eV

[10] In the absence of a periodic potential, and hence Bragg diffraction at the edge of the Brillouin zone.

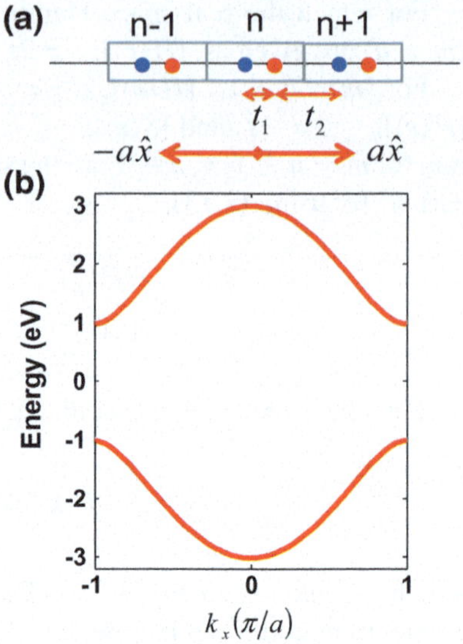

Fig. 3.23 Band structure for 1D crystal. **a** Two atoms per unit cell. **b** $E(k_x)$ band structure showing two bands with $t_1 = 2$ eV, $t_2 = 1$ eV

spacing a. Let us assume the onsite energy to be zero, whereas the hopping parameter is $-t$. By using (3.13), it can be shown that,

$$[H(k_x)] = -te^{ik_x a} - te^{-ik_x a}$$

and the Eigen values are given as,

$$E(k_x) = -2t\cos(k_x a)$$

The dispersion for the 1D crystal with one atom per unit cell and one orbital per atom has a bandwidth of $4t$, whereas the band edge is at $-2t$. One may take the onsite energy as $E_c + 2t$, which shifts the band edge to E_c.

Now, consider the 1D crystal of atoms shown in Fig. 3.23a, which has two atoms per unit cell. Let us assume the onsite energy to be zero, whereas the hopping parameter between atoms within the unit cell (with smaller distance) is t_1 and the one between atoms with larger distance is t_2. The center unit cell n, and the neighboring unit cells $n + 1$ and $n - 1$ are highlighted as well.

$[H(r)]$ for the three unit cells is given as,

$$[H(\boldsymbol{r})] = \left[\begin{array}{cc|cc|cc} 0 & -t_1 & 0 & 0 & 0 & 0 \\ -t_1 & 0 & -t_2 & 0 & 0 & 0 \\ \hline 0 & -t_2 & 0 & -t_1 & 0 & 0 \\ 0 & 0 & -t_1 & 0 & -t_2 & 0 \\ \hline 0 & 0 & 0 & -t_2 & 0 & -t_1 \\ 0 & 0 & 0 & 0 & -t_1 & 0 \end{array}\right] = \left[\begin{array}{c|c|c} \cdots & H_{n-1,n} & \cdots \\ \hline \cdots & H_{nn} & \cdots \\ \hline \cdots & H_{n+1,n} & \cdots \end{array}\right]$$

For $m = n - 1, n, n + 1$, the real space Hamiltonian matrices for the unit cell n are, respectively, given as,

$$[H(\boldsymbol{r})]_{n-1,n} = \begin{bmatrix} 0 & 0 \\ -t_2 & 0 \end{bmatrix}$$

$$[H(\boldsymbol{r})]_{nn} = \begin{bmatrix} 0 & -t_1 \\ -t_1 & 0 \end{bmatrix}$$

$$[H(\boldsymbol{r})]_{n+1,n} = \begin{bmatrix} 0 & -t_2 \\ 0 & 0 \end{bmatrix}$$

The displacement vectors and phase factors for $m = n - 1, n, n + 1$ are the same as discussed earlier. The Hamiltonian in the reciprocal sapce may be calculated by using (3.13) as follows,

$$[H(k_x)] = \begin{bmatrix} 0 & -t_1 \\ -t_1 & 0 \end{bmatrix} + \begin{bmatrix} 0 & -t_2 \\ 0 & 0 \end{bmatrix} e^{ik_x a} + \begin{bmatrix} 0 & 0 \\ -t_2 & 0 \end{bmatrix} e^{-ik_x a}$$
$$= \begin{bmatrix} 0 & -t_1 - t_2 e^{ik_x a} \\ -t_1 - t_2 e^{-ik_x a} & 0 \end{bmatrix}$$

The Eigen values are given as,

$$E(k_x) = \pm \left| -t_1 - t_2 e^{ik_x a} \right| = \pm\sqrt{t_1^2 + t_2^2 + 2t_2 \cos(k_x a)}$$

which gives two bands as shown in Fig. 3.23b. The dispersion for each band has a bandwidth of,

$$\sqrt{t_1^2 + t_2^2 + 2t_2} - \sqrt{t_1^2 + t_2^2 - 2t_2}$$

whereas the valence band edge is at $-\sqrt{t_1^2 + t_2^2 - 2t_2}$, and the conduction band edge is at $\sqrt{t_1^2 + t_2^2 - 2t_2}$, and the bandgap is given as, $2\sqrt{t_1^2 + t_2^2 - 2t_2}$.

Let us now discuss the general case of 1D nanoribbons and nanotubes as shown in Fig. 3.24. Since, nanoribbons are the easiest to handle due to their flatness, we discuss their calculation in detail, which may easily be extended to other nanotubes and nanowires.

An armchair nanoribbon with an atomic width of $N = 9$ is shown in Fig. 3.25, which has eighteen atoms per unit cell. Let us assume the onsite energy to be zero, whereas the nearest neighbor hopping parameter is $-t$. We highlight the center unit cell n, and the neighboring unit cells $n+1$ and $n-1$ as well. The real space Hamiltonian matrix for $m = n$ is then given as,

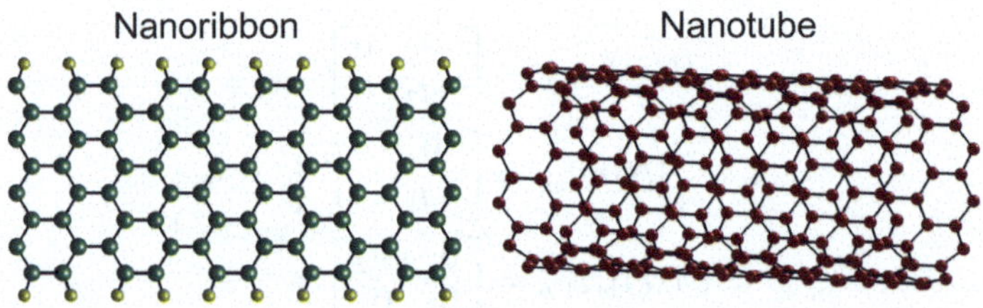

Fig. 3.24 Example of a nanoribbon and a nanotube

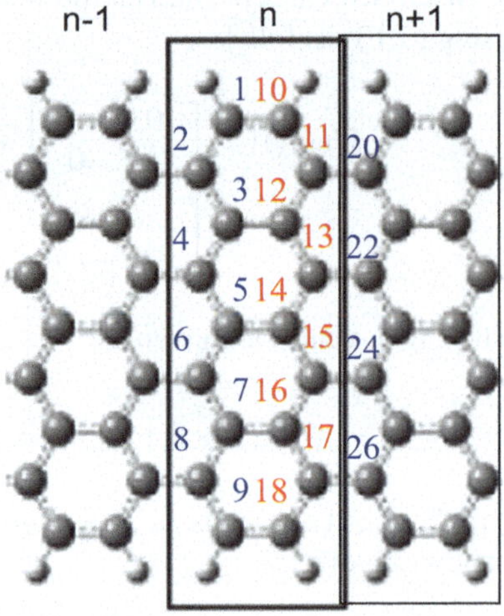

Fig. 3.25 Unit cell in an $N = 9$ armchair graphene nanoribbon. The atoms are labeled for the unit cell n, whereas the atoms for the unit cell $n+1$ are only labeled for which there is a coupling in $[H(\boldsymbol{r})]_{n+1,n}$

$$
[H(\boldsymbol{r})]_{nn} = \begin{bmatrix}
0 & -t & 0 & 0 & 0 & 0 & 0 & 0 & 0 & -t & 0 & 0 & 0 & 0 & 0 & 0 & 0 & 0 \\
-t & 0 & -t & 0 & 0 & 0 & 0 & 0 & 0 & 0 & 0 & 0 & 0 & 0 & 0 & 0 & 0 & 0 \\
0 & -t & 0 & -t & 0 & 0 & 0 & 0 & 0 & 0 & 0 & -t & 0 & 0 & 0 & 0 & 0 & 0 \\
0 & 0 & -t & 0 & -t & 0 & 0 & 0 & 0 & 0 & 0 & 0 & 0 & 0 & 0 & 0 & 0 & 0 \\
0 & 0 & 0 & -t & 0 & -t & 0 & 0 & 0 & 0 & 0 & 0 & 0 & -t & 0 & 0 & 0 & 0 \\
0 & 0 & 0 & 0 & -t & 0 & -t & 0 & 0 & 0 & 0 & 0 & 0 & 0 & 0 & 0 & 0 & 0 \\
0 & 0 & 0 & 0 & 0 & -t & 0 & -t & 0 & 0 & 0 & 0 & 0 & 0 & 0 & -t & 0 & 0 \\
0 & 0 & 0 & 0 & 0 & 0 & -t & 0 & -t & 0 & 0 & 0 & 0 & 0 & 0 & 0 & 0 & 0 \\
0 & 0 & 0 & 0 & 0 & 0 & 0 & -t & 0 & 0 & 0 & 0 & 0 & 0 & 0 & 0 & 0 & -t \\
-t & 0 & 0 & 0 & 0 & 0 & 0 & 0 & 0 & 0 & -t & 0 & 0 & 0 & 0 & 0 & 0 & 0 \\
0 & 0 & 0 & 0 & 0 & 0 & 0 & 0 & 0 & -t & 0 & -t & 0 & 0 & 0 & 0 & 0 & 0 \\
0 & 0 & -t & 0 & 0 & 0 & 0 & 0 & 0 & 0 & -t & 0 & -t & 0 & 0 & 0 & 0 & 0 \\
0 & 0 & 0 & 0 & 0 & 0 & 0 & 0 & 0 & 0 & 0 & -t & 0 & -t & 0 & 0 & 0 & 0 \\
0 & 0 & 0 & 0 & -t & 0 & 0 & 0 & 0 & 0 & 0 & 0 & -t & 0 & -t & 0 & 0 & 0 \\
0 & 0 & 0 & 0 & 0 & 0 & 0 & 0 & 0 & 0 & 0 & 0 & 0 & -t & 0 & -t & 0 & 0 \\
0 & 0 & 0 & 0 & 0 & 0 & -t & 0 & 0 & 0 & 0 & 0 & 0 & 0 & -t & 0 & -t & 0 \\
0 & 0 & 0 & 0 & 0 & 0 & 0 & 0 & 0 & 0 & 0 & 0 & 0 & 0 & 0 & -t & 0 & -t \\
0 & 0 & 0 & 0 & 0 & 0 & 0 & 0 & -t & 0 & 0 & 0 & 0 & 0 & 0 & 0 & -t & 0
\end{bmatrix}
$$

For $m = n + 1$,

$$
[H(\boldsymbol{r})]_{n+1,n} = \begin{bmatrix}
0 & 0 & 0 & 0 & 0 & 0 & 0 & 0 & 0 & 0 & 0 & 0 & 0 & 0 & 0 & 0 & 0 & 0 \\
0 & 0 & 0 & 0 & 0 & 0 & 0 & 0 & 0 & 0 & -t & 0 & 0 & 0 & 0 & 0 & 0 & 0 \\
0 & 0 & 0 & 0 & 0 & 0 & 0 & 0 & 0 & 0 & 0 & 0 & 0 & 0 & 0 & 0 & 0 & 0 \\
0 & 0 & 0 & 0 & 0 & 0 & 0 & 0 & 0 & 0 & 0 & 0 & -t & 0 & 0 & 0 & 0 & 0 \\
0 & 0 & 0 & 0 & 0 & 0 & 0 & 0 & 0 & 0 & 0 & 0 & 0 & 0 & 0 & 0 & 0 & 0 \\
0 & 0 & 0 & 0 & 0 & 0 & 0 & 0 & 0 & 0 & 0 & 0 & 0 & 0 & -t & 0 & 0 & 0 \\
0 & 0 & 0 & 0 & 0 & 0 & 0 & 0 & 0 & 0 & 0 & 0 & 0 & 0 & 0 & 0 & 0 & 0 \\
0 & 0 & 0 & 0 & 0 & 0 & 0 & 0 & 0 & 0 & 0 & 0 & 0 & 0 & 0 & 0 & -t & 0 \\
0 & 0 & 0 & 0 & 0 & 0 & 0 & 0 & 0 & 0 & 0 & 0 & 0 & 0 & 0 & 0 & 0 & 0 \\
0 & 0 & 0 & 0 & 0 & 0 & 0 & 0 & 0 & 0 & 0 & 0 & 0 & 0 & 0 & 0 & 0 & 0 \\
0 & 0 & 0 & 0 & 0 & 0 & 0 & 0 & 0 & 0 & 0 & 0 & 0 & 0 & 0 & 0 & 0 & 0 \\
0 & 0 & 0 & 0 & 0 & 0 & 0 & 0 & 0 & 0 & 0 & 0 & 0 & 0 & 0 & 0 & 0 & 0 \\
0 & 0 & 0 & 0 & 0 & 0 & 0 & 0 & 0 & 0 & 0 & 0 & 0 & 0 & 0 & 0 & 0 & 0 \\
0 & 0 & 0 & 0 & 0 & 0 & 0 & 0 & 0 & 0 & 0 & 0 & 0 & 0 & 0 & 0 & 0 & 0 \\
0 & 0 & 0 & 0 & 0 & 0 & 0 & 0 & 0 & 0 & 0 & 0 & 0 & 0 & 0 & 0 & 0 & 0 \\
0 & 0 & 0 & 0 & 0 & 0 & 0 & 0 & 0 & 0 & 0 & 0 & 0 & 0 & 0 & 0 & 0 & 0 \\
0 & 0 & 0 & 0 & 0 & 0 & 0 & 0 & 0 & 0 & 0 & 0 & 0 & 0 & 0 & 0 & 0 & 0 \\
0 & 0 & 0 & 0 & 0 & 0 & 0 & 0 & 0 & 0 & 0 & 0 & 0 & 0 & 0 & 0 & 0 & 0
\end{bmatrix}
$$

For $m = n - 1$, $[H]_{n-1,n} = [H]^{\dagger}_{n+1,n}$, where † represents the conjugate transpose. Since the tight binding hopping parameter t is real in this case, hence $t^* = t$. By using (3.13), the Hamiltonian in the reciprocal space $[H(\boldsymbol{k})]$ is given as,

$$[H(\boldsymbol{k})] = \begin{bmatrix}
0 & -t & 0 & 0 & 0 & 0 & 0 & 0 & 0 & -t & 0 & 0 & 0 & 0 & 0 & 0 & 0 & 0 \\
-t & 0 & -t & 0 & 0 & 0 & 0 & 0 & 0 & 0 & -te^{ik_xa} & 0 & 0 & 0 & 0 & 0 & 0 & 0 \\
0 & -t & 0 & -t & 0 & 0 & 0 & 0 & 0 & 0 & 0 & -t & 0 & 0 & 0 & 0 & 0 & 0 \\
0 & 0 & -t & 0 & -t & 0 & 0 & 0 & 0 & 0 & 0 & 0 & -te^{ik_xa} & 0 & 0 & 0 & 0 & 0 \\
0 & 0 & 0 & -t & 0 & -t & 0 & 0 & 0 & 0 & 0 & 0 & 0 & -t & 0 & 0 & 0 & 0 \\
0 & 0 & 0 & 0 & -t & 0 & -t & 0 & 0 & 0 & 0 & 0 & 0 & 0 & -te^{ik_xa} & 0 & 0 & 0 \\
0 & 0 & 0 & 0 & 0 & -t & 0 & -t & 0 & 0 & 0 & 0 & 0 & 0 & 0 & -t & 0 & 0 \\
0 & 0 & 0 & 0 & 0 & 0 & -t & 0 & -t & 0 & 0 & 0 & 0 & 0 & 0 & 0 & -te^{ik_xa} & 0 \\
0 & 0 & 0 & 0 & 0 & 0 & 0 & -t & 0 & 0 & 0 & 0 & 0 & 0 & 0 & 0 & 0 & -t \\
-t & 0 & 0 & 0 & 0 & 0 & 0 & 0 & 0 & 0 & -t & 0 & 0 & 0 & 0 & 0 & 0 & 0 \\
0 & -te^{-ik_xa} & 0 & 0 & 0 & 0 & 0 & 0 & 0 & -t & 0 & -t & 0 & 0 & 0 & 0 & 0 & 0 \\
0 & 0 & -t & 0 & 0 & 0 & 0 & 0 & 0 & 0 & -t & 0 & -t & 0 & 0 & 0 & 0 & 0 \\
0 & 0 & 0 & -te^{-ik_xa} & 0 & 0 & 0 & 0 & 0 & 0 & 0 & -t & 0 & -t & 0 & 0 & 0 & 0 \\
0 & 0 & 0 & 0 & -t & 0 & 0 & 0 & 0 & 0 & 0 & 0 & -t & 0 & -t & 0 & 0 & 0 \\
0 & 0 & 0 & 0 & 0 & -te^{-ik_xa} & 0 & 0 & 0 & 0 & 0 & 0 & 0 & -t & 0 & -t & 0 & 0 \\
0 & 0 & 0 & 0 & 0 & 0 & -t & 0 & 0 & 0 & 0 & 0 & 0 & 0 & -t & 0 & -t & 0 \\
0 & 0 & 0 & 0 & 0 & 0 & 0 & -te^{-ik_xa} & 0 & 0 & 0 & 0 & 0 & 0 & 0 & -t & 0 & -t \\
0 & 0 & 0 & 0 & 0 & 0 & 0 & & -t & 0 & 0 & 0 & 0 & 0 & 0 & 0 & -t & 0
\end{bmatrix}$$

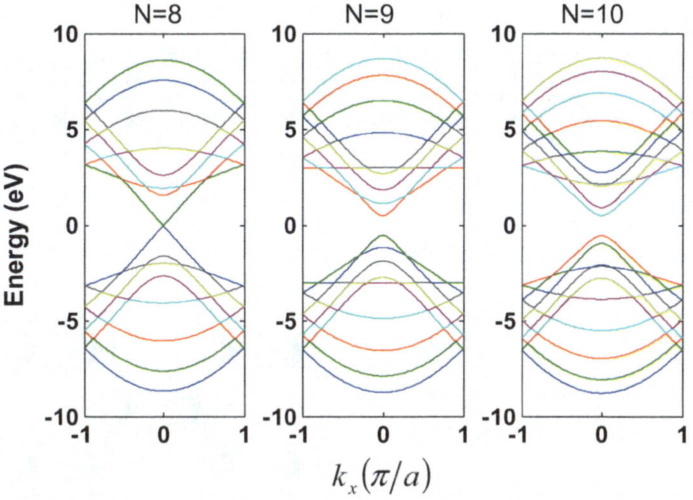

Fig. 3.26 Band structure of graphene nanoribbons with atomic widths $N = 8, 9, 10$

The Eigen values for this nanoribbon may be calculated by using a computational software like Matlab. For an armchair nanoribbon of atomic width N, one gets $2N$ bands, since there are $2N$ atoms in the unit cell and the basis set consists of one p_z-orbital per atom. The band structures for $N = 8$, $N = 9$ and $N = 10$ armchair graphene nanoribbons are shown in Fig. 3.26. One obtains semimetallic band structure for $N = 8$ and semiconducting band structure for the rest. High lying bands (i.e. bands above the conduction band) and low lying bands (i.e. bands below the valence band) are called sub bands, which are shown as well.

Two Dimensional Materials

For the two dimensional structure, the key difference from the one dimensional materials is that the wavevector $\boldsymbol{k} = (k_x, k_y)$ is now two dimensional as well as all the unit vectors, which make the displacement vector $(\boldsymbol{d_m} - \boldsymbol{d_n})$ two dimensional. The phase factor $i\boldsymbol{k} \cdot (\boldsymbol{d_m} - \boldsymbol{d_n})$ between two unit cells is the dot product of the vectors in two dimensions now.

Let us consider the graphene crystal shown in Fig. 3.27a, which has already been discussed in Chap. 2. Choosing two atoms (labeled A and B) as the unit cell, the real space lattice is shown in Fig. 3.27b, for which the unit vectors are given in (2.2). The corresponding reciprocal space lattice is shown in Fig. 3.27c. With this choice of the unit cell and the choice of the p_z-orbital basis set, the Hamiltonian for the atomic labeling shown in Fig. 3.27d is given as,

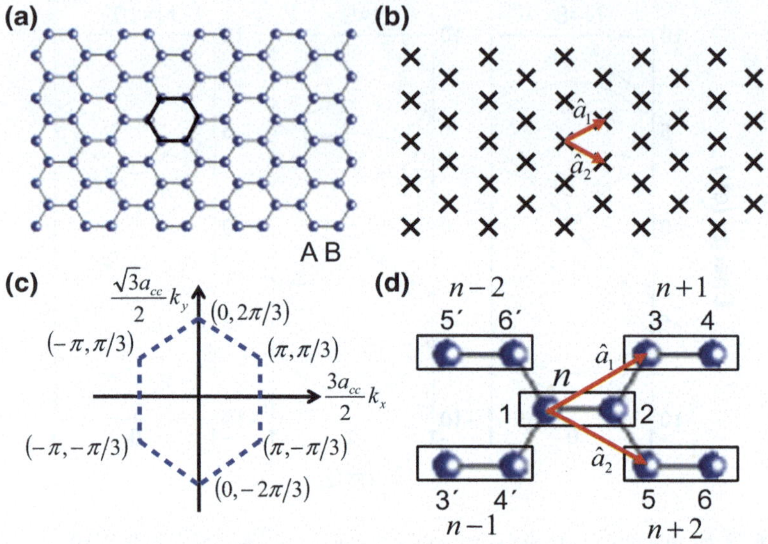

Fig. 3.27 Graphene. **a** Crystal. **b** Real space lattice. **c** Reciprocal space and Brillouin zone. **d** Nearest neighbor unit cells

$$[H(\boldsymbol{r})] = \left[\begin{array}{cccc|cc|cccc} 0 & -t & 0 & 0 & 0 & 0 & 0 & 0 & 0 & 0 \\ -t & 0 & 0 & 0 & -t & 0 & 0 & 0 & 0 & 0 \\ \hline 0 & 0 & 0 & -t & 0 & 0 & 0 & 0 & 0 & 0 \\ 0 & 0 & -t & 0 & -t & 0 & 0 & 0 & 0 & 0 \\ \hline 0 & -t & 0 & -t & 0 & -t & 0 & 0 & 0 & 0 \\ 0 & 0 & 0 & 0 & -t & 0 & -t & 0 & -t & 0 \\ \hline 0 & 0 & 0 & 0 & 0 & -t & 0 & -t & 0 & 0 \\ 0 & 0 & 0 & 0 & 0 & 0 & -t & 0 & 0 & 0 \\ \hline 0 & 0 & 0 & 0 & 0 & -t & 0 & 0 & 0 & -t \\ 0 & 0 & 0 & 0 & 0 & 0 & 0 & 0 & -t & 0 \end{array}\right] = \left[\begin{array}{c|c|c} \cdots & H_{n-2,n} & \cdots \\ \hline \cdots & H_{n-1,n} & \cdots \\ \hline \cdots & H_{nn} & \cdots \\ \hline \cdots & H_{n+1,n} & \cdots \\ \hline \cdots & H_{n+2,n} & \cdots \end{array}\right]$$

For $m = n$, the real space Hamiltonian matrix for the unit cell n is given as,

$$[H(\boldsymbol{r})]_{nn} = \begin{bmatrix} 0 & -t \\ -t & 0 \end{bmatrix}$$

The corresponding displacement and phase factor respectively are given as, $(\boldsymbol{d}_n - \boldsymbol{d}_n) = \boldsymbol{0}$ and $i\boldsymbol{k} \cdot (\boldsymbol{d}_n - \boldsymbol{d}_n) = \boldsymbol{0}$, where $\boldsymbol{k} = k_x\hat{x} + k_y\hat{y}$.

For $m = n + 1$,

$$[H(\boldsymbol{r})]_{n+1,n} = \begin{bmatrix} 0 & -t \\ 0 & 0 \end{bmatrix}$$

The corresponding displacement and phase factor respectively are given as, $\boldsymbol{d}_{n+1} - \boldsymbol{d}_n = \hat{a}_1 = \frac{3a_{cc}}{2}\hat{x} + \frac{\sqrt{3}a_{cc}}{2}\hat{y}$ and $i\boldsymbol{k}\cdot(\boldsymbol{d}_{n+1} - \boldsymbol{d}_n) = i\boldsymbol{k}\cdot\hat{a}_1 = \frac{3a_{cc}}{2}k_x + \frac{\sqrt{3}a_{cc}}{2}k_y$.

For $m = n + 2$,

$$[H(\boldsymbol{r})]_{n+2,n} = \begin{bmatrix} 0 & -t \\ 0 & 0 \end{bmatrix}$$

The corresponding displacement and phase factor respectively are given as, $\boldsymbol{d}_{n+2} - \boldsymbol{d}_n = \hat{a}_2 = \frac{3a_{cc}}{2}\hat{x} - \frac{\sqrt{3}a_{cc}}{2}\hat{y}$ and $i\boldsymbol{k}\cdot(\boldsymbol{d}_{n+2} - \boldsymbol{d}_n) = i\boldsymbol{k}\cdot\hat{a}_2 = \frac{3a_{cc}}{2}k_x - \frac{\sqrt{3}a_{cc}}{2}k_y$.

For $m = n - 1$,

$$[H(\boldsymbol{r})]_{n-1,n} = \begin{bmatrix} 0 & 0 \\ -t & 0 \end{bmatrix}$$

The corresponding displacement and phase factor respectively are given as, $\boldsymbol{d}_{n-1} - \boldsymbol{d}_n = -\hat{a}_1 = -\left(\frac{3a_{cc}}{2}\hat{x} + \frac{\sqrt{3}a_{cc}}{2}\hat{y}\right)$ and $i\boldsymbol{k}\cdot(\boldsymbol{d}_{n-1} - \boldsymbol{d}_n) = -i\boldsymbol{k}\cdot\hat{a}_1 = -\left(\frac{3a_{cc}}{2}k_x + \frac{\sqrt{3}a_{cc}}{2}k_y\right)$.

For $m = n - 2$,

$$[H(\boldsymbol{r})]_{n-2,n} = \begin{bmatrix} 0 & 0 \\ -t & 0 \end{bmatrix}$$

The corresponding displacement and phase factor respectively are given as, $\boldsymbol{d}_{n-2} - \boldsymbol{d}_n = -\hat{a}_2 = -\left(\frac{3a_{cc}}{2}\hat{x} - \frac{\sqrt{3}a_{cc}}{2}\hat{y}\right)$ and $i\boldsymbol{k}\cdot(\boldsymbol{d}_{n-2} - \boldsymbol{d}_n) = -i\boldsymbol{k}\cdot\hat{a}_2 = -\left(\frac{3a_{cc}}{2}k_x - \frac{\sqrt{3}a_{cc}}{2}k_y\right)$. By using (3.13), the Hamiltonian is given as,

$$[H(\boldsymbol{k})] = \begin{bmatrix} 0 & -t \\ -t & 0 \end{bmatrix} + \begin{bmatrix} 0 & -t \\ 0 & 0 \end{bmatrix} e^{i\boldsymbol{k}\cdot\boldsymbol{a}_1} + \begin{bmatrix} 0 & -t \\ 0 & 0 \end{bmatrix} e^{i\boldsymbol{k}\cdot\boldsymbol{a}_2} + \begin{bmatrix} 0 & 0 \\ -t & 0 \end{bmatrix} e^{-i\boldsymbol{k}\cdot\boldsymbol{a}_1} + \begin{bmatrix} 0 & 0 \\ -t & 0 \end{bmatrix} e^{-i\boldsymbol{k}\cdot\boldsymbol{a}_2}$$

$$= \begin{bmatrix} 0 & -t - te^{i\boldsymbol{k}\cdot\boldsymbol{a}_1} - te^{i\boldsymbol{k}\cdot\boldsymbol{a}_2} \\ -t - te^{-i\boldsymbol{k}\cdot\boldsymbol{a}_1} - te^{-i\boldsymbol{k}\cdot\boldsymbol{a}_2} & 0 \end{bmatrix}$$

Thus, the band structure for the two dimensional graphene is given as,

$$E(k_x, k_y) = \pm\left|-t - te^{i\boldsymbol{k}\cdot\boldsymbol{a}_1} - te^{i\boldsymbol{k}\cdot\boldsymbol{a}_2}\right|$$

$$= \pm t\sqrt{\left[1 + 2\cos\left(\frac{3a_{cc}}{2}k_x\right)\cos\left(\frac{\sqrt{3}a_{cc}}{2}k_y\right)\right]^2 + \left[2\sin\left(\frac{3a_{cc}}{2}k_x\right)\cos\left(\frac{\sqrt{3}a_{cc}}{2}k_y\right)\right]^2}$$

$$= \pm t\sqrt{1 + 4\cos\left(\frac{3a_{cc}}{2}k_x\right)\cos\left(\frac{\sqrt{3}a_{cc}}{2}k_y\right) + 4\cos^2\left(\frac{\sqrt{3}a_{cc}}{2}k_y\right)}$$

which gives two bands as shown in Fig. 3.28. The dispersion for each band has a bandwidth of $3t$. It may also be shown that there are six combinations of (k_x, k_y), for

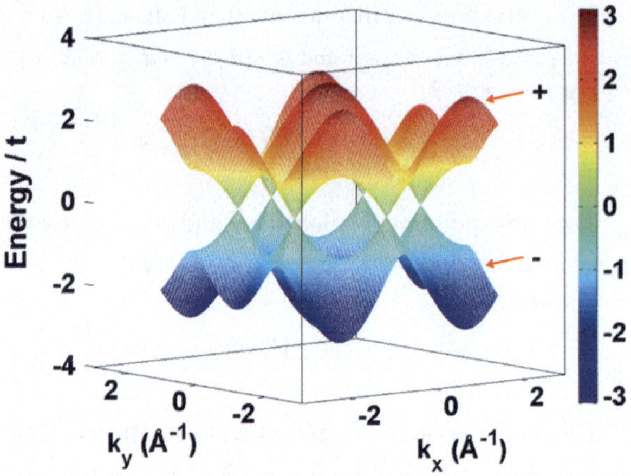

Fig. 3.28 Graphene band structure

which the conduction band and the valence band meet each other, which results in zero bandgap. These points are given as follows and also highlighted in Fig. 3.27c,

$$\left(\frac{3a_{cc}}{2}k_x, \frac{\sqrt{3}a_{cc}}{2}k_y\right) = \left(0, \frac{2\pi}{3}\right), \left(0, \frac{-2\pi}{3}\right), \left(\pi, \frac{\pi}{3}\right), \left(\pi, \frac{-\pi}{3}\right), \left(-\pi, \frac{\pi}{3}\right), \left(-\pi, \frac{-\pi}{3}\right)$$

It may appear that there are six points at which the bandgap is zero in the Brillouin zone. However only one third of each point is within the first Brillouin zone. Hence there are really two points at which the conduction band and the valence band meet to give zero bandgap. These two points are called the Dirac points and are related by the time reversal symmetry.

Apart from this, the band structure around these Dirac points is linear, which may be approximated by,

$$E(\boldsymbol{k}) = \hbar v\,|\boldsymbol{k}| \tag{3.14}$$

where v is the velocity of charge carriers in graphene, which is about 10^6 m/s≈ 1/300th the speed of light in free space. This linear dispersion is quite different from the parabolic dispersion which we have discussed so far, and it gives rise to novel photonic, magnetic, electronic and transport characteristics.

Graphene to Ribbons, Tubes and Dots

Let us do a thought experiment by taking the two dimensional graphene and cutting it along x-axis (dash-dotted line) into a 1D armchair graphene nanoribbon as shown in Fig. 3.29, which may be either rolled to form nanotubes or further cut along the y-axis to form nanodots. Similarly, one obtains 1D zigzag graphene nanoribbon by cutting the two dimensional graphene along y-axis (dashed line).

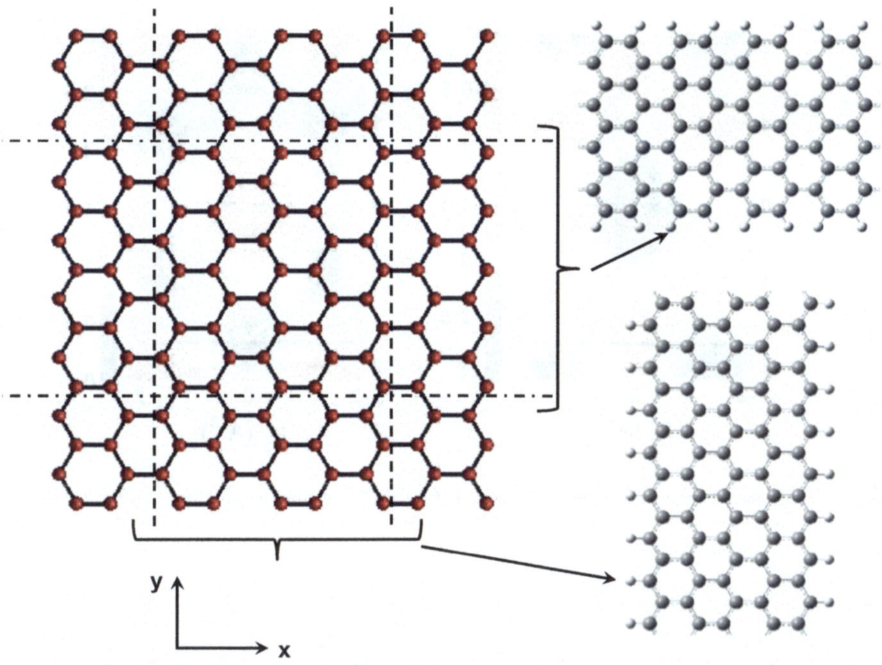

Fig. 3.29 Quantization of graphene band structure. The cut along x-axis (dash-dotted line) and y-axis (dashed line) give armchair and zigzag graphene nanoribbons, respectively. The nanoribbon edges are hydrogen passivated for clarity

Since the physical structure is constrained in y-direction for the armchair ribbon, the wavevector in this direction becomes quantized according to the following quantization condition,

$$k_y^n = \frac{n\pi}{W}$$

where n is the sub band index, and W is the width of the armchair nanoribbon, given as,

$$W = (N+1)\,\frac{\sqrt{3}a_{cc}}{2}$$

where N is at the atomic width of the nanoribbon. With the quantization of k_y^n, the two dimensional band structure of graphene becomes a one dimensional band structure, which is schematically shown in Fig. 3.30a, b for the valence band and the conduction band, respectively.

The resulting 1D bandstructure is obtained by substituting k_y^n in (3.14), and is given as,

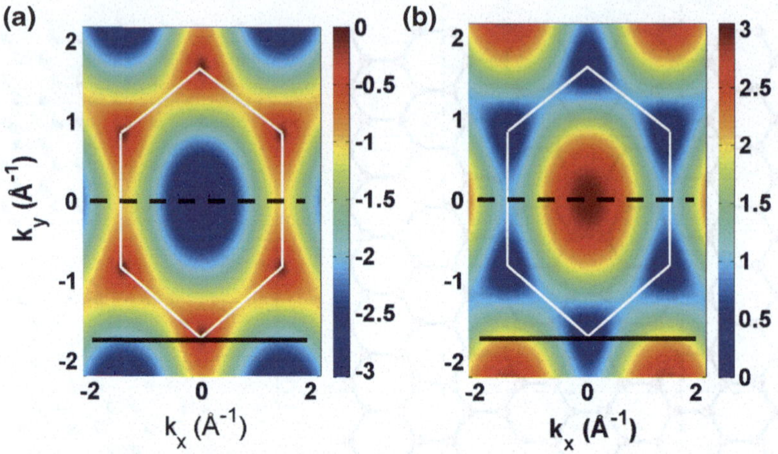

Fig. 3.30 Quantization of 2D band structure. The crossing of quantized wavevector k_y^n with one of the Dirac point gives zero bandgap shown by a solid line, otherwise a bandgap is observed for **a** valence band, and **b** conduction band

$$E(k_x, k_y^n) == \pm t\sqrt{1 + 4\cos\left(\frac{3a_{cc}}{2}k_x\right)\cos\left(\frac{n\pi}{N+1}\right) + 4\cos^2\left(\frac{n\pi}{N+1}\right)}$$

Consider $N = 8$, which gives a semimetallic band structure as shown in Fig. 3.26, and is given as,

$$\begin{aligned} E(k_x, k_y^6) &= \pm t\sqrt{1 + 4\cos\left(\frac{3a_{cc}}{2}k_x\right)\cos\left(\frac{2\pi}{3}\right) + 4\cos^2\left(\frac{2\pi}{3}\right)} \\ &= \pm\sqrt{2}t\sqrt{1 - \cos\left(\frac{3a_{cc}}{2}k_x\right)} \end{aligned}$$

Clearly at $k_x = 0$, the valence and conduction band edges are at zero and hence the bandgap is zero. The band structure obtained by using the above relationship is shown in Fig. 3.31, which matches the conduction and valence band structure in Fig. 3.26 for N = 8.

Apart from this, if one expands the cosine function by using Taylor series expansion around $k_x = 0$ and ignore the higher order terms, one obtains,

$$E(k_x, k_y^n) = \pm t\,\frac{3a_{cc}}{2}k_x$$

which shows a linear dispersion. Comparing with $\hbar v_{gx}k_x$, one obtains group velocity as,

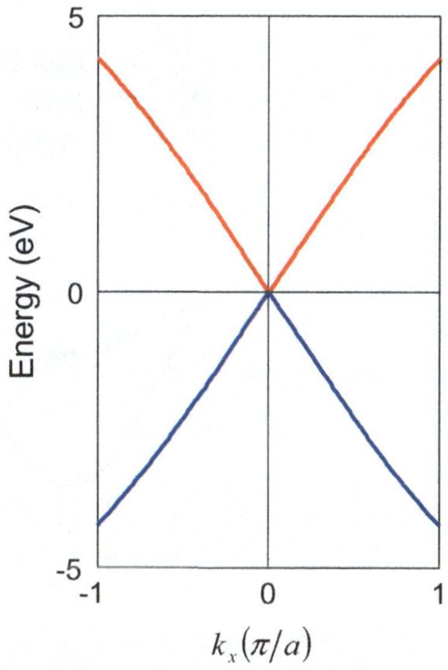

Fig. 3.31 Mapping 2D graphene bands on 1D armchair graphene nanoribbon. $N = 8$ armchair graphene nanoribbon's conduction and valence bands look the same as that of Fig. 3.26

$$v_{gx} = \frac{3a_{cc}}{2}\frac{t}{\hbar}$$

Similarly, it may be shown that for $N = 9, 10$, the armchair nanoribbon's quantized k_y wavevectors do not cross one of the two Dirac points and hence a bandgap is observed. Statistically, one third armchair graphene nanoribbons are semimetallic.

Three Dimensional Materials

In this section, we discuss three dimensional materials. Since the wavevectors have three components now i.e. $\boldsymbol{k} = (k_x, k_y, k_z)$, and the plotting becomes little cumbersome. The band structure is therefore plotted along certain symmetry directions only in the Brillouin zone (e.g. $\Gamma - X$, $\Gamma - K$, etc.). The general procedure for calculating the band structure however remains the same as discussed above. The 3D band structure may also be projected onto 2D or even 1D band structure.

We show two examples of the 3D band structure in this section, namely magnesium oxide and spin polarized iron. The band structure for magnesium oxide (with cubic lattice) is shown in Fig. 3.32 along various symmetry directions. By using LDA-DFT (in SIESTA program), one obtains a bandgap of 5.2 eV. A semi-empirical theory like EHT (extended Hückel theory) can be benchmarked with LDA-DFT. However, the experimental bandgap of magnesium oxide is 7.8 eV. In order to obtain this value, the transferable parameters of EHT can be optimized to match the experimental bandgap as shown by the dotted line. For semiconductors and insulators, getting the right bandgap is indeed important. Getting the correct experimental bandgap in

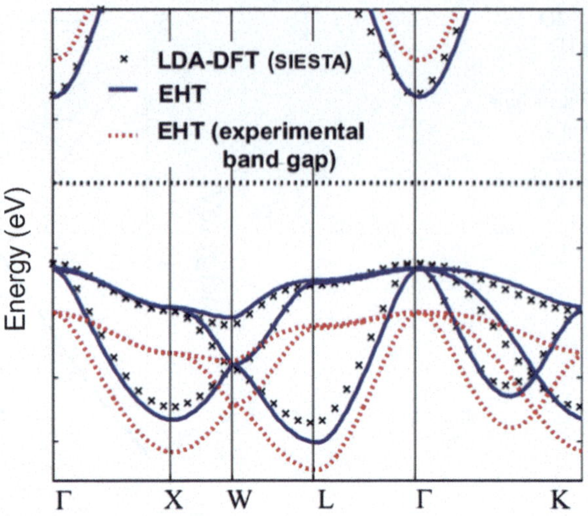

Fig. 3.32 MgO band structure along various symmetry directions [2]

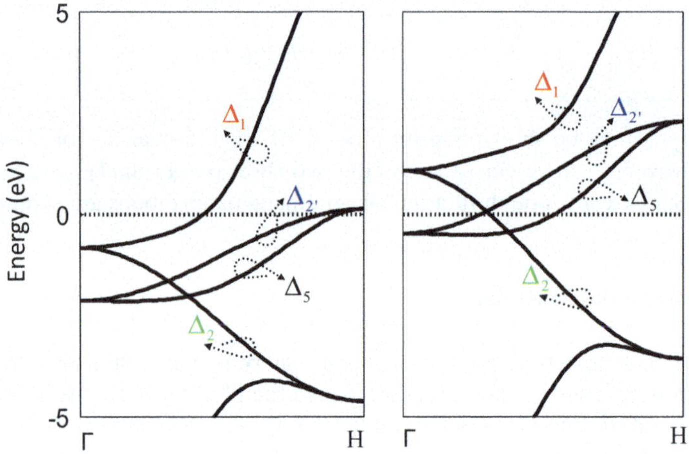

Fig. 3.33 Fe band structure along $\Gamma - H$ symmetry direction for the two spin orientations [3]

various theoretical methods like EHT, DFT (and its various approximations), tight binding theories, is an active area of research.

The band structure of ferromagnetic iron (with BCC lattice) is shown in Fig. 3.33a, b for the two spin orientations in $\Gamma - H$ symmetry direction, which is [100]. One should note that the two band structures look alike except for a constant energy shift, which is called exchange split. This leads to different concentration of the two spin orientations in ferromagnetic materials, leading to magnetism. Furthermore, the four bands exhibit orbital symmetries and are labelled as Δ_1, Δ_5, Δ_2, and $\Delta_{2'}$. We discuss these orbital symmetry effects in Chap. 6.

Problems

3.1 Using analytical solution, plot wavefunction probabilities, $|\phi(x)|^2$ for particle in a box with $n = 1$. You may use any computational software. $L = 1$ nm.

3.2 Using analytical solution, plot wavefunction probabilities, $|\phi(x)|^2$ for particle in a box with $n = 2$. You may use any computational software. $L = 1$ nm.

3.3 Using analytical solution, plot wavefunction probabilities, $|\phi(x)|^2$ for particle in a box with $n = 3$. You may use any computational software. $L = 1$ nm.

3.4 Take $a = 1$ Å with $N = 201$, where $N = 51$ corresponds to $x = 0$ and $N = 151$ to $x =$ L. With boundary conditions of $U_o = 100$ eV at the edges of box for which $N \leq 51$ and $N \geq 151$ have 100 eV potential, plot $|\phi(x)|^2$ for $n = 1$ using finite difference. Take mass equal to 3.8079 times that of the free electron mass. Comment on the tunneling into the barrier. Remember to convert the units from J to eV.

3.5 Take $a = 1$ Å with $N = 201$, where $N = 51$ corresponds to $x = 0$ and $N = 151$ to $x = L$. With boundary conditions of $U_o = 0.01$ eV at the edges of box for which $N \leq 51$ and $N \geq 151$ have 0.01 eV potential, plot $|\phi(x)|^2$ for $n = 1$ using finite difference. Take mass equal to 3.8079 times the free electron mass. Comment on the tunneling into the barrier.

3.6 For finite difference method, calculate group velocity and comment if it depends on the lattice constant a. If it does, explain why?

3.7 For finite element method. (a) Write down the equations for the basis set (ϕ_i,ϕ_j) shown in Fig. 3.12 as piecewise linear functions. (b) Calculate S_{ii}. (c) Calculate $S_{ij} = S_{ij}$. (d) Calculate H_{ii}. (e) Calculate $H_{ij} = H_{ij}$.

3.8 For a 1D linear crystal with atomic distance a, calculate $E(k)$ for (a) primitive unit cell, (b) unit cell containing two atoms, (c) unit cell containing three atoms, and (d) unit cell containing four atoms.

3.9 Using principle of Brillouin zone folding and mini band(s), connect the band structure obtained in problem 3.8a to the band structures obtained in problems 3.8b, c, d.

3.10 For an armchair graphene nanoribbon with atomic width of $N = 2$, calculate and plot $E(k)$. $t = -3$ eV.

3.11 For an armchair graphene nanoribbon with atomic width of $N = 3$, calculate and plot $E(k)$. $t = -3$ eV.

3.12 For an armchair graphene nanoribbon with atomic width of $N = 4$, calculate and plot $E(k)$. $t = -3$ eV.

3.13 For a zigzag graphene nanoribbon with atomic width of $N = 2$, calculate and plot $E(k)$. $t = -3$ eV.

3.14 For a zigzag graphene nanoribbon with atomic width of $N = 4$, calculate and plot $E(k)$. $t = -3$ eV.

3.15 For a zigzag graphene nanoribbon with atomic width of $N = 6$, calculate and plot $E(k)$. $t = -3$ eV.

3.16 Using the analytical $E(k)$ dispersion of graphene, draw a two dimensional $E(k)$ plot for graphene. Comments on how many points have zero bandgap in the Brillouin zone. $t = -3$ eV.

3.17 By projecting graphene's bandstructure on 1D for armchair graphene nanoribbons, analytically prove that one third nanoribbons are semimetallic.

Research Assignment

R3.1 One obtains a bilayer graphene by stacking two graphene monolayers. Write a one page summary of how many different ways a bilayer graphene may be stacked. Comment on their band structures. Also, look into stacking misalignment.

References

1. L.C. Oetting et al., J. Nano Educ. **8**, 47–51 (2016), arXiv:1508.03305
2. T.Z. Raza et al., J. Appl. Phys. **109**, 023–705 (2011)
3. T.Z. Raza et al., IEEE Trans. Nanotechnol. **10**, 237 (2011)

Part II
Device Characteristics and Analysis

Chapter 4
Quantum Transport

Quantum transport is broadly divided into two regimes, namely *Coherent* and *Incoherent*. Coherent transport is the norm in the absence of any scattering process, where the phase of the wavefunction is preserved. Additionally, coherent transport is always an elastic process[1] due to the absence of any scattering events in the channel region. However, if the electron interacts with some other degree of freedom within the channel, the phase of the wavefunction may be relaxed, a process called *Dephasing* or *Incoherent Scattering*.

Dephasing may occur in an elastic or an inelastic[2] scattering event, resulting in *Incoherent transport* with elastic or inelastic dephasing, respectively. In other words, if no energy is exchanged but the phase is lost, this process leads to elastic dephasing. On the other hand, if energy is exchanged during the scattering process within the channel region alongwith phase relaxation, the scattering process is called *Inelastic Dephasing*. If the energy exchange is small, one may still approximate the scattering process as *Elastic Dephasing*. Apart from this, if the momentum of the particle is exchanged with the environment without the phase relaxation, the process is called *Momentum Relaxation*.

In this chapter, we start the discussion with coherent transport for the device structure shown in Fig. 4.1. Perhaps the most important aspect of this chapter compared to the last three is that we study nonequilibrium quantum transport by solving Schrödinger equation with *open boundary conditions* in the form of contacts. Such contacts may inject electrons into the channel region and take them out, leading to current flow under *far from equilibrium* conditions. However, in some conditions, the transport may be approximated to be *close to equilibrium* condition for low bias, where one may calculate *zero bias conductance* or *low bias conductance*. In *close to equilibrium* conditions, one may very conveniently use the wavefunction approach.

---

[1] An elastic process does not involve any energy exchange.

[2] An inelastic process involves energy exchange within the law of the conservation of energy.

H. Raza, *Nanoelectronics Fundamentals*, NanoScience and Technology,
https://doi.org/10.1007/978-3-030-32573-2_4

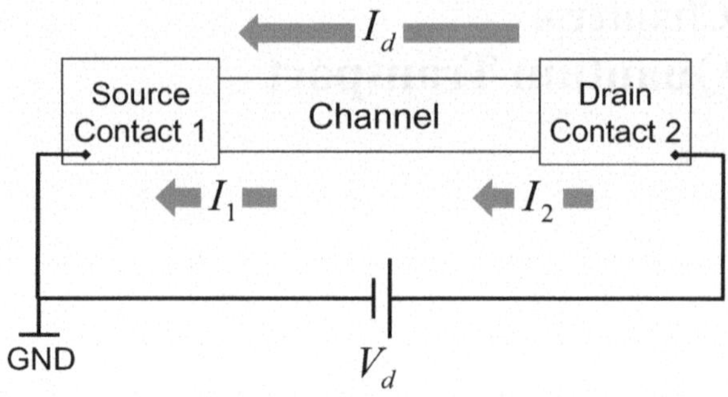

Fig. 4.1 Device schematics. Source and drain are labeled as contact 1 and contact 2, respectively for convenience. $I_d = I_1 = I_2$

However, this approach becomes difficult to keep track of in *far from equilibrium* situations, in particular the ones involving scattering. In this case, the use of *Green's function* is favored, in particular the method called *NEGF (nonequilibrium Green's function)* formalism [1, 2].

4.1 Wavefunction Approach

Starting from the 1D Schrödinger equation, the probability current[3] is given as,

$$I_d = -q\frac{i\hbar}{2m}\left(\Psi\frac{\mathrm{d}\Psi^*}{\mathrm{d}x} - \Psi^*\frac{\mathrm{d}\Psi}{\mathrm{d}x}\right) \tag{4.1}$$

which is really a definition of charge flow per unit time. If one assumes matched contacts with no reflections at the channel-contact boundary, an electron coherently entering the channel from source (contact 1) simply travels to the drain (contact 2) without any reflections, in which case, one may approximate the wavefunction to be a traveling wave. For a channel of length L_x, the normalized wavefunction for a particle in a box traveling in the positive direction is given as, $\Psi(x) = \sqrt{1/L_x}e^{ik_x x}$. Substituting in (4.1), the probability current for a certain value of k_x is then given as,

$$\tilde{I}_d(k_x) = -\frac{q}{L_x}\frac{\hbar k_x}{m}$$

For a parabolic band in 1D, with $E = \hbar^2 k_x^2/2m$, one obtains the group velocity[4] of $v_{gx} = \hbar k_x/m$, which is shown in Fig. 4.2. The probability current in terms of group velocity for a certain value of k_x thus becomes,

[3]Negative sign is due to the conventional current. Electron current flows in opposite direction to that of the conventional current.

[4]The group velocity is defined as, $v_{gx} = \mathrm{d}\omega/\mathrm{d}k_x = \mathrm{d}E/\hbar\mathrm{d}k_x$.

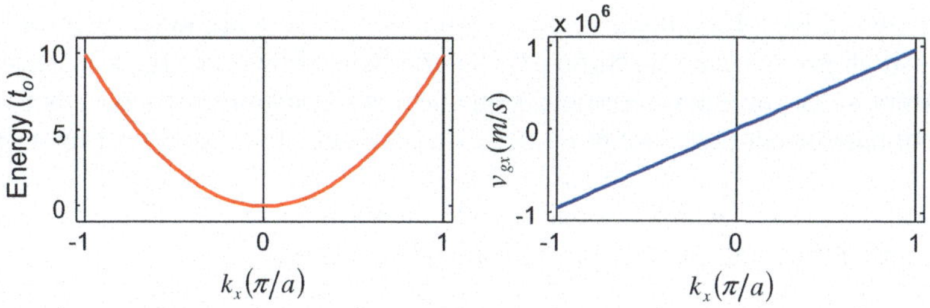

Fig. 4.2 Group velocity for a 1D parabolic band

$$\tilde{I_d}(k_x) = -\frac{q}{L_x} v_{gx} = -\frac{q}{L_x / v_{gx}} \tag{4.2}$$

For an electron traveling in the positive direction (with positive sign of the group velocity), the conventional current has a negative sign and vice versa. Furthermore, L_x / v_{gx} gives the channel transit time with [s] dimension and therefore (4.2) is really the basic definition of current flow of charge per unit time.

The total current over all occupied states k_x is given as,

$$I_d = \sum_{\text{occupied } k_x} \tilde{I_d}(k_x) = -\frac{q}{L_x} \sum_{\text{occupied } k_x} v_{gx}$$

Using the above equation, the current associated with an empty band is zero, since no states are occupied and hence no velocity contribution. What is not so straight forward is that even for a completely filled parabolic band as shown in Fig. 4.3a, the net current is zero because there are equal number of electron states with positive group velocity and electron states with negative group velocity. Whereas, for the

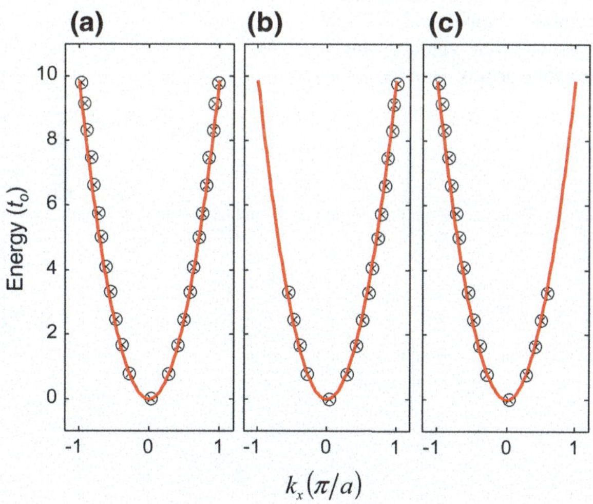

Fig. 4.3 Equilibrium and nonequilibrium conditions. **a** Filled band under equilibrium condition. **b** Partially filled band. Net electron flow is in positive x-direction. **c** Partially filled band. Net electron flow is in negative x-direction

band shown in Fig. 4.3b, there is a net current since there are more electron states with positive group velocity. Similarly, for the band shown in Fig. 4.3c, there is a net current since there are more electron states with negative group velocity. The conventional current therefore has a negative and a positive value for Fig. 4.3b and Fig. 4.3c, respectively. It is also important to note that the unoccupied states are called the hole states, which basically represent voids or absence of electrons.

The probability current may be generalized as follows,

$$I_d = -\frac{q}{L_x}\int \frac{\mathrm{d}k_x}{2\pi/L_x}\, v_{gx} = -\frac{q}{2\pi}\int \mathrm{d}k_x\, \frac{1}{\hbar}\frac{\mathrm{d}E}{\mathrm{d}k_x} = -\frac{q}{h}\int \mathrm{d}E$$

Putting the integration limits over the accessible energy range,

$$I_d = -\frac{q}{h}\int_{E_1}^{E_2} \mathrm{d}E = \frac{q}{h}(E_1 - E_2) = \frac{q}{h}\Delta E \tag{4.3}$$

where ΔE is the conduction energy window.

The energy difference, $\Delta E = E_2 - E_1$, is also called the energy conduction window. Without any external bias, i.e. under equilibrium condition, one may define the equilibrium chemical potential energy (μ_o) for the whole device (including the channel and the contact regions), which has an important role in filling the energy states. At absolute zero, all states below the chemical potential energy μ_o are filled, whereas all states above μ_o are empty as shown in Fig. 4.4 for a half filled band. Here again, the net current is zero due to the equal number of states with positive group velocity and states with negative group velocity, which is consistent with the earlier assertion of equilibrium situation.

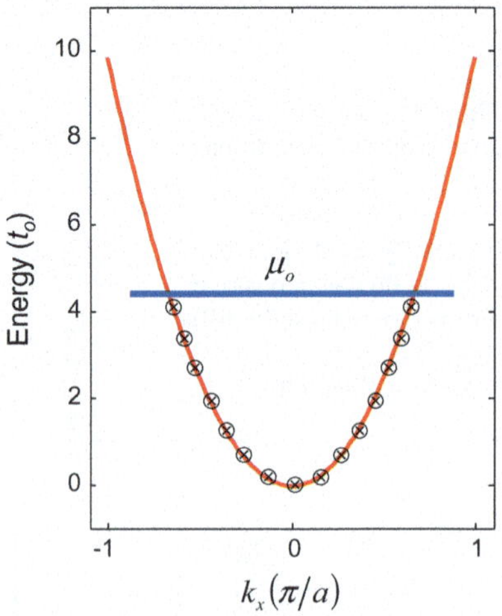

Fig. 4.4 A half filled band under equilibrium condition at absolute zero with equal number of states with positive group velocity and negative group velocity

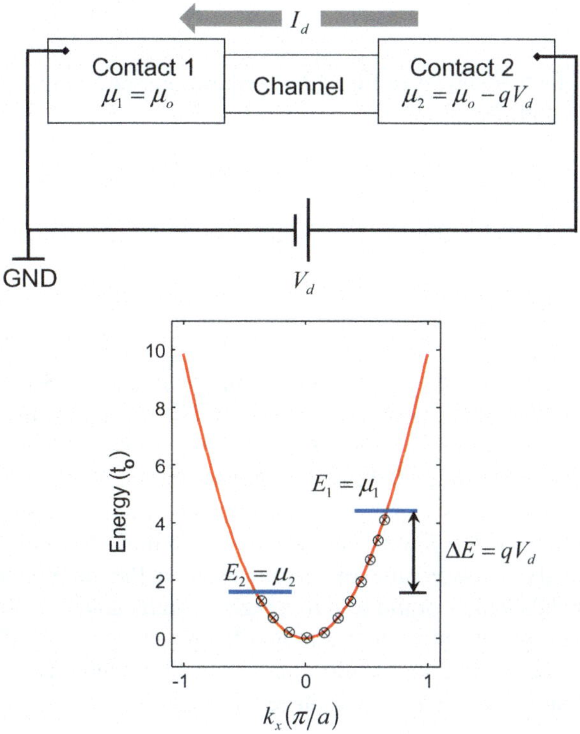

Fig. 4.5 A partially filled band at absolute zero with unequal number of states with positive group velocity and negative group velocity

However, under the situation where the source (contact 1) is grounded and drain (contact 2) has an applied bias V_d, the chemical potential energies of the two contacts are given as, $\mu_1 = \mu_o$ and $\mu_2 = \mu_o - qV_d$. Thus, a positive voltage lowers the chemical potential energy due to decreasing potential energy by qV_d and vice versa.

This condition is shown in Fig. 4.5, where the states with positive group velocity are kept filled by the source (contact 1) at $\mu_1 = \mu_o$ setting the upper limit of E_1, whereas the states with negative group velocity are emptied according to drain (contact 2) chemical potential energy $\mu_2 = \mu_o - qV_d$ due to a positive voltage V_d. The current thus becomes,

$$I_d = \frac{q}{h} \int_{\mu_2}^{\mu_1} \mathrm{d}E = \frac{q}{h} (\mu_1 - \mu_2) \tag{4.4}$$

At absolute zero, substituting the values of μ_1 and μ_2, one obtains,

$$I_d = \frac{q^2}{h} V_d$$

Taking into account the spin degeneracy of 2, the current is given as,

$$I_d = \frac{2q^2}{h} V_d$$

which is shown in Fig. 4.6. Furthermore, one may obtain the conductance per spin as follows,

$$G = \frac{q^2}{h}$$

Including the spin degeneracy of 2, the conductance is given as,

$$G = \frac{2q^2}{h}$$

which gives the conductance value of about 77 μS per state, or the resistance value of 12.9 kΩ per state. Since this quantity depends only on the universal constants of electronic charge ($q = 1.6 \times 10^{-19}$ C) and Planck's constant ($h = 6.626 \times 10^{-34}$ Js), $2q^2/h$ is also called the *quantum of conductance* and gives the maximum value of current that one may have through a single state or a single band in 1D under coherent transport. However, a device may have imperfect contacts or scattering, which may lead to a lower value of conductance. On the other hand, a device may have multiple bands which would lead to a higher conductance. To incorporate these effects at low bias or zero bias, the quantum of conductance is usually multiplied by transmission $T(\mu_o)$ around the equilibrium chemical potential energy (μ_o), giving rise to the low bias or zero bias conductance (G_o) as follows,

$$G_o = \frac{2q^2}{h} T(\mu_o)$$

Transmission may be thought of as the probability to transmit across the channel and has a range of [0,1] per band or state. Consider an incident wavefunction

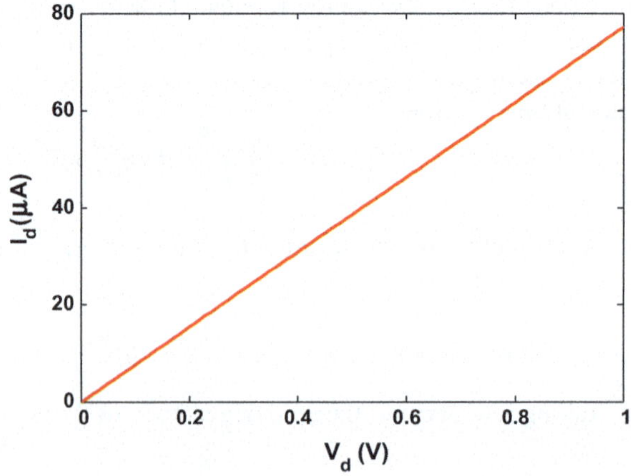

Fig. 4.6 IV characteristics for a single band including spin degeneracy showing the quantum of conductance

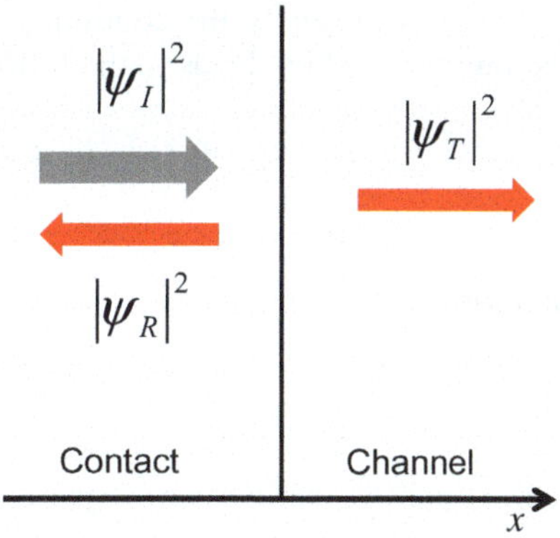

Fig. 4.7 Transmission and reflection picture of quantum transport

on a contact/channel interface as shown in Fig. 4.7. Part of this wavefunction is transmitted through this interface and the rest is reflected back. In this case, the dimensionless transmission is defined in terms of the probability densities of incident and transmitted wavefunctions as,

$$T = \frac{|\Psi_T|^2}{|\Psi_I|^2}$$

whereas, the dimensionless reflection is given as,

$$R = \frac{|\Psi_R|^2}{|\Psi_I|^2}$$

It is straight forward to show that $T + R = 1$. Furthermore, in terms of the traveling wave solutions, one may write (wavevector k_I for the incident state from the contact, wavevector k_R for the reflected wave in the contact, and wavevector k_T for the transmitted state in the channel),[5]

$$\Psi_I = Ae^{ik_I x}$$
$$\Psi_R = Be^{-ik_R x}$$
$$\Psi_T = Ce^{ik_T x}$$

which gives,

$$T = \frac{|C|^2}{|A|^2}$$
$$R = \frac{|B|^2}{|A|^2}$$

[5]Note that $k_I \equiv k_R$ being in the same material.

One has to satisfy the continuity of wavefunction and its first derivative across the interface, which leads to the following conditions,

$$\begin{aligned} \Psi_I(x=0) + \Psi_R(x=0) &= \Psi_T(x=0) \\ A + B &= C \end{aligned}$$

and for $x = 0$,

$$\begin{aligned} \frac{\mathrm{d}\Psi_I}{\mathrm{d}x} + \frac{\mathrm{d}\Psi_R}{\mathrm{d}x} &= \frac{\mathrm{d}\Psi_T}{\mathrm{d}x} \\ A - B &= \frac{k_T}{k_I} C \end{aligned}$$

since $k_I \equiv k_R$.

In general, there may be some difference in the media of the contact and the channel to have different wavevectors. For a special case when $k_I = k_T$, one may show that $B = 0$, i.e. there is no reflection. This leads to reflectionless contacts with an ideal transmission probability of unity at the interface, a very much desirable trait. However, one seldom finds such contacts and interfaces in real devices, and that is where one has to fully employ the techniques of quantum transport.

4.2 Landauer's Approach

At absolute zero, the current per spin (4.4) is given as,

$$I_d = \frac{q}{h} \int_{\mu_2}^{\mu_1} \mathrm{d}E$$

This equation has an explicit assumption that transmission is unity for the energy range between the two chemical potential energies ($\mu_{1,2}$) as discussed earlier.

In Landauer's approach towards the quantum transport, the energy dependent transmission $T(E)$ through the channel, shown in Fig. 4.8, may have an arbitrary value between zero and one per 1D band due to reflections at the contact-channel interface or other processes, which leads to,

$$I_d = \frac{q}{h} \int_{\mu_2}^{\mu_1} \mathrm{d}E \; T(E)$$

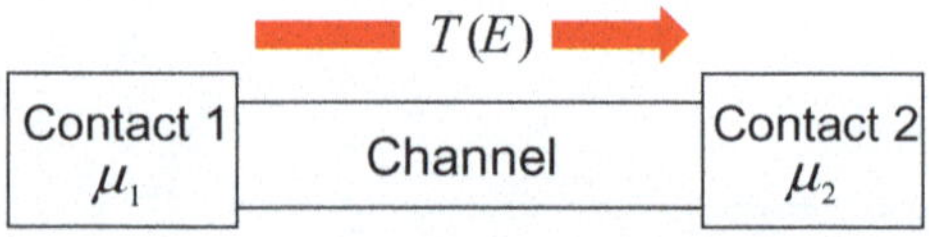

Fig. 4.8 Landauer's picture of quantum transport

Defining [consult next section and Fig. 4.10b],

$$f_1(E) - f_2(E) = \begin{cases} 1 & \mu_2 \leq E \leq \mu_1 \\ 0 & \mu_2 > E > \mu_1 \end{cases}$$

where $f_{1,2}$ are the Fermi's functions for contact 1 and contact 2 with the chemical potential energies of μ_1 and μ_2, respectively. The chemical potential energies are given as $\mu_1 = \mu_o$ and $\mu_2 = \mu_o - qV_d$ if contact 1 is connected to the ground, and voltage V_d is applied at contact 2. By using the above relation, instead of the integration limits as the chemical potential energies, one may generalize the above transport equation as follows,

$$I_d = \frac{q}{h}\int_{-\infty}^{+\infty} \mathrm{d}E\, T(E)\left[f_1(E) - f_2(E)\right]$$

Including the spin degeneracy of 2, one obtains,

$$I_d = \frac{2q}{h}\int_{-\infty}^{+\infty} \mathrm{d}E\, T(E)\left[f_1(E) - f_2(E)\right] \tag{4.5}$$

4.3 Fermi's Function

Fermi's function $f(E)$ is based on equilibrium statistical mechanics of spin-1/2 fermions, and it gives the probability of occupancy of a state by electrons. For equilibrium condition, given a certain temperature T (in K) and the equilibrium chemical potential energy (μ_o), the Fermi's function is given as follows,

$$f_o(E) = f(E, \mu_o) = \frac{1}{1 + e^{(E-\mu_o)/k_B T}}$$

where the Boltzmann constant is given as $k_B = 1.38065 \times 10^{-23}\,\mathrm{JK}^{-1} = 8.629 \times 10^{-5}\mathrm{eVK}^{-1}$. At $T = 300$ K, $k_B T = 25.8$ meV ≈ 25 meV and $1/k_B T \approx 40$ eV^{-1}. One should note that 1 eV $= 1.6 \times 10^{-19}$ J.

The Fermi's function is a Heaviside function at absolute zero, whereas at a finite temperature, it gets broadened as shown in Fig. 4.9a. This broadening is on the order of 6 $k_B T$ on a linear scale. The Fermi's function at room temperature is shown on a logarithmic scale in Fig. 4.9b. Due to the exponential dependence, the Fermi's function never really goes to zero for a well defined energy (except ∞) at a finite temperature. Furthermore, irrespective of the temperature, it may be shown that,

$$f\,(E = \mu_o) = f_o(E) = \frac{1}{2}$$

Fig. 4.9 Fermi's function for $\mu_o = 0$ on **a** linear scale, and **b** logarithmic scale

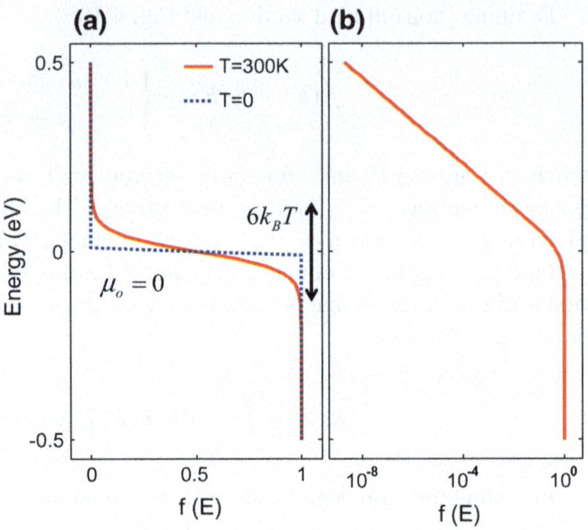

Assuming the two contacts to be in equilibrium, the Fermi's functions for the two contacts are given as,

$$f_{1,2} = \frac{1}{1 + e^{(E-\mu_{1,2})/k_B T}} \tag{4.6}$$

We next include broadening due to finite temperature and plot the Fermi's functions for contact 1 and contact 2 with an equilibrium value of $\mu_o = 0$ and an applied bias of $V_d = 0.2$ V in Fig. 4.10a. Clearly the Fermi's function of the second contact is shifted downwards by 0.2 eV due to positive voltage.

Fig. 4.10 Fermi's function. **a** At finite temperature for the two contacts with an applied bias of $V_d = 0.2$ V. Fermi's function difference $[f_1(E) - f_2(E)]$ on **b** linear scale, and **c** logarithmic scale. $\mu_o = 0$, which makes $\mu_1 = \mu_o = 0$ and $\mu_2 = \mu_o - qV_d = -0.2$ eV

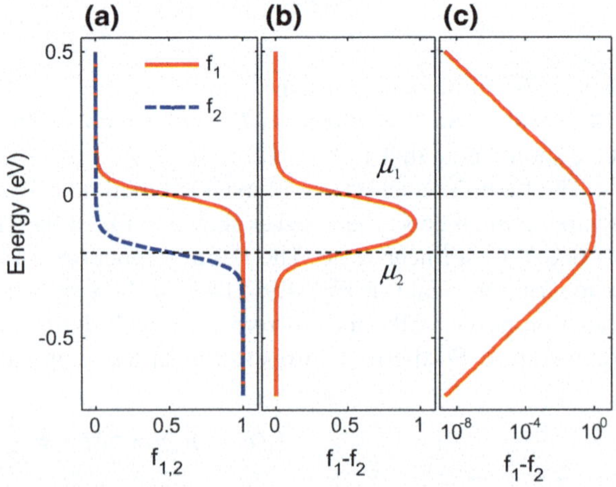

The difference of Fermi's functions $[f_1(E) - f_2(E)]$ is shown in Fig. 4.10b on a linear scale for $T = 300$ K. Within few $k_B T$ above μ_1 and few $k_B T$ below μ_2, the Fermi's function difference seems to diminish within $10k_B T$, which leads to the following limits of integration for a positive bias,

$$I_d = \frac{2q}{h} \int_{\mu_2 - 10k_B T}^{\mu_1 + 10k_B T} \mathrm{d}E \; T(E) \left[f_1(E) - f_2(E)\right]$$

Similarly for a negative bias, the limits of integration are given as,

$$I_d = \frac{2q}{h} \int_{\mu_1 - 10k_B T}^{\mu_2 + 10k_B T} \mathrm{d}E \; T(E) \left[f_1(E) - f_2(E)\right]$$

The above limits work well if within the conduction window (defined by the limits of the integral), transmission is higher or equal than the range outside the conduction window. If this is not the case, one has to increase the limits of integration to capture any peaks in the transmission, since the Fermi's function difference decays exponentially as shown in Fig. 4.10c.

Since the Fermi's function gives the probability of occupancy of a state by an electron, $[1 - f(E)]$ gives the probability of absence of an electron for a certain state or energy level. The absence or void of an electron is conveniently termed as a hole, and the probability of occupancy of a state by a hole for an equilibrium chemical potential (μ_o) is given as,

$$1 - f(E, \mu_o) = 1 - \frac{1}{1 + e^{(E-\mu_o)/k_B T}} = \frac{1}{1 + e^{-(E-\mu_o)/k_B T}}$$

It is worthwhile to emphasize that the Fermi's function may be used only under equilibrium situations, since it is derived by using the equilibrium Fermi-Dirac statistics for Fermions. For example, if microscale contacts are used for a nanoscale channel, the contacts may be assumed to remain under equilibrium, although the channel may very well be highly out of equilibrium. In such a situation, one may define chemical potential energies and hence Fermi's functions for the contacts only and not for the channel. This is, in fact, one of the founding assumptions in the Landauer's approach towards the quantum transport that the contacts remain under equilibrium to permit the use of equilibrium statistical mechanics in the form of Fermi's functions for the two contacts.

4.4 Quantum Mechanical Transmission

In this section, we further discuss the physical meaning of transmission and try to understand this quantity in terms of more fundamental concepts. Starting with the definition in terms of incident and transmitted probability densities, transmission is give as,

$$T = \frac{|\Psi_T|^2}{|\Psi_I|^2}$$

For a certain energy E within the energy window (ΔE), starting from (4.3) and after including transmission $T(E)$, current is given as,

$$I_d = \frac{q}{h} T(E) \Delta E$$

Rewriting above equation as current per unit energy, one obtains,

$$\frac{I_d}{\Delta E} = \frac{q}{h} T(E) \tag{4.7}$$

Within the basic definition of current, $I_d = v_{gx} q / L_x$, current per unit energy for a certain Eigen energy ϵ_n is given as,

$$\frac{I_d}{\Delta E} = \frac{q}{L_x} v_{gx}^n \delta\left(E - \epsilon_n\right) \tag{4.8}$$

where $\delta\left(E - \epsilon_n\right)$ is the Dirac's delta function with units of eV^{-1}. In the limit $\Delta E \rightarrow 0$, it is defined as follows,

$$\delta(E) = \begin{cases} 1/\Delta E & -\Delta E/2 \leq E \leq \Delta E/2 \\ 0 & \text{otherwise} \end{cases}$$

and shown in Fig. 4.11. Since the area under the function is unity, one obtains,

$$\int_{-\infty}^{+\infty} \mathrm{d}E \; \delta\left(E - \epsilon_n\right) = \int_{\epsilon_n - \Delta E/2}^{\epsilon_n + \Delta E/2} \mathrm{d}E \; \delta\left(E - \epsilon_n\right) = 1$$

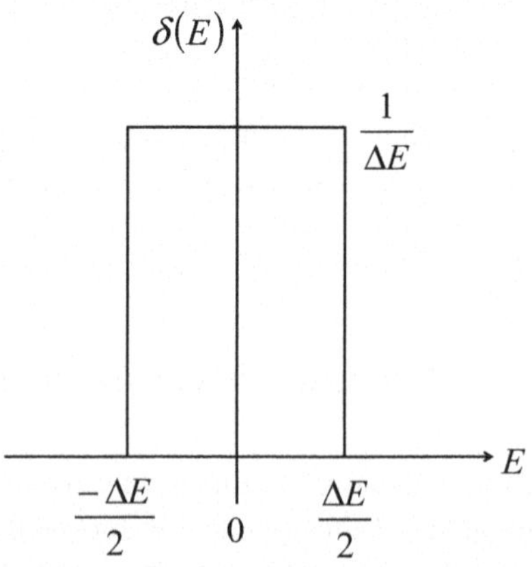

Fig. 4.11 Dirac's delta function in the limit $\Delta E \rightarrow 0$

In other words, $\delta(E-\epsilon_n)$ gives information about the presence of a state at energy ϵ_n, i.e. it is nonzero only when $E=\epsilon_n$. Comparing the right hand sides of (4.7) and (4.8), one obtains that,

$$T(\epsilon_n)=\frac{h}{L_x}v_{gx}^{n}\delta(E-\epsilon_n)$$

while summing over all the Eigen states gives the dimensionless energy resolved transmission function as follows,

$$T(E)=\frac{h}{L_x}\sum_n v_{gx}^{n}\delta(E-\epsilon_n)$$

The following quantity is called the density of states (DOS) of an individual state n and has the units of eV^{-1},

$$D^n(E)=\delta(E-\epsilon_n)$$

The transmission may thus be rewritten as,

$$T(E)=\frac{h}{L_x}\sum_n v_{gx}^{n}D^n(E) \tag{4.9}$$

The density of states gives information about the presence of states at a finite energy. For the state n, multiplying the density of states at a certain energy E with the group velocity v_{gx}^{n} for that particular state and h/L_x gives the transmission at that energy. One should note that transmission is always a positive quantity, since it is probability. Given that the density of states is also a positive quantity as it gives the number of states, the reference in the above equation is taken such that the velocity always has a positive sign.

Information about the states at a certain physical location may be written in the form of local density of states as follows,

$$LDOS(r,E)=\sum_n \Psi_n^*(r)\delta(E-\epsilon_n)\Psi_n(r)=\sum_n \delta(E-\epsilon_n)|\Psi_n(r)|^2 \tag{4.10}$$

which has units of $eV^{-1}m^{-1}$, $eV^{-1}m^{-2}$, and $eV^{-1}m^{-3}$ in 1D, 2D, and 3D, respectively.

4.5 Density of States

Irrespective of the dimensionality, the density of states is defined as,

$$D(E)=\sum_n \delta(E-\epsilon_n) \tag{4.11}$$

Although it gives information about the energy resolved states, it does not give any information whether these states are filled or not. This information has to come from the Fermi's function. The total number of electrons under equilibrium condition is given as follows,

$$N = \int_{-\infty}^{+\infty} \mathrm{d}E \, D(E) f(E)$$

Similarly, the total number of holes under equilibrium condition is given as,

$$P = \int_{-\infty}^{+\infty} \mathrm{d}E \, D(E) \left[1 - f(E)\right]$$

This computation is schematically shown in Fig. 4.12 for a certain density of states of a semiconductor, where the conduction and the valence bands are highlighted. Here it is assumed that the density of states for the two bands are symmetric (i.e. electron-hole symmetry is preserved). Since, the total number of electrons and holes are the same at equilibrium, the chemical potential energy (μ_o) is in the middle of the bandgap for an intrinsic semiconductor. However, electron-hole symmetry is broken in materials like silicon, i.e. the density of states for the conduction and valence bands are not the same due to nonidentical electronic structure of the two bands. Under these conditions, the intrinsic chemical potential energy is not at the midgap, rather slightly offset in order to achieve the equal number of electrons and holes—a necessary condition for intrinsic semiconductors at equilibrium.

For silicon (group IV element), the interstitial dopant atoms from group III (e.g. boron) lead to *missing electrons* or *surplus holes* and hence called *p*-type doping, whereas the interstitial dopant atoms from group V (e.g. phosphorus) lead to *surplus electrons* and hence called *n*-type doping. After contributing holes to the crystal, *p*-type dopant atoms become negatively charged acceptors. The *n*-type dopant atoms become positively charged donor atoms after contributing electrons to the crystal. Another example is that of GaAs, where Ga belongs to group III and As to group V. In this case, if Si atoms replace Ga atoms as interstitial dopants, the crystal has

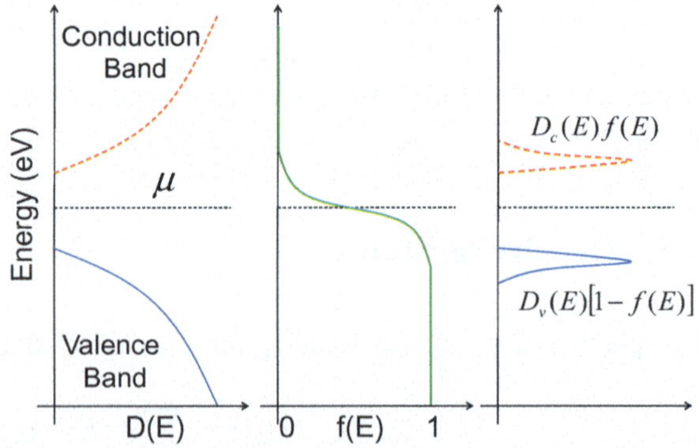

Fig. 4.12 Computing number of electrons and holes using DOS and the Fermi's function

surplus electrons that lead to *n*-type doping, whereas Si atoms replacing As atoms as interstitial dopants, the crystal has *surplus holes* leading to *p*-type doping. Dopants may be introduced either by surface diffusion process or ion implantation, discussion of both of these techniques is skipped in this chapter. However, in both processes, the foreign atoms in the crystal may lead to crystal distortion. For high end applications, yet another way to dope is neutron transmutation doping of semiconductors by using nuclear reactors. The process of nuclear transmutation changes Si atom to P, for example, which yields low-defect substrates resulting in superior device performance.

Furthermore, for *n*-type doping, the chemical potential energy is close to the conduction band, as there would be lot more electrons that holes. Similarly, for *p*-type doping, the chemical potential energy is close to the valence band, as there are more holes than electrons. Emphasizing this fact, for *n*-type and *p*-type semiconductors, the chemical potential energy is close to the conduction band and valence band, respectively.

Zero Dimensional Materials

For the 0D materials, the density of states per spin is simply given as,

$$D_{\mathrm{0D}}(E) = \sum_n \delta(E - \epsilon_n)$$

Including the spin degeneracy, the density of states is given as,

$$D_{\mathrm{0D}}(E) = 2\sum_n \delta(E - \epsilon_n)$$

which gives rise to a set of delta functions corresponding to the Eigen values ϵ_n. DOS for benzene is shown in Fig. 4.13, where the height of the delta function corresponds

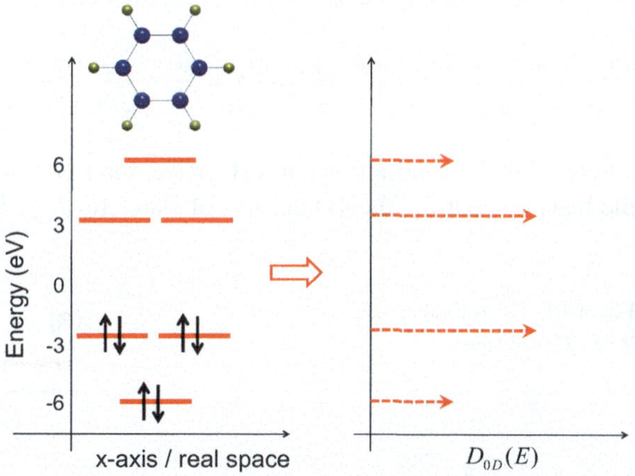

Fig. 4.13 Density of states for 0D benzene

to the degeneracy of the energy level. Since HOMO (with energy −3 eV) and LUMO (with energy +3 eV) are doubly degenerate in addition to the spin degeneracy, the DOS is double as compared to the low lying state (with energy −6 eV) and the high lying state (with energy +6 eV).

One Dimensional Materials

Consider a 1D material of length L_x as shown in Fig. 4.14a. We show 1D reciprocal space $\boldsymbol{k} = \hat{x}k_x$ in Fig. 4.14b, where the quantized spacing between the reciprocal lattice points is given as π/L_x within the 1D particle in a box picture. Since π/L_x in k_x-space gives one solution,

Number of solutions per unit length of k_x-space = $1/(\pi/L_x) = L_x/\pi$

A positive and a negative k_x-value of the same magnitude gives the same energy point within the parabolic approximation, since $E = \hbar^2 k_x^2/2m$ as shown in in Fig. 4.15. Hence, the unique energy states are half of the solutions in k_x-space given as,

Allowed energy states per unit length of k_x-space = $L_x/2\pi$

Energy states between k_x and $k_x + \mathrm{d}k_x$ = $\mathrm{d}k_x\, L_x/2\pi$

For a parabolic band with $E = \hbar^2 k_x^2/2m$, the 1D wavevector is written as, $k_x = \sqrt{2mE}/\hbar$ and $\mathrm{d}k_x = \mathrm{d}E\, m/\hbar\sqrt{2mE}$. Substituting for $\mathrm{d}k_x$, one may count states between E and $E + \mathrm{d}E$ instead of k_x and $k_x + \mathrm{d}k_x$, Hence,

The energy states between E and $E + \mathrm{d}E$ = $dE\, \frac{L_x}{h} \frac{m}{\sqrt{2mE}}$

Finally, the density of states is defined as Energy states between E and $E + \mathrm{d}E$ per unit energy $\mathrm{d}E$, given below,

$$D_{\mathrm{1D}}(E) = \frac{L_x}{h} \frac{m}{\sqrt{2mE}}$$

which is shown in Fig. 4.15. At the band edge, one obtains a singularity, known as van Hove singularity, which is a signature feature of the 1D density of states. Furthermore, the above equation is valid only within the band, i.e. for $E \geq 0$. For the conduction band edge at E_c, for $E \geq E_c$, the density of states is given as,

$$D_{\mathrm{1D}}^{c}(E) = \frac{L_x}{h} \frac{m_c^*}{\sqrt{2m_c^*\,(E - E_c)}}$$

where m_c^* is the conduction band effective mass. Similarly, for the valence band with the band edge at E_v, the 1D density of states for $E \leq E_v$ is given as,

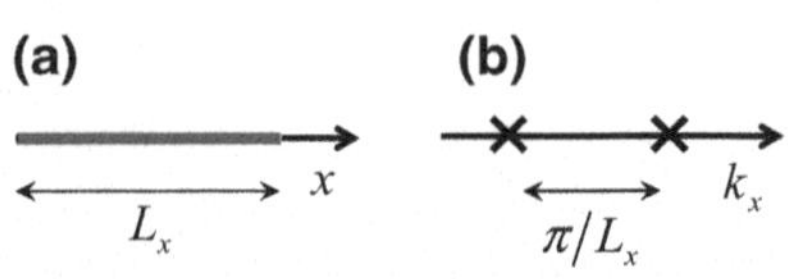

Fig. 4.14 1D. **a** Real space. **b** Reciprocal space

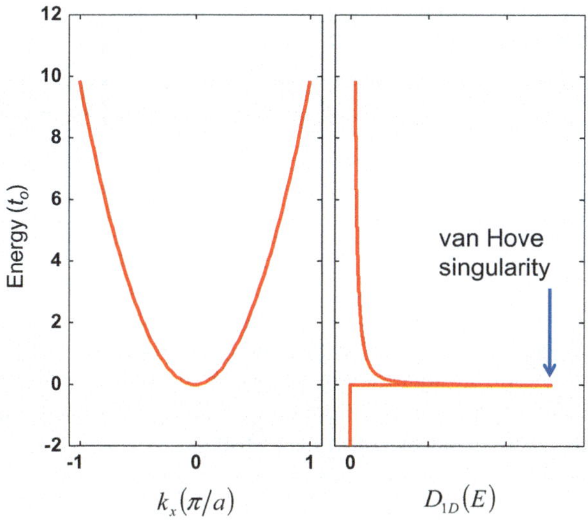

Fig. 4.15 1D density of states

$$D^{v}_{1\mathrm{D}}(E) = \frac{L_x}{h} \frac{m^*_v}{\sqrt{2m^*_v\left(E_v - E\right)}}$$

where m^*_v is the valence band effective mass.

The group velocity defined as $v_{gx} = \mathrm{d}E/\hbar \mathrm{d}k_x$ can also be calculated. Within the parabolic band approximation, one obtains, $v_{gx} = \hbar k_x/m = \sqrt{2mE}/m$. Rewriting the density of states in terms of v_{gx}, one obtains,

$$D_{1\mathrm{D}}(E) = \frac{L_x}{h} \frac{1}{v_{gx}}$$

Since, the transmission is a product of group velocity and the density of states (4.9), for one state, it is given as,

$$T(E) = \frac{h}{L_x} v_{gx} D(E)$$

It may also be shown that $T_{1\mathrm{D}}(E) = 1$. This is a very important result that the transmission is independent of energy in 1D within the energy range of the band. In fact, whenever one writes a new code or develops a new method, checking transmission to be unity in a 1D system serves as a benchmark for the validity of the code or the method.

Two Dimensional Materials

Consider a 2D material with parabolic bands having dimensions L_x and L_y as shown in Fig. 4.16a. In 2D reciprocal space $\boldsymbol{k} = \hat{x}k_x + \hat{y}k_y$, the lattice points are displaced by π/L_x and π/L_y as shown in Fig. 4.16b. Within the particle in 2D box approximation,

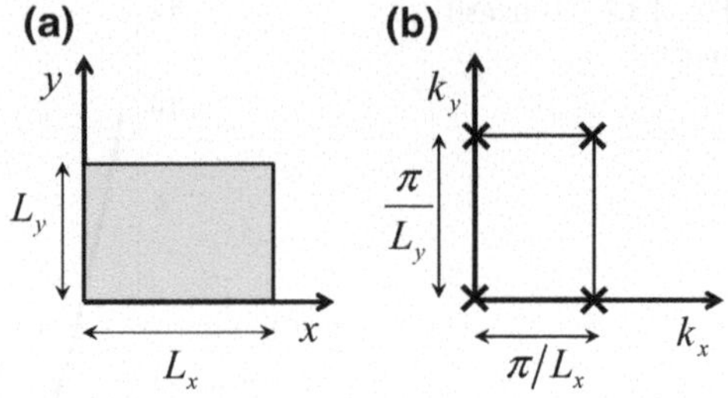

Fig. 4.16 2D. **a** Real space. **b** Reciprocal space

Number of solutions per unit area of k-space = $1/(\pi^2/L_xL_y) = L_xL_y/\pi^2$

Four combinations of points in the two dimensional reciprocal space give the same energy value. To get unique energy solutions, one has to divide by four to get the unique energy solutions as follow,

Allowed energy states per unit area of k-space = $L_xL_y/4\pi^2$

One finds the energy states in a differential area of $2\pi k dk$ bounded by the radii k and $k + dk$ as follows,

Energy states between $2\pi k^2$ and $2\pi k(k + dk)$ = $2\pi k dk\ L_xL_y/4\pi^2$ = $kdk\ L_xL_y/2\pi$

For a parabolic band, substituting $k = \sqrt{2mE}/\hbar$ and $dk = \mathrm{d}E\ m/\hbar\sqrt{2mE}$ in the allowed energy states in the k-space gives allowed states within differential energy dE as follows,

Energy states between E and $E + \mathrm{d}E$ = $\mathrm{d}E\ mL_xL_y/\pi\hbar^2$

Using the following definition of the density of states.

Energy states between E and $E + \mathrm{d}E$ per unit energy $\mathrm{d}E$ = $L_xL_y\frac{m}{2\pi\hbar^2}$

The density of states for a 2D parabolic band is therefore given as,

$$D_{2\mathrm{D}}(E) = L_xL_y\frac{m}{2\pi\hbar^2}$$

which is valid for energies greater than the band edge. Interestingly, the density of states is independent of the energy as shown in Fig. 4.17. For the conduction band edge at E_c, the 2D density of states for $E \geq E_c$ is given as,

$$D^c_{2\mathrm{D}}(E) = L_xL_y\frac{m_c^*}{2\pi\hbar^2}$$

whereas for the valence band edge at E_v, the 2D density of states for $E \leq E_v$ is given as,

$$D^v_{2\mathrm{D}}(E) = L_xL_y\frac{m_v^*}{2\pi\hbar^2}$$

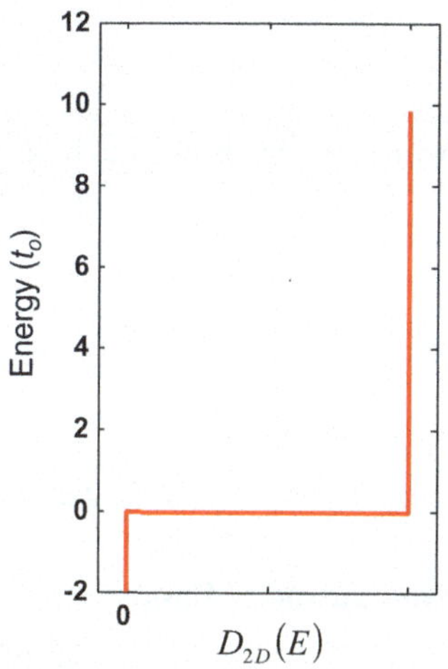

Fig. 4.17 2D density of states

Three Dimensional Materials

Consider a 3D material of dimensions L_x, L_y and L_z as shown in Fig. 4.18a. In 3D reciprocal space $\boldsymbol{k} = \hat{x}k_x + \hat{y}k_y + \hat{z}k_z$, the lattice points are displaced by π/L_x, π/L_y and π/L_z as shown in Fig. 4.18b. Hence one obtains,

Number of solutions per unit volume of k-space = $1/(\pi^3/L_xL_yL_z) = L_xL_yL_z/\pi^3$

For the eight combinations of (k_x, k_y, k_z), one gets the same energy value. Hence one has to divide the above number by eight to get the unique energy solutions, as follows,

Allowed energy states per unit volume of k-space = $L_xL_yL_z/8\pi^3$

For a differential volume $4\pi k^2 dk$ bounded by the radii k and $k + dk$, the energy states are given as,

Energy states between $4\pi k^3$ and $4\pi k^2(k + dk) = 4\pi k^2 dk\ L_xL_yL_z/8\pi^3 = k^2 dk\ L_xL_yL_z/2\pi^2$

For a parabolic band, substituting $k = \sqrt{2mE}/\hbar$ and $dk = \mathrm{d}E\ m/\hbar\sqrt{2mE}$, one obtains energy states in terms of energy E and dE as follows,

Energy states between E and $E + \mathrm{d}E = dE\ L_xL_yL_z\ m\sqrt{2mE}/2\pi^2\hbar^3$

Using the definition of the density of states, one obtains,

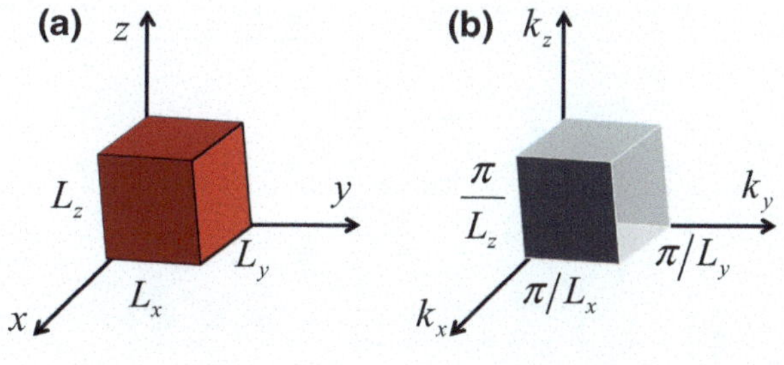

Fig. 4.18 3D. **a** Real space. **b** Reciprocal space

$$D_{3\mathrm{D}}(E) = L_x L_y L_z \frac{m\sqrt{2mE}}{2\pi^2\hbar^3}$$

which is shown in Fig. 4.19. For the band edge at zero, the above equation is valid for $E \geq 0$. For the conduction band edge at E_c and the valence band edge at E_v, one respectively obtains,

$$D^c_{3\mathrm{D}}(E) = L_x L_y L_z \frac{m_c^* \sqrt{2m_c^* (E - E_c)}}{2\pi^2\hbar^3} \qquad E \geq E_c$$

$$D^v_{3\mathrm{D}}(E) = L_x L_y L_z \frac{m_v^* \sqrt{2m_v^* (E_v - E)}}{2\pi^2\hbar^3} \qquad E \leq E_v$$

Nonparabolic Bands

So far, we have discussed the density of states for 1D, 2D and 3D materials within the parabolic band approximation. The peculiar features in the analytical expression of the density of states are not only attributed to the dimensionality, but also to the parabolic bands. If one were to use nonparabolic bands, the results of the density of states may be different from the ones obtained so far.

Although, various combinations of the dimensionality and nonparabolic bands may be applicable, we discuss a 2D material with linear bands only to highlight the importance of band dispersions and their contributions to the density of states. Following the typical procedure,

Number of solutions per unit area of k-space = $1/(\pi^2/L_x L_y) = L_x L_y/\pi^2$

Allowed energy states per unit area of k-space = $L_x L_y/4\pi^2$

Energy states between $2\pi k^2$ and $2\pi k(k + dk) = 2\pi k dk \; L_x L_y/4\pi^2 = k dk \; L_x L_y/2\pi$

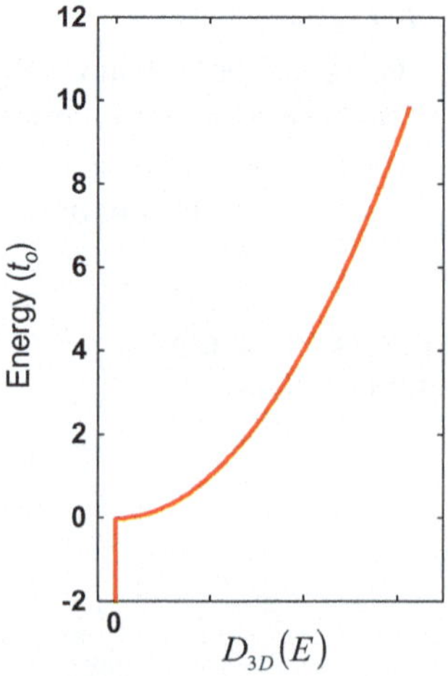

Fig. 4.19 3D density of states

Fig. 4.20 2D density of states with linear band

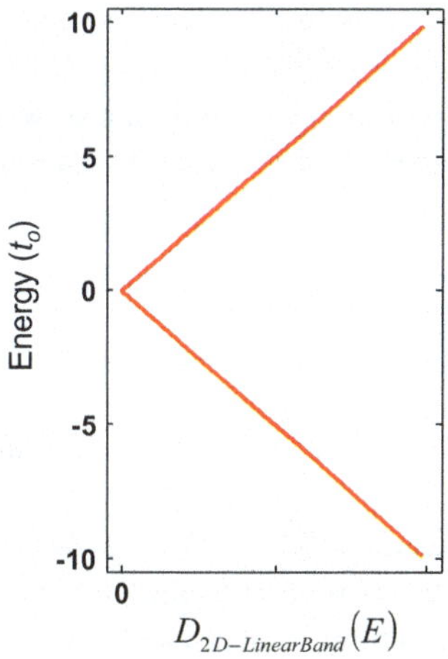

For a linear band with $E = \hbar v k$, one obtains $k = E/\hbar v$ and $dk = \mathrm{d}E/\hbar v$, which gives,

Energy states between E and $E + \mathrm{d}E$ = $dE\ EL_xL_y/2\pi\hbar^2v^2$

Finally, the density of states are given as,

$$D_{2\mathrm{D}}^{\mathrm{linear}}(E) = L_xL_y\frac{1}{2\pi\hbar^2v^2}E$$

for $E \geq 0$. where the band edge is at 0 eV. The density of states plot for 2D linear bands is shown in Fig. 4.20.

For a the conduction band edge at E_c, density of states is given as,

$$D_{2\mathrm{D}}^{\mathrm{linear}}(E) = L_xL_y\frac{1}{2\pi\hbar^2v^2}(E - E_c)$$

As discussed in the previous chapter, the 2D material graphene has linear bands and in fact has two valleys, which give the following density of states with band edge at 0 eV,

$$D_G(E) = L_xL_y\frac{1}{\pi\hbar^2v^2}E$$

4.6 Green's Function

Starting with the time dependent Schrödinger equation (3.3), one obtains,

$$\left[\frac{\partial}{\partial t} + \frac{i}{\hbar}\tilde{H}\right]\Psi(t, \boldsymbol{r}) = 0 \tag{4.12}$$

This form describes the wavefunction $\Psi(t, \boldsymbol{r})$ as the natural response of the time dependent $[\partial/\partial t + i/\hbar\,\tilde{H}]$ operator in the above equation, where $e^{-i\tilde{H}t/\hbar}$ is time evolution operator that describes the behavior of a wavefunction at time t, given a wavefunction at $t = 0$. The time dependent wavefunction solution to this equation is given as,

$$\Psi(t, \boldsymbol{r}) = e^{-i\frac{\tilde{H}}{\hbar}t}\Psi(0, \boldsymbol{r})$$

For a given wavefunction at t_o, the wavefunction at time t is given as,

$$\Psi(t, \boldsymbol{r}) = e^{-i\frac{\tilde{H}}{\hbar}(t-t_o)}\Psi(t_o, \boldsymbol{r}) \tag{4.13}$$

where $e^{-i\tilde{H}(t-t_o)/\hbar}$ describes the time evolution between the time difference $(t - t_o)$, and is related to the Green's function as we discuss next. Due to this link, Green's function is also called a propagator and has to be defined with a particular time direction.

Mathematically, Green's function is defined as an impulse response to the time dependent $[\partial/\partial t + i/\hbar\,\tilde{H}]$ operator and is given as follows for the time domain in differential form,

$$\left[\frac{\partial}{\partial t} + \frac{i}{\hbar}\tilde{H}\right] G^{r,a}(t - t_o, \boldsymbol{r}) = \delta(t - t_o) \tag{4.14}$$

where $r \equiv$ retarded[6] and $a \equiv$ advanced.[7] While wavefunction is the natural response of the operator, Green's function is a forced response of the same operator. The retarded Green's function is given as,

$$G^r(t - t_o, \boldsymbol{r}) = \begin{cases} \frac{1}{i\hbar}e^{-i\tilde{H}(t-t_o)/\hbar} & t > t_o \\ 0 & t < t_o \end{cases} \tag{4.15}$$

which is defined for a positive time and gives impulse response for future time with action at t_o. It hence satisfies causality, and therefore is widely used. On the other hand, the advanced Green's function is defined for a negative time and gives impulse response for the past time. Clearly, this Green's function does not satisfy causality, rather depicts traveling back in time.

Substituting for the retarded Green's function (4.15) in the time evolution equation (4.13), one obtains,

$$\Psi(t, \boldsymbol{r}) = i\hbar G^r(t - t_o, \boldsymbol{r})\,\Psi(t_o, \boldsymbol{r}) \tag{4.16}$$

which clearly conveys the physical action of the propagator associated with the Green's function. Transforming the retarded Green's function to the energy domain,[8] one obtains,

---

[6]Not to be confused with the real space vector $\boldsymbol{r} = (x, y, z) = (r, \theta, \phi)$.

[7]Not to be confused with the earlier use of the symbol a for the lattice constant

[8]One may also solve this integral in the complex z plane.

$$G^r(E,\mathbf{r}) = \int_{-\infty}^{+\infty} d\tau\; e^{+iE\tau/\hbar} G^r(\tau,\mathbf{r}) = \frac{1}{i\hbar}\int_{0}^{+\infty} d\tau\; e^{+i\left[E-\tilde{H}\right]\tau/\hbar}$$

where $\tau = t - t_o$. During this transformation, time and hence causality information of the retarded Green's function is lost. In addition, the integral is not well behaved due to the oscillatory nature of $e^{i\left[E-\tilde{H}\right]t/\hbar}$. To overcome these bottlenecks, one may include a factor $e^{-\eta(t-t_o)/\hbar}$ in the integral, where $\eta \approx 0$ and hence $e^{-\eta(t-t_o)/\hbar} \approx 1$ for a finite time. Also $e^{-\eta(t-t_o)/\hbar}|_{t=\infty} = 0$ makes the integral well behaved. Including the $e^{-\eta(t-t_o)/\hbar}$ factor, makes the energy resolved retarded Green's function as follows,

$$\begin{aligned} G^r(E,\mathbf{r}) &= \int_{0}^{+\infty} d\tau\; e^{i\left[E-\tilde{H}\right]\tau/\hbar} e^{-\eta\tau/\hbar} \\ &= \frac{1}{i\hbar}\int_{0}^{+\infty} d\tau\; e^{i\left[(E+i\eta)-\tilde{H}\right]\tau/\hbar} \\ &= \left[(E+i\eta)-H\right]^{-1} \end{aligned} \tag{4.17}$$

Since only the retarded Green's function is used in this text due to causality, we drop the r superscript for the sake of convenience, i.e. $G^r(E,\mathbf{r}) \equiv G(E,\mathbf{r})$. By using a basis set, since the Hamiltonian becomes a matrix $[H]$, the corresponding Green's function also becomes a matrix $[G(E)]$, which is given as follows for an orthogonal basis set,

$$[G(E)] = \left[(E+i\eta)I - [H]\right]^{-1} \tag{4.18}$$

where $(E+i\eta)$ is multiplied with an identity matrix I of the same rank as that of the Hamiltonian $[H]$ for orthogonal basis set. Furthermore, the Green's function becomes non-Hermitian due to the complex nature of the diagonal matrix elements in i. The above equation may also be written as,

$$\left[(E+i\eta)-[H]\right][G(E)] = I$$

Comparing the above matrix equation with the differential (4.14), one obtains the *impulse response* definition of Green's function in energy domain in the matrix form. Furthermore, the Hamiltonian matrix $[H]$ may be in the real space or the reciprocal space or a combination of the two. Corresponding to each choice, one obtains the Green's function as follows,

$$\begin{aligned} [H(x,y,z)] &\Leftrightarrow [G(E;x,y,z)] \\ [H(k_x,y,z)] &\Leftrightarrow [G(E;k_x,y,z)] \\ [H(k_x,k_y,z)] &\Leftrightarrow [G(E;k_x,k_y,z)] \\ [H(k_x,k_y,k_z)] &\Leftrightarrow [G(E;k_x,k_y,k_z)] \end{aligned}$$

We drop the matrix symbol [ ] henceforth for the sake of convenience.

4.7 Nonequilibrium Green's Function Formalism

Consider the two-terminal device structure in Fig. 4.21, where contact 1 and contact 2 are connected to a channel. Within the Landauer's approach, one calculates the transmission, and thereby current is given by (4.5). One method of choice for calculating the transmission is Nonequilibrium Green's Function (NEGF), where the effect of contacts is included by using self energy $\Sigma_{1,2}$ matrices for the two contacts. Furthermore, the applied bias results in a change of potential energy, which may be included by using a device potential energy U_d matrix. Hence the Green's function given in (4.18) after including the effect of $\Sigma_{1,2}$ and U_d becomes,

$$G(E) = [(E + i\eta) I - H_d - U_d - \Sigma_1 - \Sigma_2]^{-1} \tag{4.19}$$

Since, the Green's function is non Hermitian, i.e. $G \neq G^\dagger$, the non Hermitian part is defined as the Spectral function (A) and is given as,

$$A(E) = i\left(G - G^\dagger\right) \tag{4.20}$$

using which, the density of states is calculated as follows,

$$D(E) = \frac{1}{2\pi}\text{tr}\,(A) \tag{4.21}$$

Similarly, the self energies are also non Hermitian. The Hermitian part of the self energies leads to a potential energy shift, whereas the non Hermitian part gives rise to the broadening in the density of states. The Broadening functions are given as,

$$\Gamma_{1,2}(E) = i\left(\Sigma_{1,2} - \Sigma_{1,2}^\dagger\right) \tag{4.22}$$

These broadening functions give information about the coupling between the contacts and the channel and hence the lifetime of a state. NEGF is applicable only to strong coupling regime, in which case $\Gamma_{1,2}$ are quite appreciable compared to what we call as the charging energy, which we discuss later in this chapter. It may also be shown that the imaginary part of a self energy is given as,

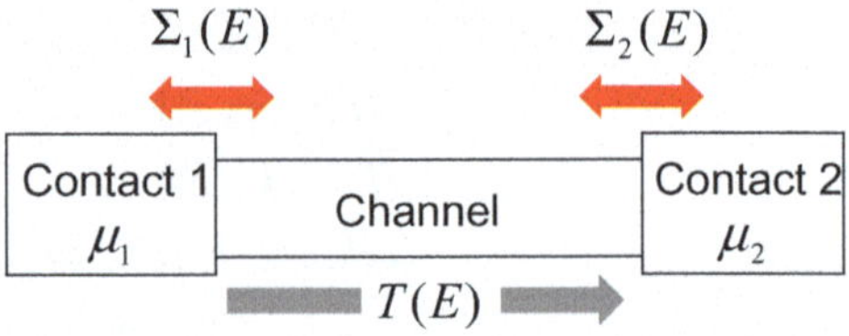

Fig. 4.21 NEGF formalism. Effect of contacts is included by using energy dependent self energies $\Sigma_{1,2}(E)$

$$\mathrm{Im}\left[\Sigma_{1,2}(E)\right] = -i\frac{\Gamma_{1,2}(E)}{2} \tag{4.23}$$

The real part may be determined by using Hilbert transform as follows,

$$\mathrm{Re}\left[\Sigma_{1,2}(E)\right] = \frac{-1}{\pi}\int_{-\infty}^{+\infty}\frac{dy}{E-y}\frac{\Gamma_{1,2}(E)}{2}$$

Given $\Gamma_{1,2}$ describes the coupling, multiplying it with the contact's Fermi functions gives the inflow functions as follows,

$$\Sigma_{1,2}^{\mathrm{in}} = \Gamma_{1,2} f_{1,2} \tag{4.24}$$

Similarly, the electron outflow (hole inflow) is defined as,

$$\Sigma_{1,2}^{\mathrm{out}} = \Gamma_{1,2}\left(1 - f_{1,2}\right) \tag{4.25}$$

It is straight forward to show that $\Sigma_{1,2}^{\mathrm{in}} + \Sigma_{1,2}^{\mathrm{out}} = \Gamma_{1,2}$. The electron correlation function is then given as,

$$G^{n}(E) = G\left(\Sigma_{1}^{\mathrm{in}} + \Sigma_{2}^{\mathrm{in}}\right)G^{\dagger} \tag{4.26}$$

and the hole correlation function is given as,

$$G^{p}(E) = G\left(\Sigma_{1}^{\mathrm{out}} + \Sigma_{2}^{\mathrm{out}}\right)G^{\dagger} \tag{4.27}$$

which describe the energy resolved electron and hole distributions respectively. Since $A = G^{n} + G^{p}$, it may be shown that,

$$G\left(\Sigma_{1}^{\mathrm{in}} + \Sigma_{2}^{\mathrm{in}}\right)G^{\dagger} + G\left(\Sigma_{1}^{\mathrm{out}} + \Sigma_{2}^{\mathrm{out}}\right)G^{\dagger} = G\left(\Gamma_{1} + \Gamma_{2}\right)G^{\dagger} = A_{1} + A_{2} = A \tag{4.28}$$

where $A_1 = G\Gamma_1 G^{\dagger}$ is the spectral contribution to the density of states due to contact 1 and $A_2 = G\Gamma_2 G^{\dagger}$ is due to contact 2. Using the electron correlation function $G^{n}(E)$, the density matrix is calculated as follows,

$$\rho = \frac{1}{2\pi}\int_{-\infty}^{+\infty} \mathrm{d}E\; G^{n}(E) \tag{4.29}$$

using which, the total number of electrons is given as,

$$N = \mathrm{tr}\left(\rho\right) \tag{4.30}$$

where tr $\equiv$ trace of a matrix.

As shown in Fig. 4.22, the total electron inflow across contact 1 and device interface is given as $\mathrm{tr}\left(\Sigma_1^{\mathrm{in}}A\right)$ and the total electron outflow across contact 1 and device interface is given as $\mathrm{tr}\left(\Gamma_1 G^n\right)$.

For a two-terminal device in the absence of scattering, current flowing across terminal 1 and across terminal 2 are the same, which one may simply label as the channel current $I_d = I_1 = I_2$. By using the information about the inflow and outflow, one may calculate the current across contact 1 and device interface as follows,

$$I_d = I_1 = I_2 = \frac{q}{h}\int_{-\infty}^{+\infty} \mathrm{d}E\ \mathrm{tr}\left(\Sigma_1^{\mathrm{in}}A - \Gamma_1 G^n\right)$$

Further simplification in the above equation results in,

$$I_d = \frac{q}{h}\int_{-\infty}^{+\infty} \mathrm{d}E\ \mathrm{tr}\left[\Gamma_1 G \Gamma_2 G^{\dagger}\right]\left[f_1 - f_2\right]$$

Including the spin degeneracy, the above two equations are given as,

$$I_d = \frac{2q}{h}\int_{-\infty}^{+\infty} \mathrm{d}E\ \mathrm{tr}\left(\Sigma_1^{\mathrm{in}}A - \Gamma_1 G^n\right) \tag{4.31}$$

$$I_d = \frac{2q}{h}\int_{-\infty}^{+\infty} \mathrm{d}E\ \mathrm{tr}\left[\Gamma_1 G \Gamma_2 G^{\dagger}\right]\left[f_1 - f_2\right] \tag{4.32}$$

Comparing with (4.5) within Landauer's approach, transmission is given as,

$$T(E) = \mathrm{tr}\left[\Gamma_1 G \Gamma_2 G^{\dagger}\right] \tag{4.33}$$

The order of matrix multiplication is not important for taking the trace of such a product.

For nonorthogonal basis set, the above equations are modified according to the unitary transformation. For a matrix M that gets added to the Hamiltonian (e.g. $\Gamma_{1,2}$, etc), the unitary transformation is given as,

$$S^{1/2}MS^{1/2}$$

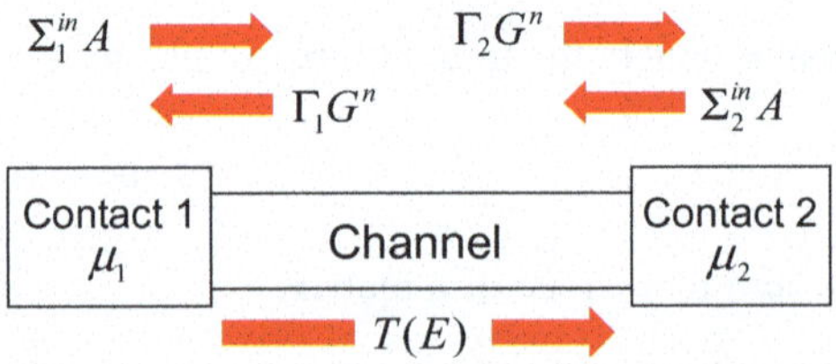

Fig. 4.22 Carrier inflow and outflow

whereas the matrix N that is inverse of matrices like M (e.g. G, etc), the unitary transformation is given as,

$$S^{-1/2}NS^{-1/2}$$

where S is the overlap matrix. For nonorthogonal basis set, the Green's function thus becomes,

$$G\,(E) = [(E + i\eta)\,S_d - H_d - U_d - \Sigma_1 - \Sigma_2]^{-1} \tag{4.34}$$

The density of states, total number of electrons, and transmission are respectively given as,

$$D(E) = \frac{1}{2\pi}\mathrm{tr}\left(S^{1/2}AS^{1/2}\right) = \frac{1}{2\pi}\mathrm{tr}\,(AS) \tag{4.35}$$

$$N = \mathrm{tr}\left(S^{1/2}\rho S^{1/2}\right) = \frac{1}{2\pi}\mathrm{tr}\,(\rho S) \tag{4.36}$$

$$T(E) = \mathrm{tr}\left(S^{1/2}\Gamma_1 S^{1/2}S^{-1/2}GS^{-1/2}S^{1/2}\Gamma_2 S^{1/2}S^{-1/2}G^{\dagger}S^{-1/2}\right) = \mathrm{tr}\left(\Gamma_1 G\Gamma_2 G^{\dagger}\right) \tag{4.37}$$

$T(E)$ equation turns out to be the same as that for the orthogonal basis set.

4.8 Self Energy

Assuming orthogonal basis set, the total Hamiltonian for contact 1, device and contact 2 in terms of the corresponding matrices (ranks listed in brackets) is given as,

$$H_{\text{total}} = \begin{bmatrix} H_{c_1}(\infty\times\infty) & H_{c_1d}(\infty\times N) & 0 \\ H_{dc_1}(N\times\infty) & H_d(N\times N) & H_{dc_2}(N\times\infty) \\ 0 & H_{c_2d}(\infty\times N) & H_{c_2}(\infty\times\infty) \end{bmatrix}$$

for which, the Green's function is given as,

$$G_{\text{total}} = \begin{bmatrix} (E+i\eta)I_{c_1} - H_{c_1} & -H_{c_1d} & 0 \\ -H_{dc_1} & (E+i\eta)I - H_d & -H_{dc_2} \\ 0 & -H_{c_2d} & (E+i\eta)I_{c_2} - H_{c_2} \end{bmatrix}^{-1} = \begin{bmatrix} g_{c_1} & g_{c_1d} & g_{c_1c_2} \\ g_{dc_1} & G_d & g_{dc_2} \\ g_{c_2c_1} & g_{c_2d} & g_{c_2} \end{bmatrix}$$

Rearranging the above equation gives,

$$\begin{bmatrix} (E+i\eta)I_{c_1} - H_{c_1} & -H_{c_1d} & 0 \\ -H_{dc_1} & (E+i\eta)I - H_d & -H_{dc_2} \\ 0 & -H_{c_2d} & (E+i\eta)I_{c_2} - H_{c_2} \end{bmatrix} \begin{bmatrix} g_{c_1} & g_{c_1d} & g_{c_1c_2} \\ g_{dc_1} & G_d & g_{dc_2} \\ g_{c_2c_1} & g_{c_2d} & g_{c_2} \end{bmatrix} = \begin{bmatrix} I_{c_1} & 0 & 0 \\ 0 & I & 0 \\ 0 & 0 & I_{c_2} \end{bmatrix}$$

Multiplying the first matrix rows with the second column of the second matrix gives the following set of equations,

$$\left[(E+i\eta)I_{c_1}-H_{c_1}\right]g_{c_1d}-H_{c_1d}G_d=0$$

$$-H_{dc_1}g_{c_1d}+\left[(E+i\eta)I-H_d\right]G_d-H_{dc_2}g_{c_2d}=I$$

$$-H_{c_2d}G_d+\left[(E+i\eta)I_{c_2}-H_{c_2}\right]g_{c_2d}=0$$

In the above set of equations, rearranging the first and the third equation, one obtains,

$$g_{c_1d}=\left[(E+i\eta)I_{c_1}-H_{c_1}\right]^{-1}H_{c_1d}G_d=g_{c_1}H_{c_1d}G_d$$

and

$$g_{c_2d}=\left[(E+i\eta)I_{c_2}-H_{c_2}\right]^{-1}H_{c_2d}G_d=g_{c_2}H_{c_2d}G_d$$

respectively, which may be substituted in the second equation, resulting in,

$$\left[(E+i\eta)I-H_d-H_{dc_1}g_{c_1}H_{c_1d}-H_{dc_2}g_{c_2}H_{c_2d}\right]G_d=I$$

The device Green's function is then given as,

$$G_d(E)=\left[(E+i\eta)I-H_d-H_{dc_1}g_{c_1}H_{c_1d}-H_{dc_2}g_{c_2}H_{c_2d}\right]^{-1}$$

Comparing with

$$G_d(E)=\left[(E+i\eta)I-H_d-\Sigma_1-\Sigma_2\right]^{-1}$$

one obtains the self energies as follows,

$$\Sigma_{1,2}=H_{dc_{1,2}}g_{c_{1,2}}H_{c_{1,2}d} \tag{4.38}$$

where $g_{c_{1,2}}$ are the contact Green's functions.

Given the contact's Green's function, one may calculate the contact's self energy on the channel. One should note that $H_{dc_{1,2}}=H^{\dagger}_{c_{1,2}d}$.

For the nonorthogonal basis set, the device Green's function is given as,

$$G(E)=\left[(E+i\eta)S_d-H_d-U_d-\Sigma_1-\Sigma_2\right]^{-1}$$

whereas the contact self energies are given as follows,

$$\Sigma_{1,2}=\left[(E+i\eta)S_{dc_{1,2}}-H_{dc_{1,2}}\right]g_{c_{1,2}}\left[(E+i\eta)S_{c_{1,2}d}-H_{c_{1,2}d}\right]$$

where $g_{c_{1,2}}=\left[(E+i\eta)S_{c_{1,2}}-H_{c_{1,2}}\right]^{-1}$ are the contact Green's function. One should note that in contrast to the orthogonal basis set, the matrices on the left and right of the contact Green's function are not Hermitian of each other for a nonorthogonal basis set, i.e. $\left[(E+i\eta)S_{dc_{1,2}}-H_{dc_{1,2}}\right]\neq\left[(E+i\eta)S_{c_{1,2}d}-H_{c_{1,2}d}\right]^{\dagger}$, although $H_{dc_{1,2}}=H^{\dagger}_{c_{1,2}d}$ and $S_{dc_{1,2}}=S^{\dagger}_{c_{1,2}d}$.

Surface Green's Function

Being semi-infinite entities, the contact's Green's function ($g_{c1,2}$) are not tractable. To solve this problem, the idea is to replace the contact's Green's function ($g_{c1,2}$) in (4.38) by Surface Green's function of the contact ($g_{s1,2}$) only, where the surface is the part next to the device. This may be achieved by dividing the contact into unit cells $\{1, 2, 3, \cdots\}$ in a way that only the nearest neighboring unit cells have nonzero hopping Hamiltonian. For each unit cell, the Hamiltonian can be a matrix itself. With this arrangement, the Green's function for the right contact within an orthogonal basis set is given as,

$$g_{c_2} = \begin{bmatrix} (E+i\eta)I_1 - [H_1] & -[H_{12}] & 0 & \cdots \\ -[H_{21}] & (E+i\eta)I_2 - [H_2] & -[H_{23}] & \cdots \\ 0 & -[H_{32}] & (E+i\eta)I_3 - [H_3] & \cdots \end{bmatrix}^{-1}$$

Writing the above equation in an abstract form of $[a]$, $[b]$, $[c]$, $[d]$ matrices below, it may be shown that one may write the Green's function for the surface portion of the contact 2, whose size is the same as that of the matrix $[a]$.

$$g_{c_2} = \begin{bmatrix} [a] & [b] \\ [d] & [c] \end{bmatrix}^{-1} = \begin{bmatrix} [g_{s_2}] & \cdots\cdots\cdots \\ \cdots & \cdots\cdots\cdots \\ \cdots & \cdots\cdots\cdots \\ \cdots & \cdots\cdots\cdots \end{bmatrix}$$

It is straightforward to show by using techniques of linear algebra that,

$$\left[g_{s_2}\right]^{-1} = [a] - [b]\,[c]^{-1}\,[d]$$

whereas due to the semi-infinite contact,

$$\left[g_{c_2}\right] = [c]^{-1}$$

making,

$$\left[g_{s_2}\right]^{-1} = [a] - [b]\left[g_{c_2}\right][d]$$

Substituting the values, one gets rid of most of the components in $[g_{c_2}]$ due to zero matrices in $[b]$ and $[d]$, resulting in,

$$\left[g_{s_2}\right]^{-1} = \underbrace{(E+i\eta)I_1 - [H_1]}_{[a]} - \underbrace{[H_{12}]}_{[b]}\left[g_{s_2}\right]\underbrace{[H_{21}]}_{[d]} \tag{4.39}$$

One has to solve this equation recursively, giving rise to the term of *recursive surface Green's function*. For a 1×1 matrix, $I_1 = 1$, which yields,

$$g_{s_2}^{-1} = E + i\eta - H_1 - H_{12}g_{s_2}H_{21}$$

$$(H_{12}H_{21})g_{s_2}^2 - (E + i\eta - H_1)g_{s_2} + 1 = 0$$

This quadratic equation may be solved as follows,

$$g_{s_2} = \frac{(E + i\eta - H_1) \pm \sqrt{(E + i\eta - H_1)^2 - 4H_{12}H_{21}}}{2H_{12}H_{21}} \tag{4.40}$$

For only the nearest neighboring unit cells with nonzero hopping Hamiltonian, (4.38) for the contact self energies may be written in terms of the surface Green's function as follows,

$$\Sigma_{1,2} = H_{dc_{1,2}} g_{s_{1,2}} H_{c_{1,2}d} \tag{4.41}$$

where only those parts of $H_{dc_{1,2}}$ and $H_{c_{1,2}d}$ are included in (4.41) for which the hopping elements are non-zero.

Finite Difference

Consider a device, consisting of a single lattice point, with two contacts that are discretized by using finite difference basis set as shown in Fig. 4.23. The Hamiltonian for contact is given below emphasizing the first three unit cells (there is one lattice point per unit cell),

$$[H] = \begin{bmatrix} 2t_o & -t_o & 0 & 0 & \cdots \\ -t_o & 2t_o & -t_o & 0 & \cdots \\ 0 & -t_o & 2t_o & -t_o & \cdots \\ \cdots & \cdots & \cdots & \cdots & \cdots \end{bmatrix}$$

Substituting $H_1 = 2t_o$ and $H_{12} = -t_o$ in (4.40), gives surface Green's function for the contact as follows,

$$g_s = \frac{(E + i\eta - 2t_o) \pm \sqrt{(E + i\eta - 2t_o)^2 - 4t_o^2}}{2t_o^2}$$

For a finite difference lattice, using $(E + i\eta)$ instead of E, the $E(k)$ dispersion relationship is given as, $E + i\eta = 2t_o\,[1 - \cos(ka)]$, making the surface Green's function,

$$g_s = \frac{-2t_o\cos(ka) \pm \sqrt{(2t_o\cos(ka))^2 - 4t_o^2}}{2t_o^2} = -\frac{e^{\pm ika}}{t_o}$$

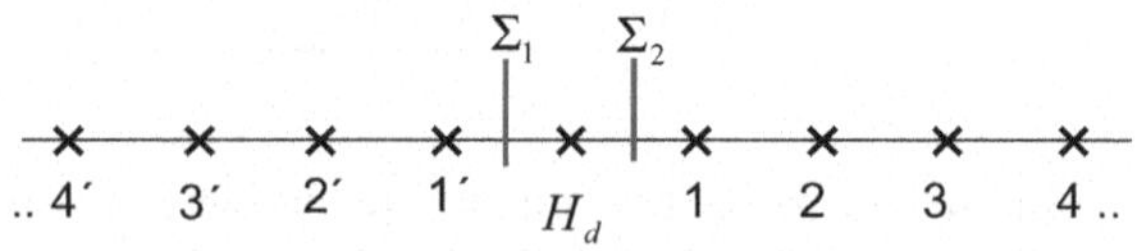

Fig. 4.23 1D finite difference lattice

The positive sign refers to a positive traveling wave and vice versa. We take the convention of a positive traveling wave, for which,

$$g_{s_{1,2}} = -\frac{e^{ik_{1,2}a}}{t_o} \tag{4.42}$$

For a given energy, the wavenumbers for contacts 1 and 2 ($k_{1,2}$) may be calculated by using the dispersion relation, $E + i\eta = U_{1,2} + 2t_o[1 - \cos(k_{1,2}a)]$, where $U_{1,2}$ is the potential energy for the two contacts due to applied bias or any other extrinsic and/or intrinsic effect. Since η is very small, it is usually ignored to get real $k_{1,2}$ values. The next step is to calculate the contact self energies. Considering the total Hamiltonian as follows,

$$[H] = \begin{bmatrix} \cdots & \cdots & \cdots & \cdots & \cdots & \cdots & \cdots \\ \cdots & 2t_o & -t_o & 0 & 0 & 0 & \cdots \\ \cdots & -t_o & 2t_o & -t_o & 0 & 0 & \cdots \\ \cdots & 0 & -t_o & 2t_o & -t_o & 0 & \cdots \\ \cdots & 0 & 0 & -t_o & 2t_o & -t_o & \cdots \\ \cdots & 0 & 0 & 0 & -t_o & 2t_o & \cdots \\ \cdots & \cdots & \cdots & \cdots & \cdots & \cdots & \cdots \end{bmatrix} = \begin{bmatrix} \cdots & \cdots & \cdots & \cdots & \cdots \\ \cdots & H_{c_1} & H_{c_1d} & & \cdots \\ \cdots & & & & \cdots \\ \cdots & H_{dc_1} & H_d & H_{dc_2} & \cdots \\ \cdots & & & & \cdots \\ \cdots & & H_{c_2d} & H_{c_2} & \cdots \\ \cdots & \cdots & \cdots & \cdots & \cdots \end{bmatrix}$$

By using (4.41) and (4.42), the contact self energies are given as,

$$\Sigma_{1,2} = t_o^2 g_{s_{1,2}} = -t_o e^{ik_{1,2}a}$$

Tight Binding

For a 1D device, consisting of a single atom, as shown in Fig. 4.24, within the tight binding approximation, the total Hamiltonian is given as,

$$[H] = \begin{bmatrix} \cdots & \cdots & \cdots & \cdots & \cdots & \cdots & \cdots \\ \cdots & 0 & -t & 0 & 0 & 0 & \cdots \\ \cdots & -t & 0 & -t & 0 & 0 & \cdots \\ \cdots & 0 & -t & 0 & -t & 0 & \cdots \\ \cdots & 0 & 0 & -t & 0 & -t & \cdots \\ \cdots & 0 & 0 & 0 & -t & 0 & \cdots \\ \cdots & \cdots & \cdots & \cdots & \cdots & \cdots & \cdots \end{bmatrix} = \begin{bmatrix} \cdots & \cdots & \cdots & \cdots & \cdots \\ \cdots & H_{c_1} & H_{c_1d} & & \cdots \\ \cdots & & & & \cdots \\ \cdots & H_{dc_1} & H_d & H_{dc_2} & \cdots \\ \cdots & & & & \cdots \\ \cdots & & H_{c_2d} & H_{c_2} & \cdots \\ \cdots & \cdots & \cdots & \cdots & \cdots \end{bmatrix}$$

where the onsite energy is at 0 (thereby making $H_d = 0$), and the hopping parameter is t. One gets, $H_1 = 0$ and $H_{1,2} = -t$. With the $E(k)$ dispersion of $E + i\eta = U_{1,2} - 2t\cos(ka)$, and by using (4.40), one obtains,

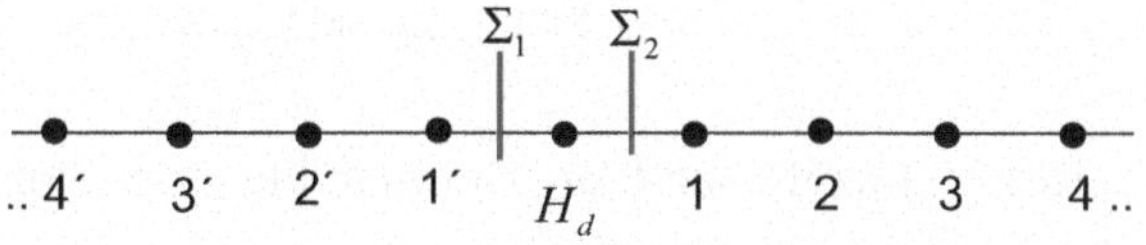

Fig. 4.24 1D crystal

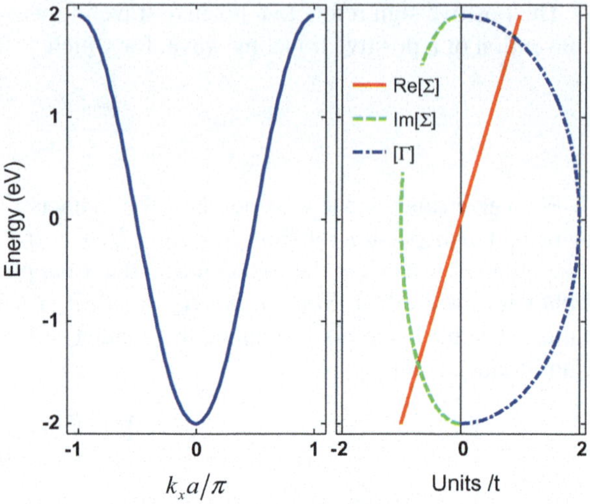

Fig. 4.25 Self energy and broadening function. With reference to the band structure, real and imaginary parts of self energy are shown along with the broadening function

$$g_{s1,2} = \frac{e^{ik_{1,2}a}}{t} \tag{4.43}$$

For the positive traveling wave, $k_{1,2}$ may be calculated (ignoring η) by using the following equation,

$$k_{1,2}a = \cos^{-1}\left[\frac{-(E - U_{1,2})}{2t}\right]$$

Next, the contact self energies (4.41) are calculated as follows,

$$\Sigma_{1,2} = H_{dc_{1,2}} g_{c_{1,2}} H_{c_{1,2}d} = t^2 g_{s_{1,2}} = -te^{ik_{1,2}a} \tag{4.44}$$

Under equilibrium condition, with $k_1 = k_2 = k_x$ (since there is no potential energy difference) and $E = -2t\cos(k_x a)$, where $k_x a = \cos^{-1}(-E/2t)$, the real and imaginary parts of the self energy are given as,

$$\text{Re}\,[\Sigma] = -t\cos(k_x a) = \frac{E}{t}$$

$$\text{Im}\,[\Sigma] = -t\sin(k_x a) = -t\sin\left[\cos^{-1}\left(\frac{-E}{2t}\right)\right]$$

The broadening function is shown as well, which is given as,

$$\Gamma = i\left(\Sigma - \Sigma^{\dagger}\right) = 2t\sin(k_x a) = 2t\sin\left[\cos^{-1}\left(\frac{-E}{2t}\right)\right]$$

Re [Σ], Im [Σ], and Γ are shown in Fig. 4.25, along with the band structure.

4.9 Coherent Transport

Let us consider the device structure in Fig. 4.24 and use the self energies in (4.44) to calculate the equilibrium density of states and transmission. This device structure depicts a single atom with 1D semi-infinite contacts (starting at atomic locations $\acute{1}$ and 1, respectively). However, this structure is equivalent to a 1D material under equilibrium. Since there is no applied voltage, $U_d = 0$. Also, $H_d = 0$. Using (4.19), the device Green's function is given as follows,

$$G(E) = \frac{1}{E + i\eta + 2te^{ik_x a}}$$

where $\Sigma_{1,2} = -te^{ik_x a}$, given $k_1 = k_2 = k_x$. Furthermore, $E + i\eta = -2t\cos(k_x a)$, which makes $G(E)$,

$$G(E) = \frac{1}{-2t\cos(k_x a) + 2te^{ik_x a}} = \frac{-i}{2t\sin(k_x a)}$$

Computing the spectral function, $A(E) = -i\left(G - G^{\dagger}\right) = 1/t\sin(k_x a)$ and the group velocity, $v_{gx} = \mathrm{d}E/\hbar\mathrm{d}k_x = 2ta\sin(k_x a)/\hbar$, one may write the spectral function in terms of the group velocity, $A(E) = 2a/\hbar v_{gx}$. Finally, the density of states is given as, $D(E) = 1/2\pi\mathrm{tr}(A) = a/\pi\hbar v_{gx}$, which has the same structure as that of 1D density of states in Sect. 4.5.

Next, one may calculate the broadening functions as follows, $\Gamma_{1,2} = i\left(\Sigma_{1,2} - \Sigma_{1,2}^{\dagger}\right) = 2t\sin(k_x a)$, which leads to the following inflow and outflow functions, $\Sigma_{1,2}^{\mathrm{in}} = \Gamma_{1,2}f_o = 2t\sin(k_x a)f_o$ and $\Sigma_{1,2}^{\mathrm{out}} = \Gamma_{1,2}\left(1 - f_o\right) = 2t\sin(k_x a)\left(1 - f_o\right)$, where f_o is the Fermi's function for the equilibrium chemical potential energy μ_o. Furthermore, the electron correlation function is given as, $G^n(E) = G\left(\Sigma_1^{\mathrm{in}} + \Sigma_2^{\mathrm{in}}\right)G^{\dagger} = f_o/t\sin(k_x a)$ and the hole correlation function is given as, $G^p(E) = A - G^n = \left(1 - f_o\right)/t\sin(k_x a)$.

Using the electron correlation function, the density matrix is calculated as,

$$\rho = \frac{1}{2\pi}\int_{-\infty}^{+\infty} \mathrm{d}E\; G^n(E)$$

where the total number of electrons (without the spin degeneracy) is given as $N = \mathrm{tr}(\rho)$ and the transmission is given as,

$$T(E) = \mathrm{tr}\left(\Gamma_1 G \Gamma_2 G^{\dagger}\right) = \mathrm{tr}\left(2t\sin(k_x a)\frac{-i}{2t\sin(k_x a)}2t\sin(k_x a)\frac{-i}{2t\sin(k_x a)}\right) = 1$$

This transmission is the signature result of 1D band structure, where the transmission is independent of energy as discussed in Sect. 4.5. As shown in Fig. 4.26, the van Hove singularities are clearly visible in the density of states. The transmission is also unity within the bandwidth and is zero outside.

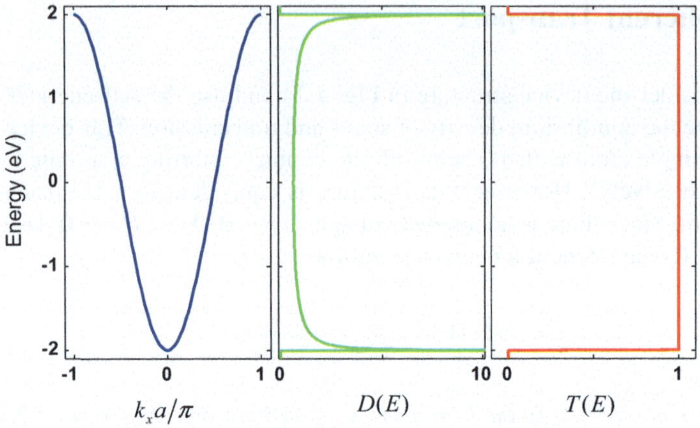

Fig. 4.26 NEGF calculation for 1D equilibrium density of states and transmission

Zero Dimensional Conductor

Let us now consider the same structure, i.e. a channel consisting of a single atom with two semi-infinite 1D contacts, but break the symmetry by applying a voltage V_d at contact 2. Under this nonequilibrium condition, the device Hamiltonian remains the same, i.e., $H_d = 0$. We assume a symmetric voltage drop as shown in Fig. 4.27, which gives the potential energy $U_d = -qV_d/2$ in the device region. The Green's function is given as,

$$G(E, V_d) = \frac{1}{E + i\eta + qV_d/2 + te^{ik_1 a} + te^{ik_2 a}}$$

where the self energies are given as $\Sigma_{1,2} = -te^{ik_{1,2}a}$. The wavevectors $k_{1,2}$ may be obtained by using, $E = U_{1,2} - 2t\cos(k_{1,2}a)$, where $U_1 = 0$ and $U_2 = -qV_d$ since contact 1 is grounded and applied bias is applied to contact 2.

Thus,

$$k_1 a = \cos^{-1}\left[\frac{-E}{2t}\right]$$

$$k_2 a = \cos^{-1}\left[\frac{-(E + qV_d)}{2t}\right]$$

The bias dependent spectral function is given as, $A(E, V_d) = i\left(G - G^{\dagger}\right)$ and hence the density of states is given as, $D(E, V_d) = \frac{1}{2\pi}\mathrm{tr}(A)$, which is shown in Fig. 4.28. One should note that the energy level for the atom is at zero, since $H_d = 0$. However, due to very strong coupling with the contacts, the level is broadened and in fact, the density of states looks like that of the 1D contact for $V_d = 0$. For a finite V_d, the van Hove singularity peaks shift and the shape is modified, but nonetheless, the features are more like 1D contact's features than the 0D channel's.

This is particularly true for nanodevices where sometimes the electronic properties are determined predominantly by the contacts than the channel itself, a stark contrast from micron scale devices, where channel characteristics are the most important.

The broadening functions are given as, $\Gamma_{1,2} = i\left(\Sigma_{1,2} - \Sigma_{1,2}^{\dagger}\right) = 2t\sin(k_{1,2}a)$. Finally, the inflow and outflow functions are, $\Sigma_{1,2}^{\text{in}} = \Gamma_{1,2}f_{1,2} = 2t\sin(k_{1,2}a)f_{1,2}$ and $\Sigma_{1,2}^{\text{out}} = \Gamma_{1,2}\left(1 - f_{1,2}\right) = 2t\sin(k_{1,2}a)\left(1 - f_{1,2}\right)$, where $f_{1,2}$ are the Fermi's function for contact 1 and contact 2. The corresponding chemical potential energies are $\mu_1 = \mu_o$ and $\mu_2 = \mu_o - qV_d$. One may further calculate the electron correlation functions, as well as the density matrix and the total number of electrons as follows,

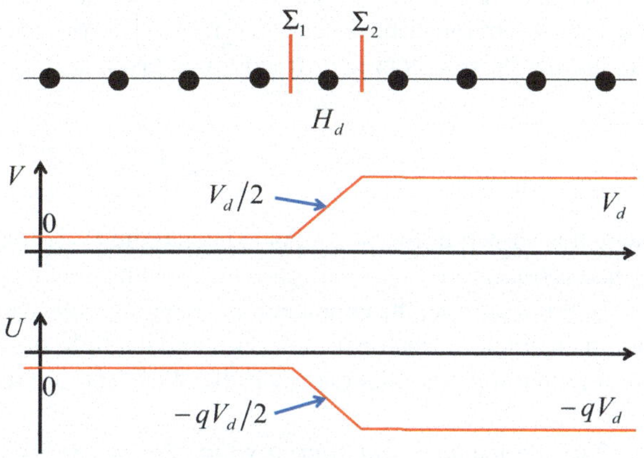

Fig. 4.27 Single atom conductor with 1D semi-infinite contacts and the corresponding potential profile

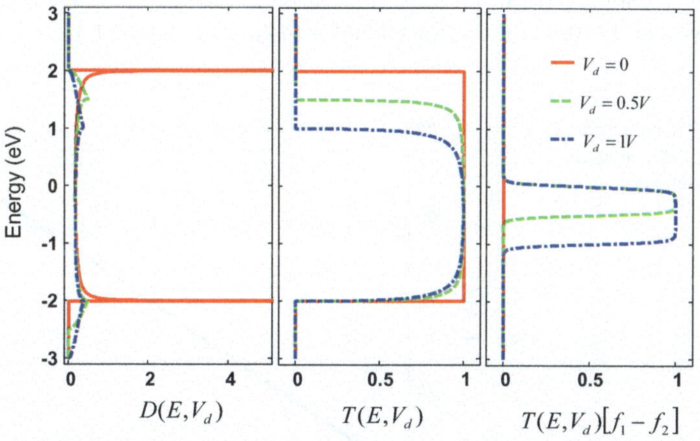

Fig. 4.28 Nonequilibrium density of states, transmission, and the product of transmission and Fermi's function difference for a single atom conductor at $T = 300$ K. Equilibrium properties for $V_d = 0$ are shown as a reference

$$G^n(E) = G\left(\Sigma_1^{\text{in}} + \Sigma_2^{\text{in}}\right)G^\dagger = 2t\, G\left[\sin(k_1 a)f_1 + \sin(k_2 a)f_2\right]G^\dagger$$

$$\rho = \frac{1}{2\pi}\int_{-\infty}^{+\infty} \mathrm{d}E\; G^n(E)$$

$$N = \mathrm{tr}(\rho)$$

Finally, the transmission is given as,

$$T(E) = \mathrm{tr}\left(\Gamma_1 G \Gamma_2 G^\dagger\right)$$

and is shown in Fig. 4.28 as a function of bias. The product of transmission and Fermi's function difference $T(E)\left[f_1 - f_2\right]$ is plotted as well. Finally, the current including the spin degree of freedom, is given as,

$$I_d = I_1 = I_2 = \frac{2q}{h}\int_{-\infty}^{+\infty} \mathrm{d}E\; T(E)\left[f_1 - f_2\right]$$

which is shown in Fig. 4.29. As expected the IV is linear and a quantum of conductance is observed.

In this example, the contacts form what we call as Ohmic contacts with the single atom channel, i.e. the onsite energies and the tight binding parameters (equivalent to saying that the potential energy or band offsets) are the same for the device and the contacts.

Next, consider a situation in which $H_d = -0.5$ eV, i.e. there is an offset in the onsite energy with a shift of -0.5 eV. Furthermore, we take the magnitude of the self energy to be $1/10$th of the value taken for Figs. 4.28 and 4.29, i.e.

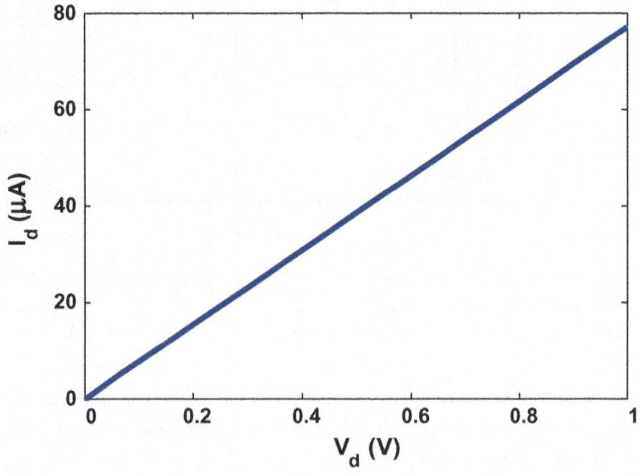

Fig. 4.29 IV Characteristics of a single atom conductor with ideal contacts. $T = 300$ K

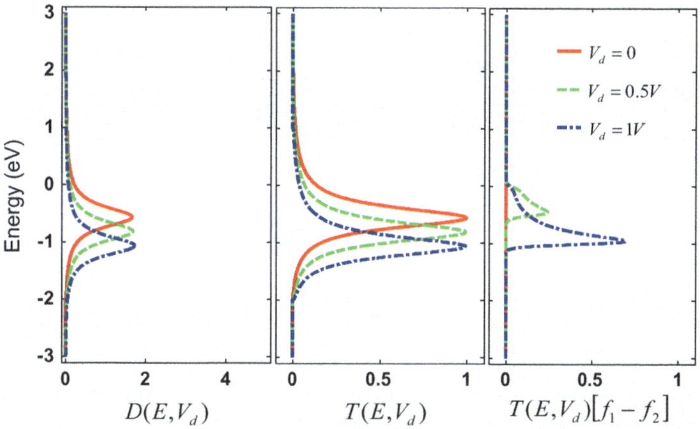

Fig. 4.30 Nonequilibrium density of states, transmission, and the product of transmission and Fermi's function difference for a single atom conductor with 1/10th self energy as in Fig. 4.28 and with onsite energy of the device atom at −0.5 eV at $T = 300$ K. Equilibrium properties for $V_d = 0$ are shown as a reference

$$\Sigma_{1,2} = -\frac{t}{10}e^{ik_{1,2}a}$$

Using these parameters, the nonequilibrium density of states is shown in Fig. 4.30. One should note that even for $V_d = 0$, the density of states is shifted from −0.5 eV due to the real part of the contact self energies. The bias dependent shift is also observed.

The shape of the density of states plot looks more like a broadened delta function, which in fact is a Lorentzian function, given as,

$$D(E, V_d) = \frac{1}{\pi}\frac{\gamma/2}{(E-\epsilon_o)^2 + (\gamma/2)^2}$$

where ϵ_o is the peak position and γ is the peak width and is given as the total broadening due to the two contacts in the absence of scattering. Since contact induced broadening is relatively smaller, the 0D features of the channel are visible in Fig. 4.30, which are reflected in the transmission and $T(E)\,(f_1 - f_2)$ plots as well. The nonlinear IV characteristics are shown in Fig. 4.31, which clearly shows less than the quantum of conductance value due to nonideal contacts leading to reflections and hence reduced transmission and nonlinear conduction.

One Dimensional Conductor

Now, let us consider the nonequilibrium properties of a 1D conductor as shown in Fig. 4.32. For the sake of convenience, we take only four atoms in the channel region. Within the nearest neighbor tight binding approximation, assuming zero onsite energy, the device Hamiltonian is given as,

$$H_d = \begin{bmatrix} 0 & -t & 0 & 0 \\ -t & 0 & -t & 0 \\ 0 & -t & 0 & -t \\ 0 & 0 & -t & 0 \end{bmatrix}$$

The electrostatic potential energy matrix due to applied voltage is give as,

$$U_d = \begin{bmatrix} U_{d1} & 0 & 0 & 0 \\ 0 & U_{d2} & 0 & 0 \\ 0 & 0 & U_{d3} & 0 \\ 0 & 0 & 0 & U_{d4} \end{bmatrix}$$

For a linear voltage drop (within linear screening approximation) in the device region, the potential energy matrix becomes,

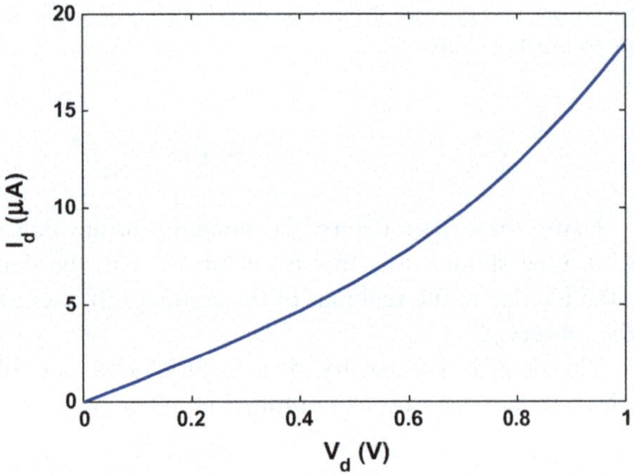

Fig. 4.31 IV Characteristics of a single atom conductor showing non-Ohmic conduction with 1/10th self energy as in Fig. 4.29 and with onsite energy of device atom at -0.5 eV. $T = 300$ K

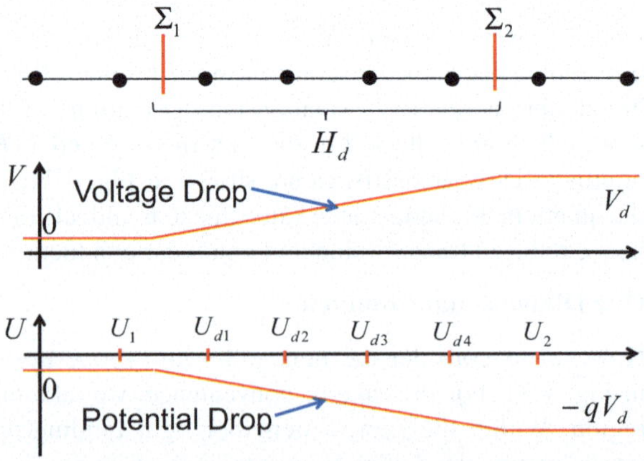

Fig. 4.32 1D conductor with 1D contacts

$$U_d = -qV_d \begin{bmatrix} 0.2 & 0 & 0 & 0 \\ 0 & 0.4 & 0 & 0 \\ 0 & 0 & 0.6 & 0 \\ 0 & 0 & 0 & 0.8 \end{bmatrix}$$

whereas the contact atoms right next to the channel have the potential energies, $U_1 = 0$ and $U_2 = -qV_d$, which gets added to the $E(k)$ dispersion of the two contacts as follows, $E = U_{1,2} - 2t\cos(k_{1,2}a)$. The self energies for the two contacts are calculated as follows,

$$\Sigma_1 = \begin{bmatrix} -te^{ik_1 a} & 0\,0\,0 \\ 0 & 0\,0\,0 \\ 0 & 0\,0\,0 \\ 0 & 0\,0\,0 \end{bmatrix}$$

$$\Sigma_2 = \begin{bmatrix} 0\,0\,0 & 0 \\ 0\,0\,0 & 0 \\ 0\,0\,0 & 0 \\ 0\,0\,0 & -te^{ik_2 a} \end{bmatrix}$$

The nonequilibrium Green's function is then calculated as,

$$G(E, V_d) = \begin{bmatrix} E + i\eta - U_{d1} + te^{ik_1 a} & -t & 0 & 0 \\ -t & E + i\eta - U_{d2} & -t & 0 \\ 0 & -t & E + i\eta - U_{d3} & -t \\ 0 & 0 & -t & E + i\eta - U_{d4} + te^{ik_2 a} \end{bmatrix}^{-1}$$

using which, the nonequilibrium spectral function is given as, $A(E, V_d) = i\left(G - G^\dagger\right)$. One obtains the nonequilibrium density of states as, $D(E, V_d) = \frac{1}{2\pi}\mathrm{tr}(A)$, which is shown in Fig. 4.33, which clearly show the 1D van Hove singularities, as expected. The broadening functions are given as,

$$\Gamma_1 = i\left(\Sigma_1 - \Sigma_1^\dagger\right) = \begin{bmatrix} 2t\sin(k_1 a) & 0\,0\,0 \\ 0 & 0\,0\,0 \\ 0 & 0\,0\,0 \\ 0 & 0\,0\,0 \end{bmatrix}$$

$$\Gamma_2 = i\left(\Sigma_1 - \Sigma_1^\dagger\right) = \begin{bmatrix} 0\,0\,0 & 0 \\ 0\,0\,0 & 0 \\ 0\,0\,0 & 0 \\ 0\,0\,0 & 2t\sin(k_2 a) \end{bmatrix}$$

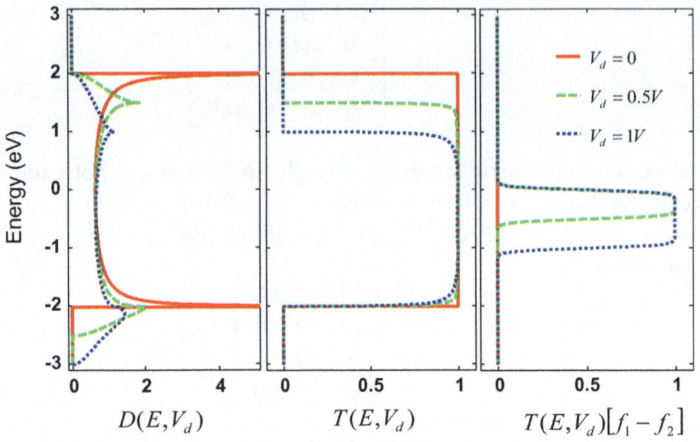

Fig. 4.33 Nonequilibrium density of states, transmission, and the product of transmission and Fermi's function difference for 1D conductor. Equilibrium properties for $V_d = 0$ are shown as a reference

Finally, the inflow functions are,

$$\Sigma_1^i n = \Gamma_1 f_1 = \begin{bmatrix} 2t\sin(k_1 a)f_1 & 0\,0\,0 \\ 0 & 0\,0\,0 \\ 0 & 0\,0\,0 \\ 0 & 0\,0\,0 \end{bmatrix}$$

$$\Sigma_2^i n = \Gamma_2 f_2 = \begin{bmatrix} 0\,0\,0 & 0 \\ 0\,0\,0 & 0 \\ 0\,0\,0 & 0 \\ 0\,0\,0 & 2t\sin(k_2 a)f_2 \end{bmatrix}$$

where $f_{1,2}$ are the Fermi's function for the two contacts with the chemical potential energies $\mu_1 = \mu_o$ and $\mu_2 = \mu_o - qV_d$. Taking the equilibrium chemical potential energy (μ_o) at zero ensures that the chemical potential energy is in the middle of the band. One may further calculate the electron correlation functions, as well as the density matrix and the total number of electrons as follows,

$$G^n(E) = G\left(\Sigma_1^{\text{in}} + \Sigma_2^{\text{in}}\right)G^\dagger$$

$$\rho = \frac{1}{2\pi}\int_{-\infty}^{+\infty} \mathrm{d}E\; G^n(E)$$

$$N = \mathrm{tr}(\rho)$$

Finally, the nonequilibrium transmission is given as,

$$T(E) = \mathrm{tr}\left(\Gamma_1 G \Gamma_2 G^{\dagger}\right)$$

and is shown in Fig. 4.33 as a function of bias. Consistent with the 1D transmission, unity transmission within the conduction window is observed. Note that the conduction window is shifting due to a shift in the band edges due to applied bias. The product of transmission and Fermi's function difference $T(E)\,(f_1 - f_2)$ is shown as well, which reflects unity product with some additional broadening due to the contact Fermi's functions.

Including spin degree of freedom, current is then given as,

$$I_d = I_1 = I_2 = \frac{2q}{h} \int_{-\infty}^{+\infty} \mathrm{d}E\; T(E)\left[f_1 - f_2\right]$$

which is shown in Fig. 4.34. As expected the IV characteristics are linear and a quantum of conductance is observed. Here, we use Ohmic contacts, where the self energies are taken in a way that there are no reflections at the contact-channel interface, i.e. the contacts are ideal. On the other hand, in case of the nonideal contacts, one observes multiple reflections, not only due to the electron waves incident at the channel from either of the two contacts, but also within the device due to the electron waves impinging on the contacts from channel.

This bouncing back and forth leads to resonances inside the channel region, which lead to additional resonance peaks[9] in the density of states as shown in Fig. 4.35. Here, the contact self energies are taken as 1/10th of the ideal values, i.e.

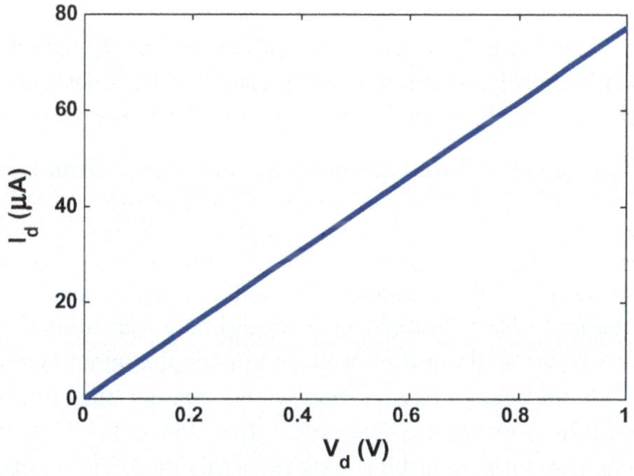

Fig. 4.34 IV characteristics of 1D conductor. $T = 300$ K

[9]For metallic contacts, these are also called MIGS (metal induced gap states).

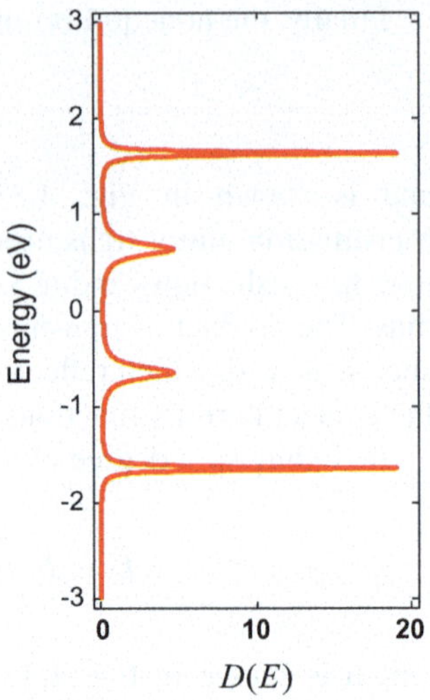

Fig. 4.35 Density of states plot for the 1D conductor with nonideal contacts

$$\Sigma_1 = \begin{bmatrix} -te^{ik_1 a} & 0\,0\,0 \\ 0 & 0\,0\,0 \\ 0 & 0\,0\,0 \\ 0 & 0\,0\,0 \end{bmatrix} \frac{1}{10}$$

$$\Sigma_2 = \begin{bmatrix} 0\,0\,0 & 0 \\ 0\,0\,0 & 0 \\ 0\,0\,0 & 0 \\ 0\,0\,0 & -te^{ik_2 a} \end{bmatrix} \frac{1}{10}$$

Such nonideal contacts lead to unique transport characteristics, again a key difference from microscale devices. In nanodevices, contacts are as important as the channel itself and may very well determine the final transport characteristics.

Transport in Two Dimensional and Three Dimensional Channels

So far we have discussed 1D transport only, where the Hamiltonian is written as, $[H(x)]$. The corresponding Green's function is $[G(E; x)]$. Irrespective of the dimensionality of the contacts, the self energy is cast into the same dimension as the channel's Hamiltonian. One should note that following the recipe within NEGF, no information about the multi dimensional contacts would be lost.

In the case of a two dimensional channel, the Hamiltonian is written as, $[H(x, y)]$ and the corresponding Green's function is $[G(E; x, y)]$. If the transport is in only one direction, say the x-axis only (also known as longitudinal direction) and if the symmetry is not broken in the channel width direction, i.e. y-axis (also known as transverse direction), one may transform the Hamiltonian in the y-direction, provided the width is much greater than the electron's wavelength, yielding $[H(x, k_y)]$. The corresponding Green's function is then given as $[G(E; x, k_y)]$. The rest of the calculations may be done independently for each value of k_y in the absence of scattering. This leads to a tremendous simplification, since a two dimensional transport problem may be reduced to a set of one dimensional transport equation. To further clarify, if in 1D, the Hamiltonian size is 1000 × 1000. For the same dimensions in 2D, the Hamiltonian size would be 1M × 1M, where M stands for million. Since the Green's function involves taking inverse of matrices of this size, which becomes computationally very expensive with the increasing matrix size. However, with the Fourier transformation in k_y, one has to solve only one thousand independent transport problems of 1000 × 1000 Hamiltonian size.

Similarly for a three dimensional Hamiltonian with $[H(x, y, z)]$, if the transport is in x-direction only, one may transform the other two spatial directions (the transverse directions) to the reciprocal (k_y, k_z) space, as follows $[H(x, k_y, , k_z)]$, which gives the Green's function as $[G(E; x, k_y, k_z)]$ essentially reducing a 3D problem to a 1D one. However, one should note that if the symmetry is broken in the transverse direction, e.g. by a varying potential due to an external or internal source or due to scattering, one may not use Fourier transform and hence has to solve the complete problem in real space or orbital space, which is usually quite challenging.

Ballistic Transport

As shown in Fig. 4.29 for a one atom conductor and in Fig. 4.34 for a 1D conductor, the current is the same with ideal contacts by using coherent NEGF formalism. It may be further shown that within the assumption of coherent quantum transport, the current is independent of the channel length as shown in Fig. 4.36. This phenomenon is also called Ballistic transport.

However, it is in contrast to Ohm's law, which states that,

$$R = \frac{1}{\sigma}\frac{l}{S}$$

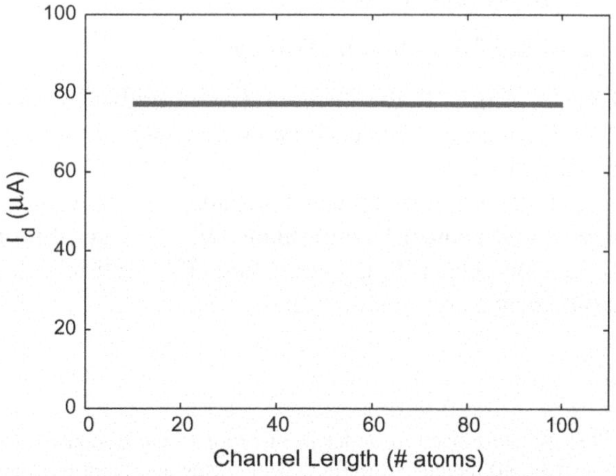

Fig. 4.36 Ballistic transport versus channel length at 300K for $V_d = 1$ V

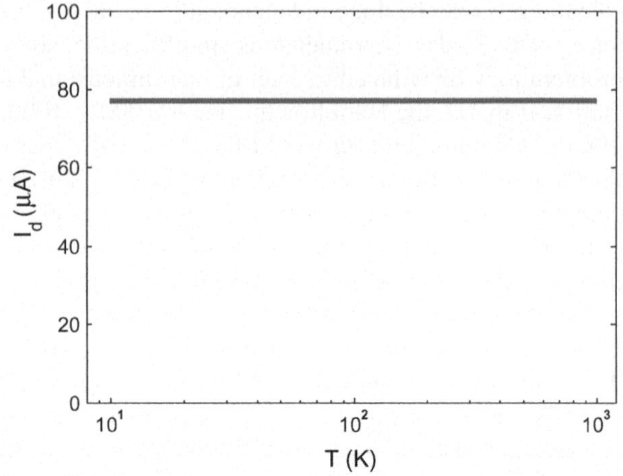

Fig. 4.37 Coherent transport versus temperature for $V_d = 1$ V

where, σ is the conductivity with the dimensions of $[S/m]$, l is the conductor length, and S is the cross sectional area. To emphasize, Ohm's law is not always applicable at the nanoscale. In fact, experimentally, one does observe a length independent transport within a certain length scale, called the mean free path (λ). However, if the device length is larger than the mean free path, Ohm's law becomes applicable and the transport becomes incoherent, which is further discussed in the next section.

One should note that the trend of decreasing resistance with the increasing area is consistent with NEGF. In 3D, for each wavevector (k_y, k_z) in the transverse direction, one gets a 1D transport channel. With increasing area, the number of transverse wavevectors increase and hence the current increases, which leads to a decreasing resistance.

It may also be shown that the cohrent transport is independent of temperature as well [shown in Fig. 4.37], apart from the minuscule dependence of the Fermi's function broadening on temperature.[10]

Algorithm for Coherent Transport

Consider Fig. 4.38 for the numerical algorithm of the coherent transport by using the NEGF equations. For each value of applied bias, the step-by-step implementation usually involves,

(1) The input parameters, variables and constants are specified in the form of device and contact Hamiltonians (H_{d,c_1,c_2}), equilibrium chemical potential energy (μ_o), applied bias (V_d), temperature (T), electronic charge (q), Planck's constant (h), Boltzmann's constant (k_B), etc.

[10]On the other hand, in order to incorporate incoherent effects, we need to go beyond coherent transport and include scattering (both elastic and inelastic) in this transport model to capture any length dependence or temperature dependence, which we discuss in the next section.

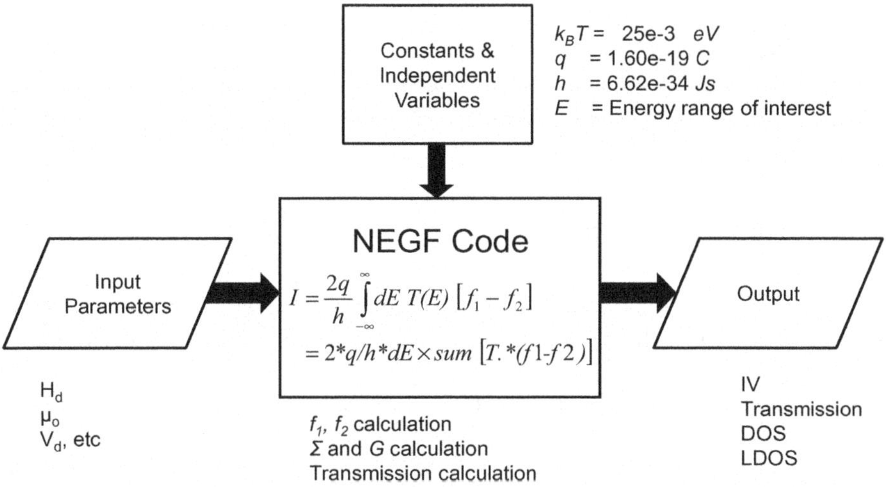

Fig. 4.38 Algorithm for calculating coherent transport properties by using NEGF equations

(2) Using the information about the applied bias, contact chemical potential energies ($\mu_{1,2}$) are calculated.

(3) The range of energy is established over which the transport calculation is to be performed. Usually, the minimum value of the energy grid is $10k_\mathrm{B}T$ below the minimum chemical potential energy and the maximum value is $10k_\mathrm{B}T$ above the maximum chemical potential energy, but this is not always the case. In principle, this integration is from $-\infty$ to $+\infty$ energy values.

(4) The energy grid spacing is a fraction of $k_\mathrm{B}T$. Usually, $0.1k_\mathrm{B}T$ is adequate for most calculations.

(5) For the established energy array, Fermi's functions $(f_{1,2})$ are calculated and the difference $(f_1 - f_2)$ is stored in an array.

(6) Within the energy FOR loop for each energy value, self energies $\Sigma_{1,2}$ are calculated, using which, Green's function (G) and the broadening functions $(\Gamma_{1,2})$ are calculated. Spectral function (A) and electron/hole correlation functions $(G^{n,p})$ may also be calculated. Thus, for each value of energy, transmission (T) is calculated as $\mathrm{tr}\left(\Gamma_1 G \Gamma_2 G^{\dagger}\right)$.

(7) Finally, the current is calculated by using, $2 * q/\hbar * \mathrm{d}E * \mathrm{sum}\left[T \cdot * (f_1 - f_2)\right]$, which is the numerical form of the Landauer's equation. One should note that the symbol $\cdot *$ represents element by element multiplication (the same as in Matlab), i.e. multiplication at the same energy point and finally adding the product. One should

note that current may also be calculated by using the following numerical form, $2 * q/\hbar * \mathrm{d}E * \mathrm{sum}\left[\mathrm{trace}\left(\Sigma_1^{\mathrm{in}} A - \Gamma_1 G\right)\right]$.

(8) Finally, the output quantities like IV characteristics, nonequilibrium transmission, nonequilibrium density of states, etc, are calculated.

4.10 Incoherent Transport

Within NEGF, scattering is included by employing a scattering self energy Σ_s as shown in Fig. 4.39. The nonequilibrium Green's function including scattering is given as,

$$G(E) = [(E + i\eta) I - H_d - U_d - \Sigma_1 - \Sigma_2 - \Sigma_s]^{-1} \tag{4.45}$$

for which, the spectral function is given as, $A(E) = i\left(G - G^{\dagger}\right)$, which gives the density of states as,

$$D(E) = \frac{1}{2\pi} \mathrm{tr}\,(A)$$

The broadening, inflow, and outflow functions are given as,

$$\Gamma_{1,2}(E) = i\left(\Sigma_{1,2} - \Sigma_{1,2}^{\dagger}\right)$$

$$\Sigma_{1,2}^{\mathrm{in}} = \Gamma_{1,2} f_{1,2}$$

$$\Sigma_{1,2}^{\mathrm{out}} = \Gamma_{1,2}\left(1 - f_{1,2}\right)$$

where, $\Gamma_{1,2} = \Sigma_{1,2}^{\mathrm{in}} + \Sigma_{1,2}^{\mathrm{out}}$. The electron spectral function is given as,

$$G^n(E) = G\left(\Sigma_1^{\mathrm{in}} + \Sigma_2^{\mathrm{in}}\right) G^{\dagger}$$

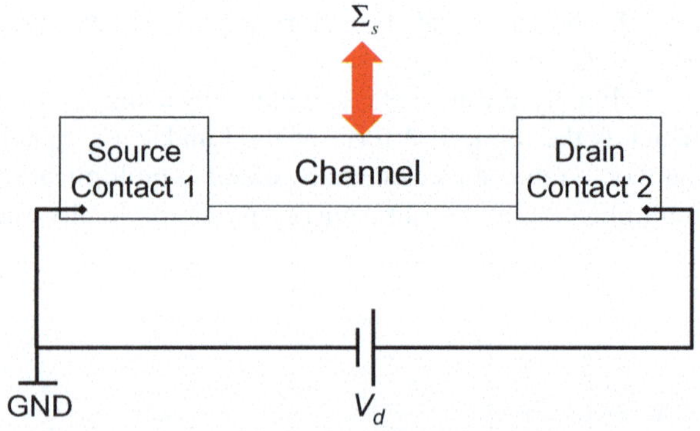

Fig. 4.39 Scattering self energy within NEGF

using which, one finds the density matrix as follows,

$$\rho = \frac{1}{2\pi} \int_{-\infty}^{+\infty} \mathrm{d}E\ G^n(E)$$

The current for the two contacts including spin degeneracy is given as follows,

$$I_{1,2} = \frac{2q}{h} \int_{-\infty}^{+\infty} \mathrm{d}E\ \mathrm{tr}\left(\Sigma_{1,2}^{\mathrm{in}} A - \Gamma_{1,2} G^n\right)$$

where it should be emphasized that after including scattering, the above equation CANNOT be reduced to the following,

$$I_{1,2} \neq \frac{q}{h} \int_{-\infty}^{+\infty} \mathrm{d}E\ \mathrm{tr}\left(\Gamma_1 G \Gamma_2 G^{\dagger}\right) [f_1 - f_2]$$

However, the concept of transmission is still valid within the Landauer's approach,

$$I_1 = \frac{q}{h} \int_{-\infty}^{+\infty} \mathrm{d}E\ T'(E) [f_1 - f_2]$$

which gives the transmission function for a nonzero Fermi's function difference as follows,

$$T'(E) = \frac{\mathrm{tr}\left(\Sigma_{1,2}^{\mathrm{in}} A - \Gamma_{1,2} G^n\right)}{f_1 - f_2} \tag{4.46}$$

It should be emphasized that one may define a current through the scattering self energy and it should be zero, explicitly given as,

$$I_s = \frac{2q}{h} \int_{-\infty}^{+\infty} \mathrm{d}E\ \mathrm{tr}\left(\Sigma_s^{\mathrm{in}} A - \Gamma_s G^n\right) = 0 \tag{4.47}$$

Since the scattering may be thought of as a *virtual* contact, and the number of electrons across this *virtual* contact should be conserved.

Various scattering processes may be broadly divided into two classes, namely inelastic (with energy exchange) and elastic (without energy exchange) scattering. Within elastic scattering, there may further be two scattering processes, namely Phase Breaking Scattering or Dephasing and Momentum Relaxation Scattering. Although one may describe these processes independently, in reality, a single event may involve various scattering processes concurrently. Consider the vertical transition in Fig. 4.40a between the initial state i in the conduction band and the final state f in the valence band. Clearly this transition indicates an inelastic process involving an energy exchange of $E_f - E_i$. One may not deduce from the $E(k)$ diagram, whether dephasing is present. In the absence of dephasing, this process is simply Inelastic Scattering, whereas in the presence of dephasing, the process is Inelastic Dephasing.

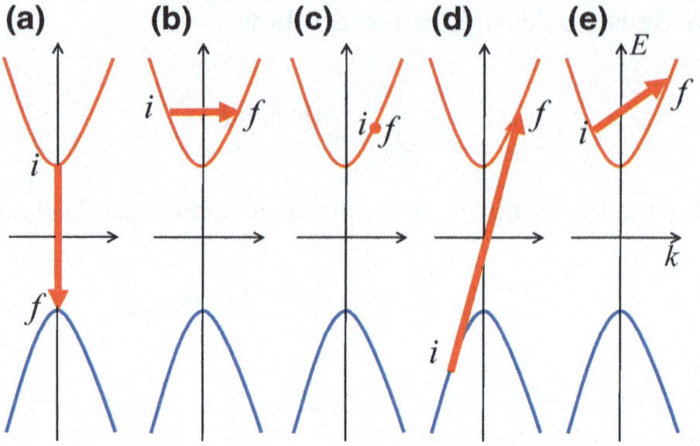

Fig. 4.40 Scattering processes. **a** Inelastic scattering—interband, **b** momentum relaxation scattering—intraband, **c** dephasing or phase breaking scattering indicating no change in momentum and energy, **d** inelastic and momentum relaxation scattering—interband, **e** inelastic and momentum relaxation scattering—intraband. i = initial state, f = final state

Furthermore, this scattering is also called Interband Scattering, since there are two bands involved.

In Fig. 4.40b, we show an Intraband Scattering event with horizontal transition. Clearly, this event is Elastic Scattering involving Momentum Relaxation only with an exchange of $\hbar\left(k_f - k_i\right)$ momentum. In the absence of dephasing, this scattering is Momentum Relaxation Scattering, whereas with dephasing, the scattering is Momentum Relaxation Dephasing. In Fig. 4.40c, the process has no energy or momentum exchange. In the case of phase breaking scattering, this process simply leads to Dephasing. In Fig. 4.40d, the process is inelastic interband scattering with momentum relaxation, whereas Fig. 4.40e represents the same scattering process, albeit intraband. In Figs. 4.40d,e, if phase breaking scattering were to accompany with the energy and momentum exchange, the processes would be interband and intraband inelastic dephasing with momentum relaxation respectively.

While we discuss various scattering processes in abstract form in this section, it is important to associate them with various physical mechanisms. Consider a photon that has both energy and momentum. If electrons interact with such a photon, this interaction may lead to both energy exchange and momentum exchange, thus leading to inelastic scattering with momentum relaxation. In light matter interaction, usually electrons do not loose phase information, in which case, dephasing is absent. Furthermore, since momentum is given as $p = h/\lambda = hf/c$, photons have minimum momentum for a given energy, since they have maximum possible speed (c) in free space. Although, the photon momentum is small, it is nonetheless nonzero, hence this transition is not truly vertical in $E(k)$ diagram. However, for such a transition, one may approximate the momentum exchange to be zero and hence assume the transition to be vertical. Photon absorption, e.g. in a photodiode, takes an electron from low energy state (e.g. valence band) to high energy state (e.g. conduction band).

Similarly, photon emission, e.g. in an LED (light emitting diode), brings an electron from high energy state (e.g. conduction band) to low energy state (e.g. valence band).

Similarly, for phonon scattering, one may have both the energy exchange and the momentum exchange. However, since phonons are mere atomic vibrations, their energy is usually very close to the thermal energy scale, i.e. $k_B T$, which is only about 25 meV at the room temperature. On the other hand, a phonon may have a large momentum. For low energy phonons, one may approximate the energy exchange to be small and hence assume the transition to be horizontal. It is understandable that phonon absorption takes an electron from low momentum state to high momentum state and *vice versa* for phonon emission. Furthermore, interaction with the phonon degree of freedom usually leads to phase breaking scattering or dephasing.

Inelastic Scattering

Within the Born approximation, the scattering inflow and outflow functions due to emission and absorption process is given as,

$$\Sigma_s^{\text{in}}(E; i, j) = D^{\text{em}}(\hbar\omega; i, k; j, l)\, G^n(E + \hbar\omega; k, l) + D^{\text{ab}}(\hbar\omega; i, k; j, l)\, G^n(E - \hbar\omega; k, l)$$
$$\Sigma_s^{\text{out}}(E; i, j) = D^{\text{em}}(\hbar\omega; i, k; j, l)\, G^p(E - \hbar\omega; k, l) + D^{\text{ab}}(\hbar\omega; i, k; j, l)\, G^p(E + \hbar\omega; k, l)$$

where $D^{em,ab}(\hbar\omega; i, k; j, l)$ are the emission and the absorption rank-4 tensor functions, and $\hbar\omega$ is the energy quantum for a phonon or a photon of angular frequency ω. This equation physically states that an electron from the state with indices (k, l) may be scattered into the state with indices (i, j). Henceforth, we drop these indices for further discussion, which simplifies the above equation to,

$$\Sigma_s^{\text{in}}(E) = D^{\text{em}}(\hbar\omega)\, G^n(E + \hbar\omega) + D^{\text{ab}}(\hbar\omega)\, G^n(E - \hbar\omega)$$
$$\Sigma_s^{\text{out}}(E) = D^{\text{em}}(\hbar\omega)\, G^p(E - \hbar\omega) + D^{\text{ab}}(\hbar\omega)\, G^p(E + \hbar\omega)$$

The broadening function due to scattering is given as $\Gamma_s = \Sigma_s^{\text{in}} + \Sigma_s^{\text{out}}$, where, the scattering self energy is given as,

$$\Sigma_s = \text{Re}\,[\Sigma_s] - i\frac{\Gamma_s}{2}$$

The real part of the self energy is given as the Hilbert transform of the imaginary part of the self energy as follows,

$$\text{Re}\,[\Sigma_s(E)] = \frac{-1}{\pi}\int_{-\infty}^{+\infty} \frac{dy}{E - y}\frac{\Gamma_s(E)}{2}$$

Furthermore, the emission and absorption functions are given as,

$$D^{\text{em}}(\hbar\omega) = (N + 1)D_o(\hbar\omega)$$
$$D^{\text{ab}}(\hbar\omega) = N\, D_o(\hbar\omega)$$

where $D_o\,(\hbar\omega)$ is the scattering function, which is a temperature independent rank-4 tensor, and is given as,

$$D_o\,(\hbar\omega) = \int_{-\infty}^{+\infty} \mathrm{d}\boldsymbol{r}\; \phi_i^* \phi_j^* U U^* \phi_k \phi_l$$

where U is the scattering potential energy. N is the Bose factor that gives the number of Bosons with spin-1 at a finite temperature, and is given as,

$$N = \frac{1}{e^{\hbar\omega/k_\mathrm{B}T} - 1}$$
$$N + 1 = \frac{1}{1 - e^{-\hbar\omega/k_\mathrm{B}T}}$$

In high frequency/energy limit, for $\hbar\omega \gg k_\mathrm{B}T$ or $\hbar\omega/k_\mathrm{B}T \gg 1$, one obtains,

$$N = \frac{1}{e^{\hbar\omega/k_\mathrm{B}T} - 1} \approx 0$$
$$N + 1 = \frac{1}{1 - e^{-\hbar\omega/k_\mathrm{B}T}} \approx 1$$

which leads to,

$$D^{\mathrm{em}}(\hbar\omega \gg k_\mathrm{B}T) \approx D_o(\hbar\omega)$$
$$D^{\mathrm{ab}}(\hbar\omega \gg k_\mathrm{B}T) \approx 0$$

Thus, the scattering inflow, outflow and broadening functions are given as,

$$\Sigma_s^{\mathrm{in}} \approx D_o G^n(E + \hbar\omega)$$
$$\Sigma_s^{\mathrm{out}} \approx D_o G^p(E - \hbar\omega)$$
$$\Gamma_s = \Sigma_s^{\mathrm{in}} + \Sigma_s^{\mathrm{out}} = D_o\left[G^n(E + \hbar\omega) + G^p(E - \hbar\omega)\right]$$

Scattering due to high energy phonons is weekly temperature dependent. Since D_o is temperature independent, the temperature dependent Fermi's functions do affect the electron and hole spectral functions $G^{n,p}$.

Inelastic scattering with high energy phonons is schematically shown in Fig. 4.41a. In this process, an electron absorbs a phonon of energy $\hbar\omega$ and ends up at energy E, where in the initial state, the energy is $E - \hbar\omega$. Similarly, an electron emits a phonon of energy $\hbar\omega$ and ends up at energy E, where in the initial state, the energy is $E + \hbar\omega$. These subtle transitions are not visible in the IV characteristics. However, they show up as kinks in the first derivative $(\mathrm{d}I_d/\mathrm{d}V_d)$, and as peaks or dips in the second derivative $(\mathrm{d}^2I_d/\mathrm{d}V_d^2)$ as shown in Fig. 4.41b. This technique forms the basis of what is known as InElastic Tunneling Spectroscopy (IETS) to probe electron-phonon interactions in the quantum transport.

In the low frequency/energy limit, for $\hbar\omega \ll k_B T$ or $\hbar\omega/k_B T \ll 1$, ignoring the higher order terms in the Taylor series, one obtains, $e^{\hbar\omega/k_B T} \approx 1 + \hbar\omega/k_B T$, which leads to,

$$N \approx N + 1 \approx \frac{k_B T}{\hbar\omega}$$

The emission and absorption dephasing functions take the form,

$$D^{\text{em}}(\hbar\omega \ll k_B T) \approx \frac{k_B T}{\hbar\omega} D_o(\hbar\omega)$$
$$D^{\text{ab}}(\hbar\omega \ll k_B T) \approx \frac{k_B T}{\hbar\omega} D_o(\hbar\omega)$$

Approximating $E + \hbar\omega \approx E - \hbar\omega \approx E$, the scattering inflow, outflow and broadening functions become,

$$\Sigma_s^{\text{in}}(E) \approx \frac{2k_B T}{\hbar\omega} D_o G^n(E)$$
$$\Sigma_s^{\text{out}}(E) \approx \frac{2k_B T}{\hbar\omega} D_o G^p(E)$$
$$\Gamma_s(E) \approx \frac{2k_B T}{\hbar\omega} D_o A(E)$$

where $A(E) = G^n + G^p$. Thus, the scattering due to low energy phonons are strongly temperature dependent.

Elastic Scattering

Using the structure of scattering inflow and outflow functions, the inflow, outflow, and dephasing functions for elastic scattering are defined as,

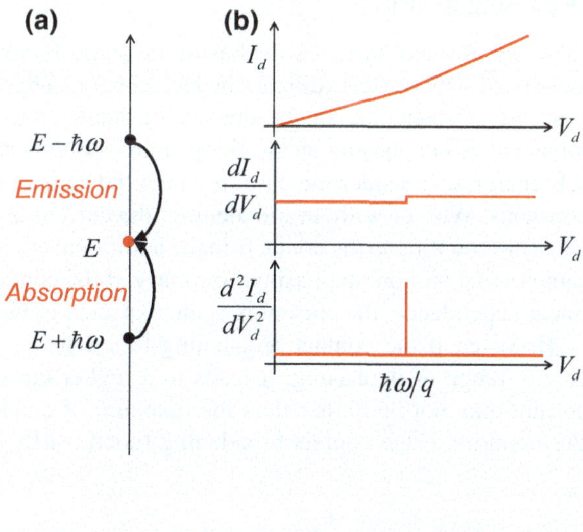

Fig. 4.41 Inelastic scattering with high energy phonons. **a** Scattering process in the energy domain, **b** dI_d/dV_d, and d^2I_d/dV_d^2 characteristics showing high frequency inelastic scattering feature

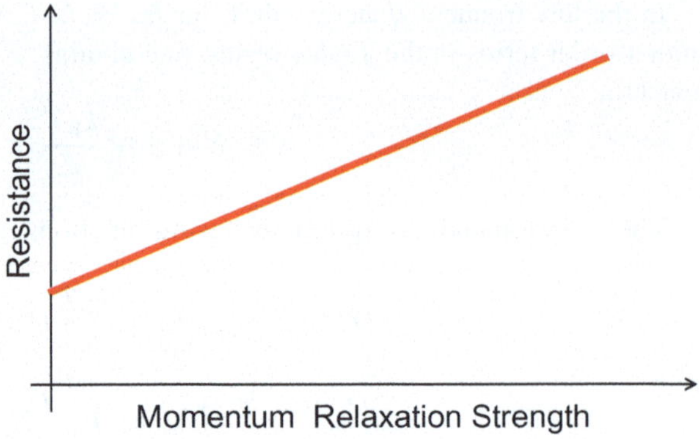

Fig. 4.42 Momentum relaxation scattering. Resistance increases with the increasing scattering strength

$$\Sigma_s^{\text{in}}(E; i, j) \approx D^{el}(i, k, j, l)G^n(E; k, l)$$
$$\Sigma_s^{\text{out}}(E; i, j) \approx D^{el}(i, k, j, l)G^p(E; k, l)$$
$$\Gamma_s^{(}E; i, j) \approx D^{el}(i, k, j, l)A(E; k, l)$$

where $D^{\text{el}}(i, k, j, l)$ is the elastic scattering function. While the inelastic scattering is included by coupling electron and hole correlation functions $G^{n,p}$ at energies $E \pm \hbar\omega$ with the scattering inflow and outflow functions at energies E, the details of momentum relaxation and phase relaxation scattering is included in the details of the rank-4 tensor of scattering function D^{el}, which we discuss later.

Since current is a directional quantity, instead of depending on the speed, it depends on the velocity of charge carriers and hence the momentum. While the effect of phase relaxation on the charge transport requires a more detailed discussion, it is quite clear that the momentum relaxation always leads to an increase in resistance, which is schematically shown in Fig. 4.42 with an increasing value of the momentum relaxation strength.

Dephasing Revisited

Like any physical process, dephasing or phase breaking scattering has a life time associated with it. According to the Heisenberg's uncertainty principle, this life time leads to a broadening in the density of states $D(E)$ as shown in Fig. 4.43. The impact of this dephasing on the charge transport depends on the details of the contact self energies. Consider the case in which the contact broadening functions $\Gamma_{1,2}$ are constants. With or without broadening, the current is not expected to change much since the coupling to the contacts in the additional energy range with dephasing is the same as that without dephasing. Similarly, if the contact broadening functions show linear dependence, the current is again expected to be independent of dephasing.

However, if the contact broadening functions $\Gamma_{1,2}$ are higher in the additional energy range of dephasing, it leads to a higher current. One should note that the current may not be higher than the quantum of conductance per band or channel. Furthermore, if the contact broadening functions $\Gamma_{1,2}$ are smaller in the additional

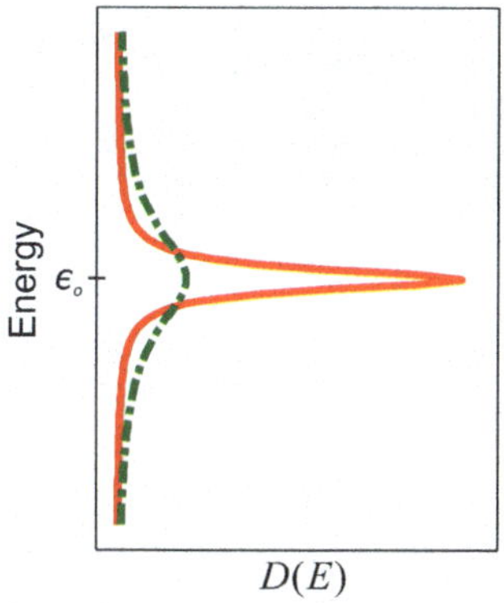

Fig. 4.43 Dephasing. The solid (red) curve has the contact induced broadening only, whereas the dashed (green) curve has additional broadening due to dephasing. Both plots are peaked at the energy level ϵ_o

energy range of dephasing, the current decreases with dephasing. In short, while with momentum scattering, resistance always increases, the behavior of resistance dependence on the scattering strength is rather complex for dephasing.

At this point, we should clarify a terminology about coherent transport and ballistic transport. For coherent transport, resistance is independent of the device length, such a transport behavior is called ballistic transport. However, the resistance sometimes does not change with increasing length even in the presence of dephasing, which also qualifies as ballistic transport.

Butticker Probe

One of the simplest ways to include dephasing in Landauer's approach is by using Butticker probe, also known as Landauer-Butticker approach. For such a probe, the scattering self energy is given as,

$$\Sigma_s = -i\eta_{\mathrm{B}}[I]$$

and the scattering broadening function is given as,

$$\Gamma_s = 2\eta_{\mathrm{B}}[I]$$

where η_{B} is the Butticker probe strength. With the inclusion of such dephasing, the density of states may be written as a Lorentzian function for a single energy level as follows (a good approximation to the actual density of states as shown in Fig. 4.30),

$$D = \frac{1}{\pi} \frac{\gamma/2}{(E - \epsilon_o)^2 + (\gamma/2)^2}$$

where γ is the net broadening function and ϵ_o is the peak location. For a single energy level, it is the sum of the contact and the scattering/dephasing broadening functions, and is given as follows,

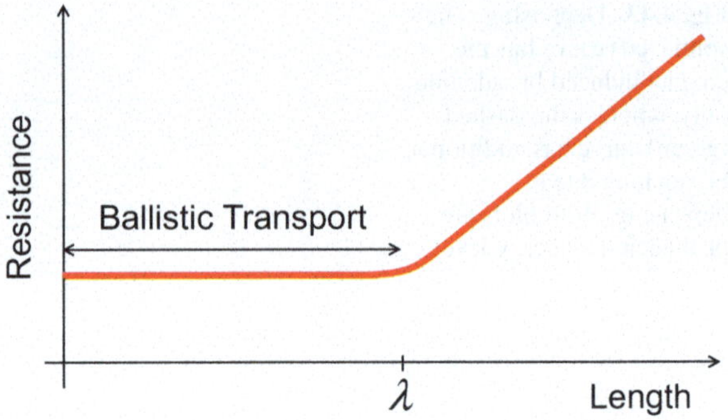

Fig. 4.44 Mean free path (λ). Transition from the ballistic transport to the diffusive transport

$$\gamma = \Gamma_1 + \Gamma_2 + \eta_B$$

where $\Gamma_{1,2}$ are scalars for a single energy level. If the contact broadening is smaller than scattering/dephasing broadening, i.e., $\Gamma_{1,2} \ll \eta_B$, the density of states is given as,

$$D = \frac{1}{\pi} \frac{\eta_B}{(E - \epsilon)^2 + \eta_B^2}$$

which results in the broadening of the density of states due to dephasing.

From Ballistic to Diffusive Transport

While ballistic transport could be coherent or incoherent, it always refers to the length independent transport. In this context, a certain length scale, called the mean free path (λ), refers to a mean distance over which scattering events happen. If the device length l is less than the mean free path λ, the resistance is length independent in contrast to the Ohm's law, whereas, for the device length l greater than the mean free path λ, the resistance increases with device length in a linear fashion—similar to Ohm's law.

This behavior is shown in Fig. 4.44 and may be expressed as follows,

$$R = \frac{1}{\sigma S}(l + \lambda)$$

For $l \ll \lambda$, $l + \lambda \approx \lambda$, the resistance in the ballistic limit is given as,

$$R = \frac{\lambda}{\sigma S}$$

For $l \gg \lambda$, $l + \lambda \approx l$, one recovers Ohm's law in the diffusive limit as follows,

$$R = \frac{l}{\sigma S}$$

Finally, by including appropriate scattering processes in the coherent transport, one may simulate transport from ballistic to diffusive regime.

4.11 Selfconsistent Mean Field Transport

Under equilibrium conditions, one may calculate the equilibrium Green's function and hence the density matrix,

$$\rho_{\text{eq}} = \frac{1}{2\pi} \int \mathrm{d}E \; G^n_{\text{eq}}(E)$$

using which, one may calculate the equilibrium number of electrons as follows, $N_{\text{eq}} = \text{tr}(\rho_{\text{eq}})$. Under nonequilibrium conditions, the density matrix changes (ρ) and the number of electrons (N) change from the equilibrium condition.

Starting from Gauss's law,

$$\nabla \cdot \mathcal{D} = \rho$$

where $\mathcal{D}$ is the electric flux density. With $\mathcal{D} = \varepsilon \mathcal{E}$, where $\mathcal{E}$ is the electric field intensity and ε is the permittivity, one obtains,

$$\nabla \cdot (\varepsilon \mathcal{E}) = \rho$$

Since $\mathcal{E} = -\nabla V$ and $U = -qV$, substituting

$$\mathcal{E} = \frac{1}{q} \nabla U$$

one obtains $\nabla \cdot (\varepsilon \nabla U) = q\rho$. If the dielectric constant ε is spatially independent, above relation is simplified to,

$$\nabla^2 U = q \frac{\rho}{\varepsilon}$$

which is a familiar form of Poisson's equation. U is also called Hartree potential energy.[11]

With a change in the density matrix ρ, the potential energy changes, which should be included in the nonequilibrium Green's function calculation. Consequently, one may calculate the density matrix ρ by using NEGF, which changes the Poisson's or Hartree potential energy as discussed above. Thus one needs to solve the Poisson's equation and NEGF equations selfconsistently with each other, also known as Hartree or Electrostatic selfconsistent loop.

[11] In the case of the charge free $\rho = 0$ region, one obtains Laplace's equation as follows, $\nabla^2 U = 0$, which results in a potential energy with linear screening.

In the simplest picture of capacitive charging, the selfconsistent potential energy may be approximated as,

$$U = U_{\text{ch}} \left(N - N_{\text{eq}}\right)$$

where N and N_{eq} are the number of electrons under nonequilibrium and equilibrium conditions, respectively. U_{ch} is the charging energy, which is defined as,

$$U_{\text{ch}} = \frac{\mathrm{d}U}{\mathrm{d}N}$$

In a simple model with capacitance C, the potential energy (U) is given as,

$$U = \frac{Q^2}{2C}$$

where Q is the total charge given as $Q = Nq$ for N electrons, where q is the electronic charge. Thus the potential energy becomes,

$$U = \frac{q^2N^2}{2C}$$

and the charging energy for adding N-electrons to the channel (or removing) is given as,

$$U_{\text{ch}}^{N} = \frac{\mathrm{d}U}{\mathrm{d}N} = \frac{q^2}{C}N$$

For $N = 1$, the single electron charging energy becomes,

$$U_{\text{ch}} = \frac{\mathrm{d}U}{\mathrm{d}N} = \frac{q^2}{C}$$

One should note that if the number of electrons increase, the change in the self-consistent potential energy U is positive and hence the density of states floats up due to the additional negative charge in the device region and hence a positive potential energy change. With a decrease in the number of electrons, the change in the selfconsistent potential energy U is negative, which results in a downward shift in the density of states due to the *missing negative charge*, which is equivalent to an additional positive charge and hence negative potential energy.

An additional selfconsistent loop may exist for scattering. Consider the scattering self energy, which depends on the Green's function and vice versa, which needs to be solved selfconsistently. For elastic scattering, the self consistency may be solved for each energy point independently, whereas for the inelastic scattering, the transport quantities at various energy points are coupled and hence the selfconsistent loop involves the complete energy range of interest. With this selfconsistent loop, the Born approximation for scattering becomes the selfconsistent Born approximation (SCBA). These selfconsistent loops are schematically shown in Fig. 4.45 along with a flow chart in Fig. 4.46.

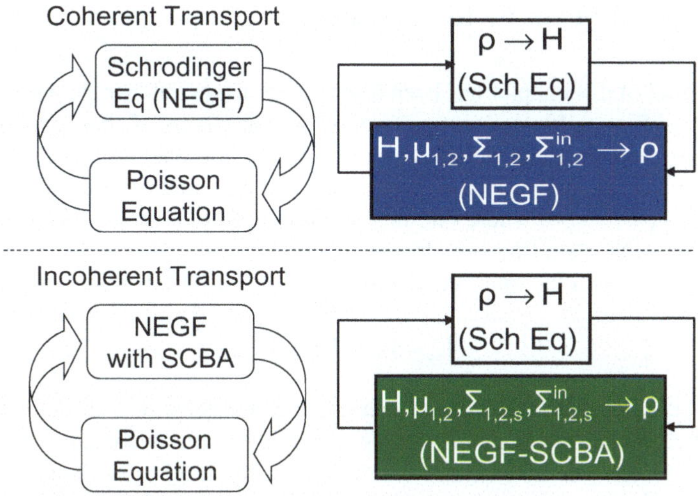

Fig. 4.45 Selfconsistent solution for coherent and incoherent transport

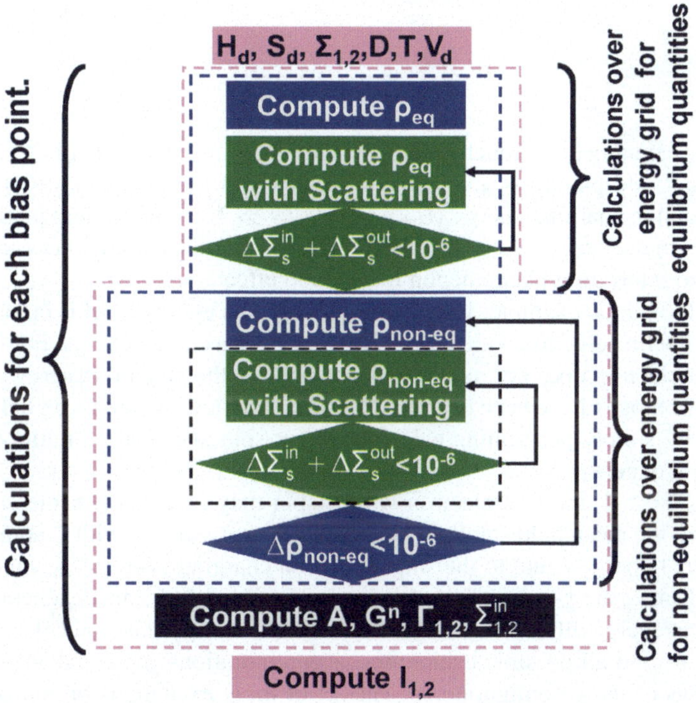

Fig. 4.46 Algorithm for selfconsistent solution. Different colors reflect various blocks in Fig. 4.45

4.12 Beyond Mean Field Transport

The main goal of this chapter has been to introduce mean field single particle NEGF formalism within a selfconsistent picture, where the effects of charge perturbations due to an external bias are included in a mean field approach for a single electron. It may be understood that NEGF breaks down if the contact couplings are weak in the limit of $\Gamma_{1,2} \approx 0$, where one may obtain unphysical results. For the sake of completeness, the scattering broadening functions should also be brought into this discussion, for which the total broadening function for NEGF-SCBA is given as,

$$\Gamma = \Gamma_1 + \Gamma_2 + \Gamma_s$$

However, since the broadening functions are matrices, it is difficult to decide whether the broadening is small or large. For this reason, one usually looks at the broadening in the density of states for evaluating the total broadening in the peak width (γ) due to the contacts and the scattering. For simplicity, let us consider a single energy level ϵ for which $\Gamma_{1,2,s}$ are scalars and the energy spectrum may be described by a Lorentzian, and is given as,

$$D = \frac{1}{\pi} \frac{\gamma/2}{(E - \epsilon)^2 + (\gamma/2)^2}$$

If the broadening (γ) is much greater than the charging energy (U_{ch}), i.e. $U_{\text{ch}}/\gamma \ll 1$, the mean field picture is applicable.[12] However, if the broadening (γ) is much smaller than the charging energy (U_{ch}), i.e. $U_{\text{ch}}/\gamma \gg 1$, the mean field picture is not applicable. In the later case, the transport regime is called Coulomb blockade, which may lead to interesting phenomenon like Kondo effect, etc.

Let us look at an example, where to begin with, an energy level is empty. While putting in the first electron with $\uparrow$-spin, one may need a certain energy. For $U_{\text{ch}}/\gamma \ll 1$, the next electron does not require much change in the potential energy since U_{ch} is small. We mostly have been dealing with such situations within mean field picture of NEGF by simply performing calculations per spin and finally multiplying with two for spin degeneracy.

For $U_{\text{ch}}/\gamma \gg 1$, the first electron with $\uparrow$-spin may require the same amount of energy as in the mean field solution. However, the next electron with $\downarrow$-spin requires an additional energy equal to the single electron charging energy U_{ch}. Since U_{ch} is appreciable now, the $\uparrow$-spin and the $\downarrow$-spin electronic levels are not degenerate, rather split with an energy difference of U_{ch}. With multiple energy states and electrons, one may have to find all possible excitations or configurations and eventually evaluate probabilities of these configurations. The set of these excitations or configurations are also known as Configuration space or Fock space, the analysis of which is beyond the scope of this book.

---

[12]Usually a factor of 10 is used.

For practical applications, usually, one wishes to have strong contact couplings to have higher drive current, which requires mean field selfconsistent NEGF transport analysis. Devices with weak contacts do have some niche applications, like single electron detectors, etc.

In this chapter, we have discussed the electron behavior as a Fermi Gas, which obeys Fermi-Dirac statistics. While this may be a valid assumption for a wide range of nanomaterials, the nanoscale devices may follow the Fermi liquid behavior. Moreover, in 1D nanostructures, electrons and holes may behave as a Tomonaga–Lüttinger Liquid. Similarly, in 2D and 3D, electron and hole degrees of freedom may exhibit properties of a Wigner Solid. Electrons and holes may also behave like a plasma in some cases—a field of study known as Plasmonics. However, such topics are beyond the scope of this book. One finds it interesting to compare these varying behaviors of charge carriers with the fact that matter does exist in four distinct forms of gas, liquid, solid and plasma.

Problems

4.1 For $T = 300$ K and 600 K, plot Fermi function. Assume $\mu_o = 0$. Comment on temperature broadening.

4.2 For $E = ak^3$, calculate and plot density of states for one dimension.

4.3 For $E = ak^3$, calculate and plot density of states for two dimension.

4.4 For $E = ak^3$, calculate and plot density of states for three dimension.

4.5 For $E = ak^4$, calculate and plot density of states for one dimension.

4.6 For $E = ak^4$, calculate and plot density of states for two dimension.

4.7 For $E = ak^4$, calculate and plot density of states for three dimension.

4.8 In Fig. 4.33 for 1D, the bias independent transmission is either zero (where band is absent) or one (where band is present). However, in the bias dependent transmission, the transmission has fractional values around $E = -2$ eV and some other points. Qualitatively explain the fractional nonequilibrium transmission. Note that one would get similar trends in problem 4.9.

4.9 One wishes to have parabolic band over an energy range of 1 eV. With a choice of finite difference lattice, what should be the minimum lattice constant a if the effective mass is equal to the free electron mass. Explain your answer.

4.10 Calculate the value of the Fermi's function for,

(a) $E = \mu_o$
(b) $E = \infty$
(c) $E = -\infty$

4.11 Using single orbital basis set for a 1D channel and contacts with $t = 1$ eV, for a device with nine atoms, calculate and plot self energy for $E = -3$ eV to 3 eV with energy grid spacing of 0.1 $k_B T$. Assume a linear voltage drop with $U = -qV_d$ [0.1 0.2 0.3 0.4 0.5 0.6 0.7 0.8 0.9]. Assume $T = 300$ K and $\mu = 0$.

4.12 Using single orbital basis set for a 1D channel and contacts with $t = 1$ eV, for a device with 9 atoms, calculate and plot equilibrium transmission for $E = -3$

Fig. 4.47 Problem 4.15

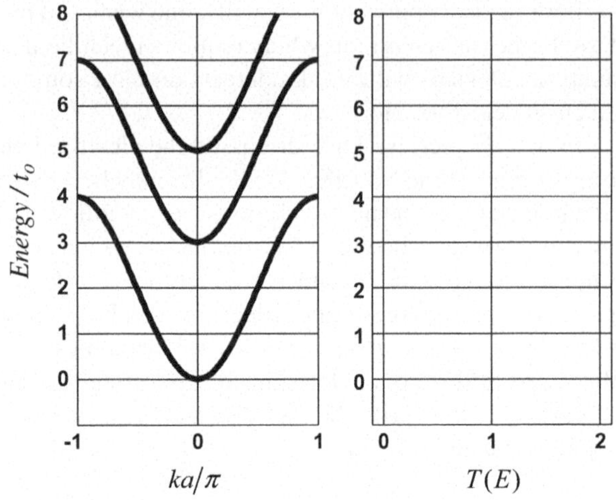

eV to 3 eV with energy grid spacing of 0.1 k_BT. Assume a linear voltage drop with $U = -qV_d$ [0.1 0.2 0.3 0.4 0.5 0.6 0.7 0.8 0.9]. Assume T = 300 K and μ = 0.

4.13 Using single orbital basis set for a 1D channel and contacts with t = 1 eV, for a device with 9 atoms, calculate and plot nonequilibrium transmission at $V_d = 0.5$V and $V_d = 1$V with energy grid spacing of 0.1 k_BT. Assume a linear voltage drop with $U = -qV_d$ [0.1 0.2 0.3 0.4 0.5 0.6 0.7 0.8 0.9]. Assume T = 300 K and μ = 0.

4.14 Using single orbital basis set for a 1D channel and contacts with t = 1 eV, for a device with 9 atoms, calculate and plot IV characteristics for V_d = zero to 1 V. Comment on the relationship between the value of current and the quantum of conductance. Assume a linear voltage drop with $U = -qV_d$ [0.1 0.2 0.3 0.4 0.5 0.6 0.7 0.8 0.9]. Assume T = 300 K and μ = 0.

4.15 For a 1D channel with ideal contacts (no reflection), the band structure is shown in Fig. 4.47, draw equilibrium transmission plot.

Research Assignment

R4.1 For silicon, use sp^3 tight binding Hamiltonian to calculate the band structure and the density of states. By using Fourier transform, calculate transmission in $\Gamma - H$ [100] direction.

References

1. S. Datta, *Quantum Transport: Atom to Transistor* (Cambridge University Press, Cambridge, UK, 2005)
2. H. Raza, *Understanding transport through molecules on silicon (doctoral dissertation)*. Purdue University, West Lafayette, IN, USA (2007)

Chapter 5
Charge Based Devices

In this chapter, we discuss ideal and nonideal conduction behavior of various two terminal and three terminal nanodevices based on semiconducting channels. The end of chapter problems encourage the enthusiastic reader to couple NEGF formalism from the previous chapter to the IV characteristics of the nanodevices discussed in this chapter. For the two terminal devices, the channel is connected to the source and the drain contacts, using which one may inject electrons or holes into the channel and extract electrons or holes out of the channel. Source and drain contacts are also referred to as cathode and anode, respectively, with reference to the electron injection and the extraction. For the three terminal devices, the third contact (called gate) is used to electrostatically control the channel to switch the device between ON and OFF states. The current through the gate contact is usually undesirable and hence termed as the gate leakage current. However, in certain devices like nonvolatile flash memory,[1] this leakage current may be made use of, which we discuss in Chap. 7.

The charge transport could be unipolar or bipolar, i.e. whether the dominant transport consists of one polarity of charges (i.e. electrons or holes) or both electrons and holes contribute to the device current. Strictly speaking, since the semiconductor has both conduction and valence bands and hence electrons and holes as the charge carriers, both electron and hole currents flow at nonzero temperature under nonequilibrium conditions. However, if either the electron current or the hole current is much smaller and may be ignored, the device is termed as unipolar. For our discussion it becomes important, since in unipolar devices, one has to calculate the transport through one of the two bands, i.e. either the conduction or the valence bands for electron or hole transport, respectively. On the other hand, in bipolar devices, the current is calculated by using a two band model. Furthermore, the conduction through the valence band and the conduction band is assumed independent of each other, which is a valid assumption in the absence of scattering.

---

[1] A nonvolatile memory keeps its memory state stored during power cycle, whereas a volatile memory looses information during power cycle.

H. Raza, *Nanoelectronics Fundamentals*, NanoScience and Technology,
https://doi.org/10.1007/978-3-030-32573-2_5

We start the discussion with a two terminal bipolar pn junction diode consisting of a bilayer of p-type and n-type semiconductors. This device is also used as a rectifier. Next, we discuss a bipolar Zener diode, which is a special kind of pn junction device with degenerately doped[2] n-type and p-type semiconductors. The last two terminal device we discuss is a unipolar resonant tunneling diode, which is a multilayer structure of metal, insulator and semiconductor materials. Due to the unique tunneling probabilities and transmission function, this diode leads to a negative differential resistance. This phenomenon has been used in bistable circuit applications, like memories and oscillators, etc.

The three terminal device we discuss is a unipolar field effect transistor (FET), where the electric field due to gate voltage controls the channel conduction. The channel is electronically isolated from the gate contact by using a dielectric, with a goal of minimal wavefunction overlap between the channel and the gate contact. If the gate contact is made of metal and the dielectric is an oxide, the transistor is also called MOSFET (metal-oxide-semiconductor FET).[3] Indeed, there are FETs where the gate is polysilicon and the dielectric is an insulator other than an oxide. While the role of gate dielectric in FET is to enable highest possible electric field with smallest leakage current, the gate stack may also be engineered in the nonvolatile flash memory to enable charge storage in the storage node housed in the gate dielectric region as discussed in Chap. 7.

5.1 pn Junction Diode

Consider the isolated p-type and n-type semiconductors in Fig. 5.1a. The band diagram $E(x)$ for the two isolated materials is shown under equilibrium conditions in Fig. 5.1b.

After joining the p-type and n-type semiconductors together, one forms a depletion region around the metallurgical junction as shown in Fig. 5.1c. The circuit symbol is shown as well. The associated band diagram for the pn junction diode is shown in Fig. 5.1d under equilibrium condition.[4] One should note that the conduction and the valence bands bend around the depletion region, which represents a change in the potential energy. At the interface, holes from the p-type region and electrons from the n-type region cross the metallurgical interface [shown by the dashed line in Fig. 5.1c, d] and eventually recombine leaving behind negatively charged ionized acceptors and positively charged ionized donors, respectively, schematically shown in Fig. 5.1d. The negative ionized acceptor charges inside the depletion region in p-type semiconductor and the positive ionized donor charges in the n-type semiconductor

[2]Doping on the order of $10^{18} - 10^{20}$ cm^{-3}.

[3]MOSFET is usually just written as MOS—a convention we also follow in this book unless otherwise necessary.

[4]At equilibrium, one would have the same equilibrium chemical potential energy (μ_o) throughout the device.

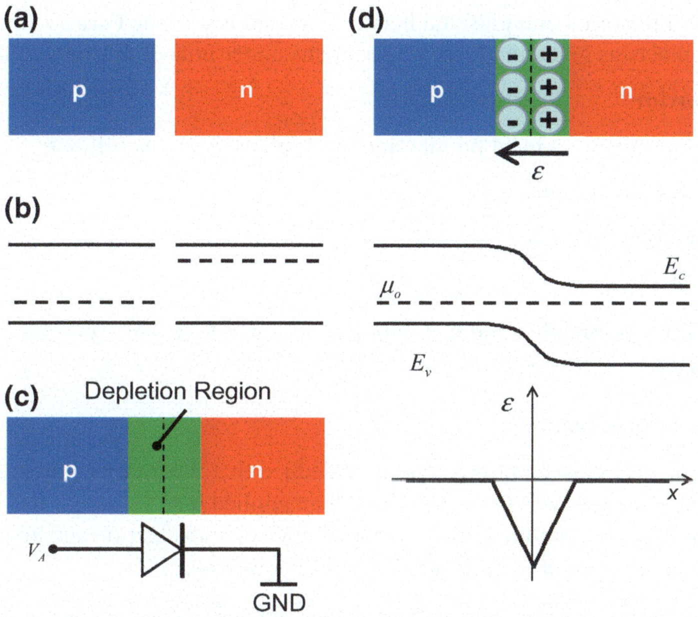

Fig. 5.1 pn junction diode. **a** Isolated p-type and n-type semiconductors. **b** Energy band diagram for the isolated p-type and n-type semiconductors. **c** Schematic device structure of the pn junction diode and the circuit symbol. **d** Depletion region charges, equilibrium energy band diagram, and electric field profile for the pn junction diode

lead to an intrinsic electric field in the depletion width, which has a negative value as shown in Fig. 5.1d.

This electric field is also manifested by the band bending in the conduction band and the valence band. Since the potential energy ($P.E.$) with respect to a constant reference energy (E_{ref}) is defined as,

$$P.E.(x) = E_c(x) - E_{\mathrm{ref}}$$

which also equals $P.E. = -qV$, where V is the potential that relates to the magnitude of the electric field intensity in $1D$ as, $\mathcal{E} = -\mathrm{d}V/\mathrm{d}x$, which leads to,

$$\mathcal{E} = \frac{1}{q}\frac{\mathrm{d}E_c}{\mathrm{d}x}$$

where $\mathrm{d}E_{\mathrm{ref}}/\mathrm{d}x = 0$. Thus, the band bending in the conduction band and the valence band at equilibrium reflects intrinsic electric field in the device region. One should also note that the pn junction diode has an intrinsic electric filed intensity, but the current is still zero under equilibrium condition. It is important to emphasize that it is not the electric field intensity which leads to a net current, rather a difference in the chemical potential energies and hence the difference in the Fermi's functions of the two contacts as given in (4.5). This happens under nonequilibrium conditions.

Ideal Behavior

The assumptions for an ideal pn junction diode analysis are as follows,

(1) 1D analysis with unity transmission.

(2) No scattering processes.

(3) No breakdown.

(4) No RG (recombination and generation of electron-hole pairs) processes.

(5) No light.

(6) No series resistance.

Using the above assumptions, one may deduce that the current flowing through the diode at finite temperature T (in K) with an applied bias V_A across the depletion width. Since the pn junction diode is a bipolar device, both electron and hole currents (I_n and I_p) are important, and the total current is given as,

$$I_D = I_n + I_p$$

Starting with Landauer's equation (4.5), one may write the electron current as,

$$I_n = \frac{2q}{h} \int_{-\infty}^{+\infty} \mathrm{d}E \; T_n(E) \left[f_1(E) - f_2(E) \right]$$

where, the transmission through the conduction band (see Fig. 5.2) is a Heaviside function, and is given as,

$$T_n(E) = \begin{cases} 1 & E_{c(p)} < E \leq \infty \\ 0 & -\infty \leq E \leq E_{c(p)} \end{cases}$$

The electron current thus becomes,

$$I_n = \frac{2q}{h} \int_{E_{c(p)}}^{+\infty} \mathrm{d}E \; \left[f_1(E) - f_2(E) \right] \tag{5.1}$$

where $E_{c(p)}$ is the value of the conduction band edge at the p-side of the junction and is bias dependent, since the conduction band edge moves with the applied bias as follows,

$$E_{c(p)} = E_{co(p)} - qV_A$$

where $E_{co(p)}$ is the equilibrium conduction band edge at the p-side.

The n-side (contact 1) is connected to the ground for which, $\mu_1 = \mu_o$, and the p-side (contact 2) is connected to the positive terminal of the voltage source for which,

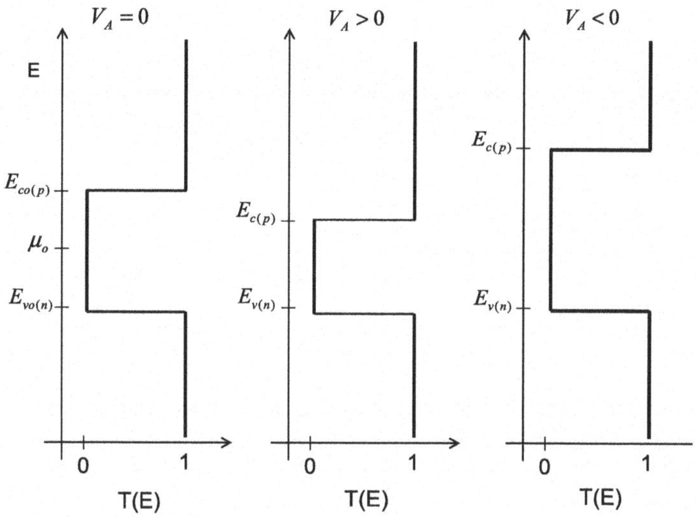

Fig. 5.2 pn junction diode. Transmission at equilibrium ($V_A = 0$), and as a function of the forward bias ($V_A > 0$) and the reverse ($V_A < 0$) bias

$\mu_2 = \mu_o - qV_A$. The Fermi's functions for the two contacts (p-side and n-side regions of the junction) are given as,

$$f_{1,2}(E) = \frac{1}{1 + e^{(E-\mu_{1,2})/k_B T}}$$

By substituting the Fermi's function of the two contacts in (5.1), it can be shown that,

$$I_n \propto \left[e^{qV_A/k_B T} - 1\right]$$

which leads to,

$$I_n = I_{no}\left[e^{qV_A/k_B T} - 1\right]$$

where I_{no} is the proportionality constant and is related to q, h, k_B, T, and material parameters.

Similarly, it may be shown that the hole current is given as,

$$I_p \propto \left[e^{qV_A/k_B T} - 1\right]$$

which leads to,

$$I_p = I_{po}\left[e^{qV_A/k_B T} - 1\right]$$

where I_{po} is the proportionality constant and is related to q, h, k_B, T, and material parameters. The total diode current is then given as,

$$I_D = I_n + I_p = (I_{no} + I_{po})\left[e^{qV_A/k_BT} - 1\right]$$

Putting $I_o = I_{no} + I_{po}$, one obtains the well known ideal diode current equation as follows,

$$I_D = I_o\left[e^{qV_A/k_BT} - 1\right] \tag{5.2}$$

which gives the rectifying IV characteristics as shown in Fig. 5.3 on a linear scale and in Fig. 5.4 on a logarithmic scale.

For the applied voltage $V_A \ll -3k_BT/q$, $e^{qV_A/k_BT} \approx 0$, hence the current is bias independent and given as,

$$I_D \approx -I_o$$

I_o is also called the reverse leakage current and is shown in Fig. 5.3.

Similarly for $V_A \gg +3k_BT/q$,

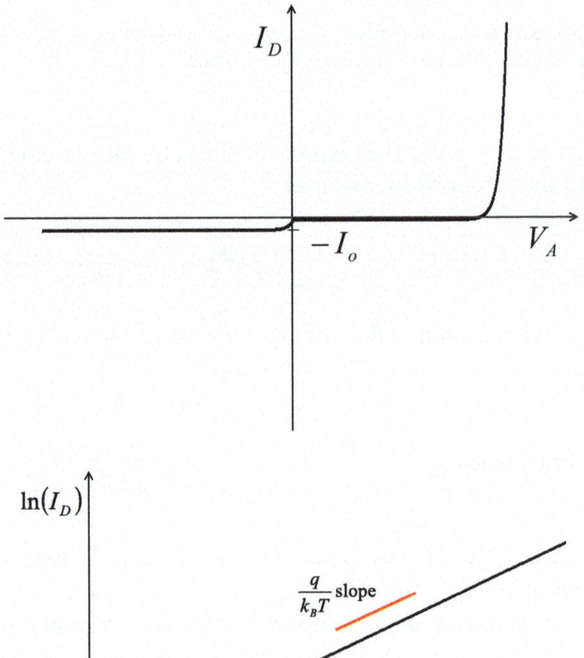

Fig. 5.3 pn junction diode. Ideal IV characteristics on a linear scale

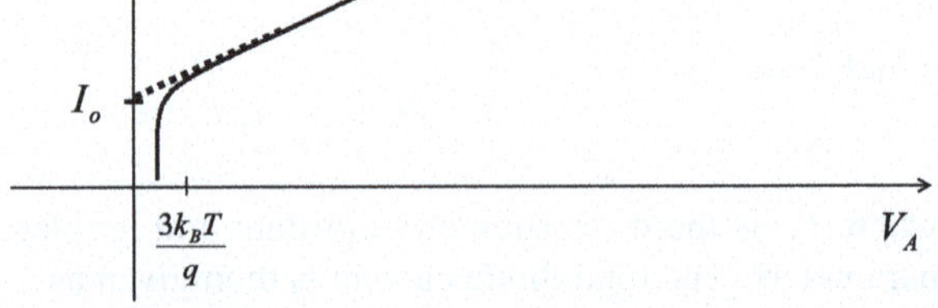

Fig. 5.4 pn junction diode. Ideal IV characteristics on a logarithmic scale

$$I_D \approx I_o e^{qV_A/k_B T}$$

$$ln(I_D) \approx ln(I_o) + \frac{qV_A}{k_B T}$$

where the slope is given as,

$$\frac{\partial\,[ln(I_D)]}{\partial V_A} \approx \frac{q}{k_B T} \tag{5.3}$$

as shown in Fig. 5.4. Since the slope is temperature dependent and involves only a universal constant and the charge of an electron, a pn junction diode is used as a highly sensitive temperature sensor.

Nonideal Behavior

Next we discuss the nonideal behavior of the pn junction diodes. Such nonideal behavior usually involves scattering, which may be included in NEGF by using a scattering self energy as discussed in the previous chapter.

By including scattering due to *RG* (recombination and generation) processes involving electron-hole pairs, an additional current flows, which is given as,

$$I_{RG} = I_{oRG}\left[e^{qV_A/2k_B T} - 1\right] \tag{5.4}$$

with a slope of,

$$\frac{\partial\,[ln(I_{RG})]}{\partial V_A} = \frac{q}{2k_B T} \tag{5.5}$$

The total current (including RG current) is thus given as,

$$I_D = I_o\left[e^{qV_A/k_B T} - 1\right] + I_{oRG}\left[e^{qV_A/2k_B T} - 1\right] \tag{5.6}$$

where, $I_o \ll I_{oRG}$ as shown in Fig. 5.5. Additionally, the current saturates at high bias due to series resistance effect (which consists of the contact resistance and the finite resistance of p-type and n-type regions). Given the total bias of V_B, the applied bias V_A across the depletion region is given as,

$$V_A = V_B - IR_s$$

Current in (5.6) is then given as,

$$I_D = I_o\left[e^{q(V_B - IR_s)/k_B T} - 1\right] + I_{oRG}\left[e^{q(V_B - IR_s)/2k_B T} - 1\right] \tag{5.7}$$

Finally, with large reverse bias, the reverse leakage current increases due to avalanche process enabled by a high electric field intensity as shown in Fig. 5.6. The avalanche breakdown is a chain reaction, where a high energy electron gives off its excess energy to an electron in the valence band and creates an electron-hole pair. The generated electrons lead to further avalanche scattering in addition to the

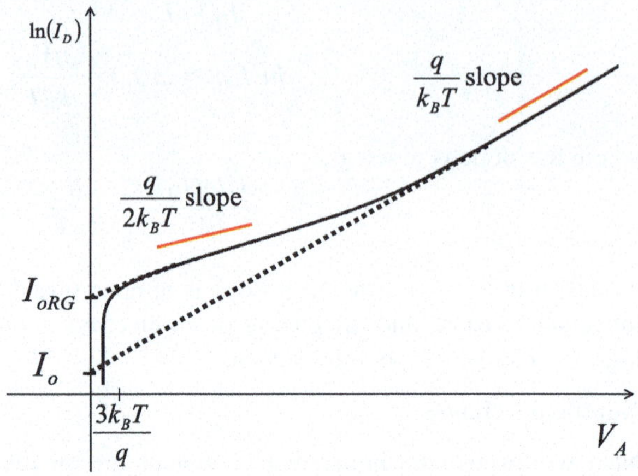

Fig. 5.5 pn junction diode. Nonideal IV characteristics on a logarithmic scale under forward bias

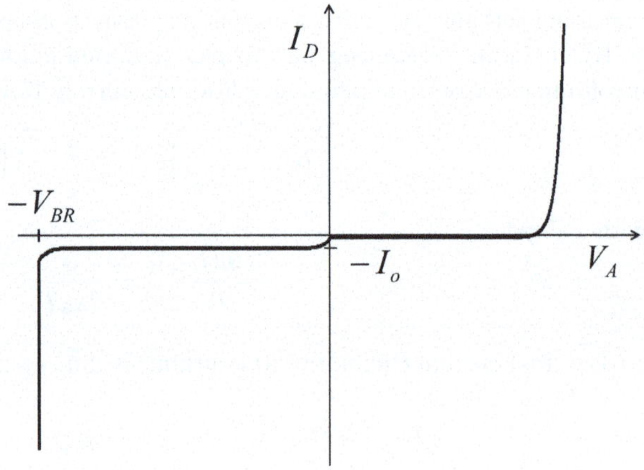

Fig. 5.6 pn junction diode. Nonideal IV characteristics on a linear scale under reverse bias

original electrons resulting in a multiplicative process. Avalanche breakdown may be included within NEGF equations by including an inelastic interband scattering self energy. Avalanche breakdown leads to defect generation as well as heat dissipation, which may lead to device degradation. Therefore, it is instructive not to operate this diode in the avalanche breakdown regime.

5.2 Zener Diode

With low or moderate doping, the depletion width is large, usually in microscale. However, with degenerate doping of the p-type and the n-type regions, the depletion width is on the order of *few tens of nanometers*. Depletion widths of up to 10 nm are

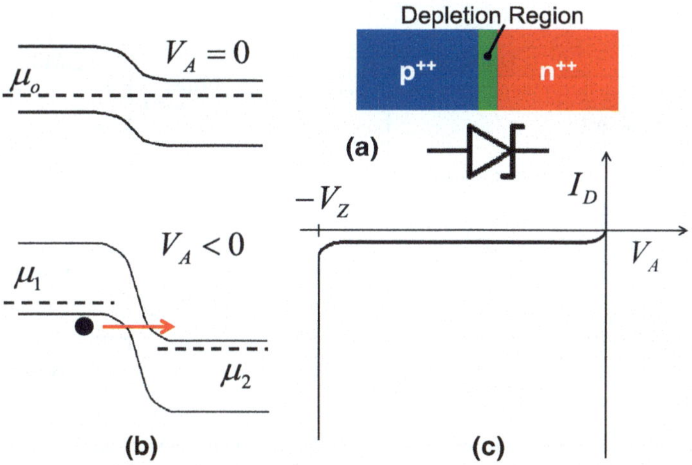

Fig. 5.7 Zener diode. **a** Device structure and circuit symbol. **b** Band diagram under equilibrium, and nonequilibrium conditions. **c** IV characteristics under reverse bias

achievable with p^{++}–n^{++} Zener diodes.[5] The device structure and circuit symbol are shown in Fig. 5.7a.

Under negative bias V_A, electrons from the valence band of p-type semiconductor quantum mechanically tunnel through the nanoscale depletion width into the conduction band of the n-type semiconductor as shown in Fig. 5.7b. This leads to a different kind of breakdown called Zener Breakdown and happens at rather low voltage called Zener voltage (V_Z) (within $4 - 6\ E_g/q$, where E_g is the semiconductor bandgap) as shown in Fig. 5.7c. This is one of the key features of the Zener breakdown that distinguishes it from the Avalanche breakdown. Furthermore, since Zener diode does not involve any scattering processes rather a gentle tunneling process, it does not lead to any defect generation and heat dissipation. The Zener diodes may thus be operated in the breakdown regime over much longer time duration than the Avalanche diodes. Zener diodes may also be gated to make a three terminal tunnel transistor, which shows very high ON/OFF current ratios under small supply voltage requirements.

5.3 Field Effect Transistor

Let us consider three terminal devices and the associated circuit symbols. The conduction through these devices may be electrostatically controlled by using an appropriate polarity and the magnitude of the gate voltage. The p-channel MOS (generally referred to as pMOS) is shown in Fig. 5.8a, where the source and the drain contacts consist of degenerately doped p^{++} regions, and the substrate consists of n-type semiconductor. The source/drain contacts and the substrate interface thus form pn

[5] The superscript $^{++}$ denotes degenerate doping.

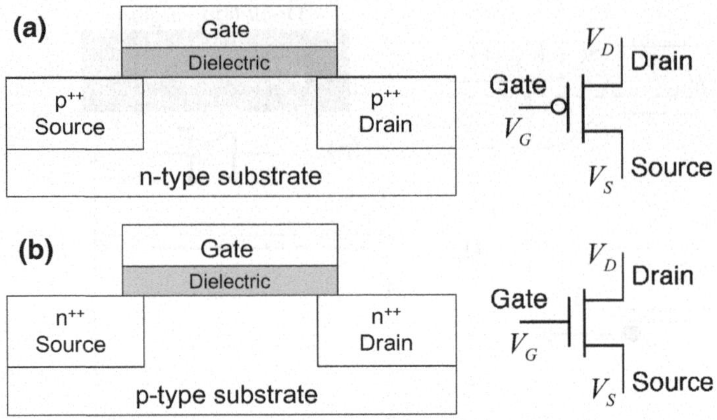

Fig. 5.8 Metal-oxide-semiconductor field effect transistor. Schematic device structure and circuit symbol for **a** pMOS, and **b** nMOS

junction diodes, which are connected back to back—thereby turning the transistor OFF in the absence of a gate voltage. However, by applying an appropriate negative gate voltage, one may accumulate enough holes at the substrate/gate interface to form an inversion layer and hence a p-channel between the source and the drain, thereby turning the transistor ON.

Similarly, an n-channel MOS (generally referred to as nMOS) is shown in Fig. 5.8b, where the source and the drain contacts consist of degenerately doped n^{++} regions, and the substrate consists of p-type semiconductor. In this case, by applying a large enough positive gate voltage, one may accumulate enough electrons at the substrate/gate interface to form an n-channel between the source and the drain, thereby turning the transistor ON.

Electrostatics

For an nMOS, isolated band diagrams for the source, channel, and drain are shown in Fig. 5.9a. nMOS with source/drain contacts shown in Fig. 5.8b may be thought of as two back-to-back connected pn^{++} junction diodes. Due to this similarity, the band diagram of the channel and the source/drain contacts looks like a series combination of the band diagrams of two pn junction diodes as shown in Fig. 5.9b. The band diagram for a shorter channel length is also shown in Fig. 5.9c.

For $V_{DS} = 0$,[6] we next discuss the band diagrams for various polarities of the gate voltages. For $V_{GS} = 0$,[7] the electron in the source contact see a barrier as shown in Fig. 5.10a. For $V_{GS} > 0$, the barrier may be lowered as shown in Fig. 5.10b leading to ON state of the transistor. Similarly, for $V_{GS} < 0$, the barrier height may be increased as shown in Fig. 5.10c, leading to the OFF state of the transistor. It should be noted that

[6] $V_{DS} = V_D - V_S$. Usually, source contact in an nMOS is connected to ground and in a pMOS is connected to the supply voltage.

[7] $V_{GS} = V_G - V_S$.

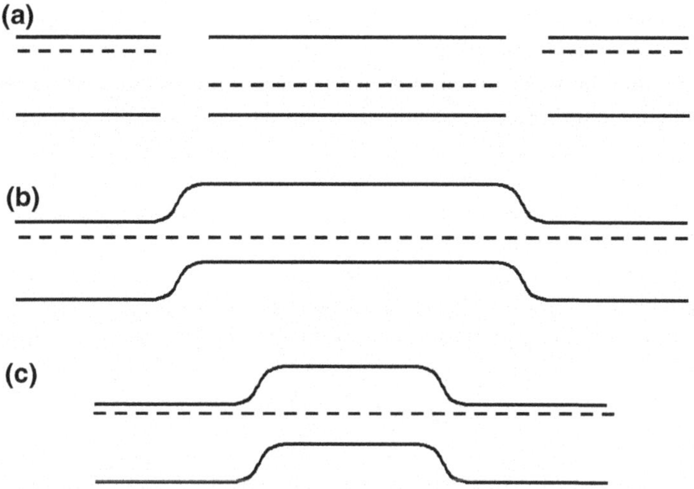

Fig. 5.9 nMOS. Equilibrium band diagram for **a** isolated source, channel and drain, **b** nMOS with a long channel, **c** nMOS with a short channel

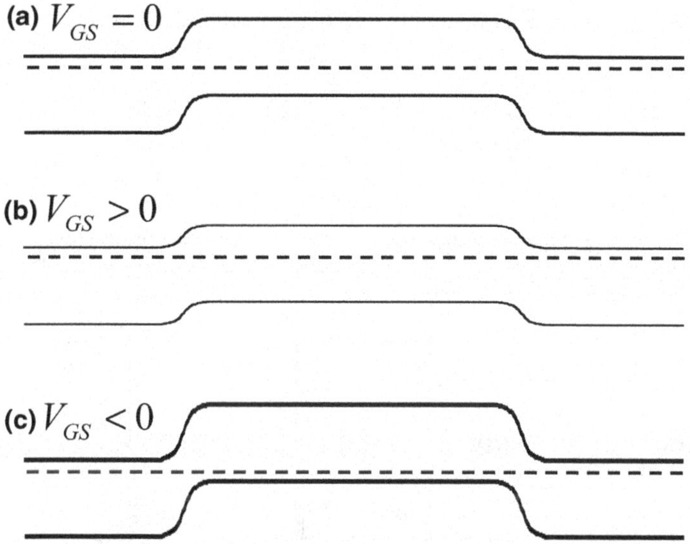

Fig. 5.10 nMOS. Energy band diagram as a function of the gate voltage for **a** $V_{GS} = 0$, **b** $V_{GS} > 0$ and **c** $V_{GS} < 0$

with a finite drain voltage, the OFF current may never be zero at a finite temperature due to the exponential nature of the Fermi's function as discussed later.

Without the degenerately doped source and drain contacts the device is called a MOS capacitor where the first terminal is the gate contact and the second terminal is usually the channel itself. In reference to the nMOS in Fig. 5.11a, MOS capacitor with p-type substrate is shown in Fig. 5.11b, the analysis of which would be helpful in understanding nMOS operation.

Assuming that the work function of the metal and the silicon match, one obtains flat bands under equilibrium conditions as shown in Fig. 5.11c. The capacitance corresponding to this flat band condition is called the flat band capacitance (C_{FB}) and is shown in Fig. 5.12 with a flat band voltage (V_{FB}) of zero in this case.[8] However, with a difference in the chemical potential energies of the metal and the semiconducting channel, the flat band voltage is nonzero.

For the p-type substrate, with $V_G < 0$, majority carrier holes are accumulated at the oxide-semiconductor interface as shown in Fig. 5.11d, resulting in a parallel plate capacitor with the accumulation capacitance equal to the oxide capacitance as follows,

$$C_{acc} = C_o = S\frac{K_{ox}\varepsilon_o}{t_{ox}} \tag{5.8}$$

and is shown in Fig. 5.12, where K_{ox} is the relative dielectric constant of oxide film[9] and t_{ox} is the oxide thickness. S is the cross sectional area of the gate contact.

With an increasing $V_G > 0$, the majority carrier holes are repelled from the oxide-semiconductor interface, giving rise to an increasing depletion region of width W as shown in Fig. 5.11e. This results in another capacitance in series with the oxide

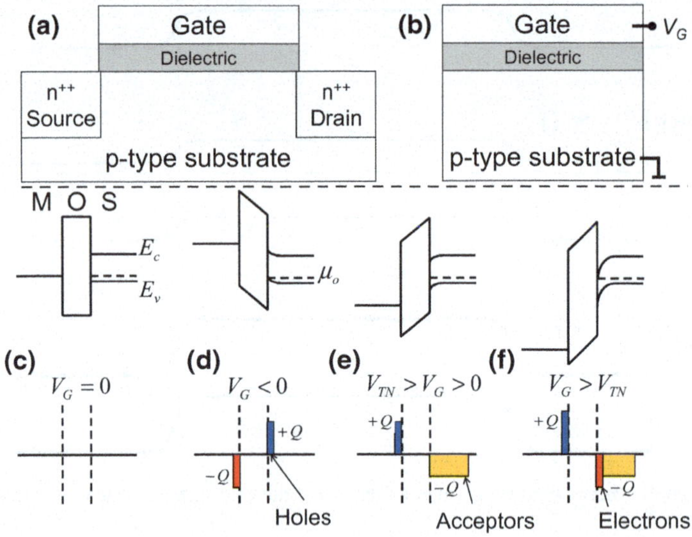

Fig. 5.11 MOS Capacitor. **a** nMOS device structure. **b** MOS capacitor device structure. Band diagrams and charge profiles under **c** flat band, **d** accumulation, **e** depletion, and **f** inversion

[8] While we discuss electrostatic capacitance here, the discussion of quantum capacitance is beyond the scope of this book.

[9] $K_{ox} = 3.9$ for silicon dioxide.

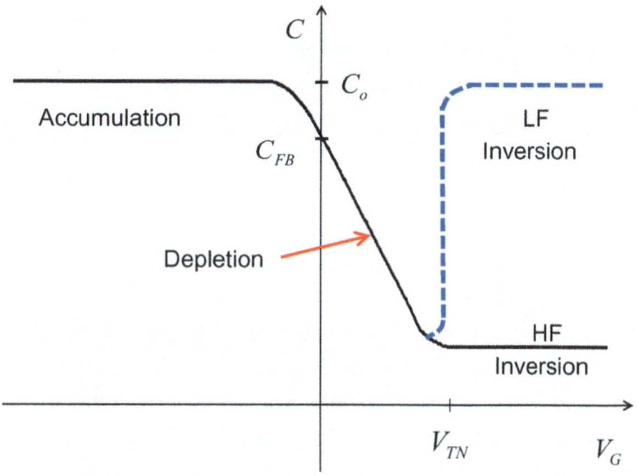

Fig. 5.12 CV characteristics of MOS capacitor with p-type substrate

capacitance, which decreases with increasing depletion width. The net depletion capacitance is then given as,

$$C_{\text{dep}} = C_o || C_s = \frac{C_o C_s}{C_o + C_s} = \frac{C_o}{1 + \frac{K_{ox} W}{K_s t_{ox}}} \tag{5.9}$$

With further increase in V_{G}, the minority carrier electrons start to collect at the oxide-semiconductor interface. At a certain gate voltage, called the threshold voltage (V_{TN}) for nMOS, the minority carrier density becomes equal to the ionized dopant density in the depletion region.[10] For $V_{\text{G}} = V_{\text{TN}}$, the depletion width is given as W_T, and the depletion capacitance is as follows,

$$C_{\text{dep}}(V_{\text{G}} = V_T) = C_o || C_s = \frac{C_o C_s}{C_o + C_s} = \frac{C_o}{1 + \frac{K_{ox} W_T}{K_s t_{ox}}}$$

For $V_{\text{G}} > V_{\text{TN}}$, the minority carrier density becomes greater than the ionized dopant density at the oxide-semiconductor interface, and the channel becomes inverted as shown in Fig. 5.11f, which in an nMOS results in the ON state of the transistor.

The minority carriers have a RG (recombination generation) lifetime on the order of microseconds. Under low frequency excitation (less than about 1 MHz), the minority carriers follow the change in the gate voltage and hence the minority carrier inversion layer follows the gate voltage. Thus, the inversion capacitance under low frequency (LF) excitation is equal to the oxide capacitance as shown in Fig. 5.12 and given as follows,

$$C_{\text{inv}}(\text{LF}) = C_o = S\frac{K_{ox}\varepsilon_o}{t_{ox}} \tag{5.10}$$

[10]The region below the threshold voltage (V_{TN}), i.e. $V_{\text{G}} < V_{\text{TN}}$ for nMOS, is called subthreshold region.

However, under high frequency (HF) gate voltage excitation, the minority carriers lag behind, which results in the depletion width edge following the gate voltage, and the inversion capacitance gets pinned at,

$$C_{\text{inv}}(\text{HF}) = \frac{C_o}{1 + \frac{K_{ox} W_T}{K_s t_{ox}}} \tag{5.11}$$

While the above discussion of CV characteristics is for MOS capacitors, it may also be applied to MOSFETs with a key difference that under inversion conditions, the minority carrier electrons do not have to come from the electron-hole pair thermal generation process, rather they may be injected from the source/drain contacts, which are highly doped n^{++}-type semiconductors. Therefore, CV characteristics of a MOS transistor looks exactly like that of a LF MOS capacitor. Apart from this, there is no speed penalty for minority carrier injection in a MOS transistor. This enables picosecond (THz) operation at the device level, which is instrumental to achieve nanosecond (GHz) operation at the circuit level.

We discuss nMOS in this section, but similar discussion applies to the pMOS with opposite polarity of the gate voltage. Indeed, the threshold voltage for pMOS is negative and is given as V_{TP}. In order to achieve inversion in a pMOS to turn it ON, $V_{\text{GS}} < V_{\text{TP}}$.

Ideal Behavior

The assumptions for an ideal MOSFET analysis are as follows,

(1) 1D analysis with unity transmission.

(2) No scattering processes.

(3) No breakdown.

(4) No RG processes.

(5) No light.

(6) No series resistance.

Starting with the Landauer's equation, the channel current is as follows,

$$I = \frac{2q}{h} \int_{-\infty}^{+\infty} \mathrm{d}E \; T(E) \left[f_1(E) - f_2(E)\right]$$

where the transmission is shown in Fig. 5.13 as a function of gate bias. Since FET is a unipolar device, we calculate the transport through the conduction band only for an nMOS, and through valence band only for pMOS.

For an nMOS, the channel current becomes,

$$I_D = \frac{2q}{h} \int_{E_{c(ch)}}^{+\infty} \mathrm{d}E \; \left[f_1(E) - f_2(E)\right]$$

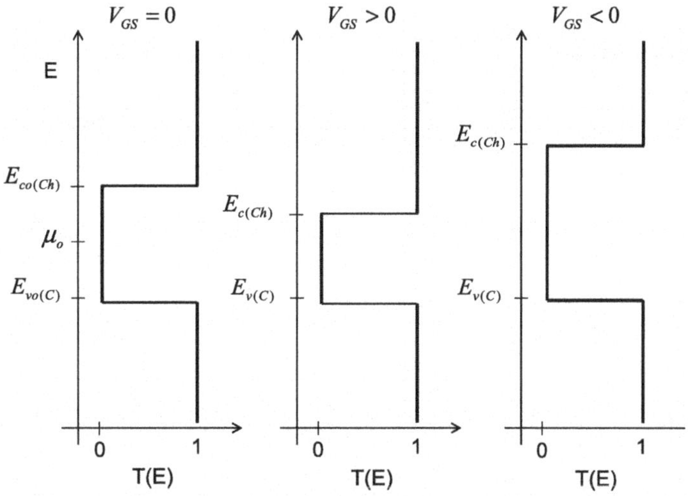

Fig. 5.13 nMOS. Transmission as a function of the gate voltage

where $E_{c(ch)}$ is the value of the conduction band edge in the channel region and is gate voltage dependent. The conduction band edge moves with the gate voltage as follows,

$$E_{c(ch)} = E_{co(ch)} - \frac{qV_{\mathrm{G}}}{m},$$

where $E_{co(ch)}$ is the equilibrium conduction band edge for the channel and m is the body factor given as,

$$m = 1 + \frac{C_{\mathrm{dep}}}{C_{ox}}. \tag{5.12}$$

where C_{dep} is the depletion capacitance. The body factor describes the electrostatic control of the gate on the channel, and is ideally unity.

While we ignore the effect of V_{DS} here, the conduction band edge is drain bias dependent too and in fact varies nonlinearly from the source to the drain under a finite source-drain voltage V_{DS}.

The channel current shown in Fig. 5.14 is thus given as,

$$I_D = \frac{2q}{h}\int_{E_{co(ch)}-qV_{\mathrm{G}}}^{+\infty} \mathrm{d}E\ \left[f_1(E) - f_2(E)\right]$$

For $(V_{\mathrm{GS}} - V_{\mathrm{TN}}) \le -3k_{\mathrm{B}}T/q$, the transfer characteristics $(I_{\mathrm{D}} - \mathrm{V_{GS}})$ are as follows,

$$I_D = I_{Do}e^{qV_{\mathrm{GS}}/k_{\mathrm{B}}T} \tag{5.13}$$

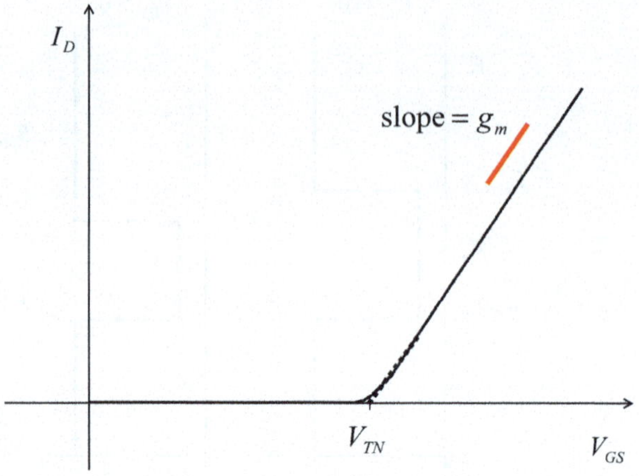

Fig. 5.14 nMOS. $I_\mathrm{D} - \mathrm{V_{GS}}$ transfer characteristics on a linear scale

where I_{Do} is drain voltage dependent and is zero under equilibrium conditions, i.e. $V_\mathrm{DS} = 0$.

Rearranging the channel current as,

$$\ln(I_D) = \ln(I_{Do}) + \frac{qV_\mathrm{GS}}{k_\mathrm{B}T}$$

$$\log_{10}(I_D) = \log_{10}(I_{Do}) + \frac{qV_\mathrm{GS}}{2.3 \times k_\mathrm{B}T}$$

gives a slope of,

$$\frac{\partial\left[\log_{10}(I_D)\right]}{\partial V_\mathrm{GS}} = \frac{q}{2.3 \times k_\mathrm{B}T}$$

The inverse subthreshold slope is then defined as follows,

$$S = \left[\frac{\partial\left[\log_{10}(I_D)\right]}{\partial V_\mathrm{GS}}\right]^{-1} = 2.3 \times \frac{k_\mathrm{B}T}{q} \tag{5.14}$$

which at room temperature is about 60 mV/decade as shown in Fig. 5.15.

For $(V_\mathrm{GS} - V_\mathrm{TN}) \geq 3k_\mathrm{B}T/q$,

$$I_D = g_m\left(V_\mathrm{GS} - V_\mathrm{TN}\right)$$

where g_m is the transconductance, which is as follows,

$$g_m = \frac{\partial I_D}{\partial V_\mathrm{GS}} \tag{5.15}$$

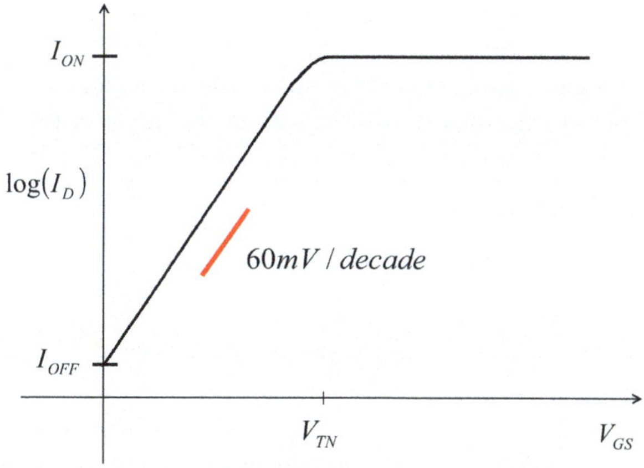

Fig. 5.15 nMOS. $I_D - V_{GS}$ transfer characteristics on a logarithmic scale at room temperature showing 60 mV/decade of inverse subthreshold slope

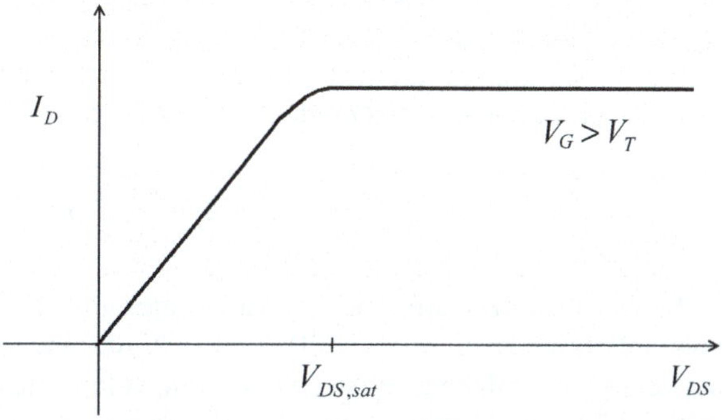

Fig. 5.16 nMOS. $I_D - V_{DS}$ output characteristics on a linear scale

and is shown in Fig. 5.14.

While it is not straightforward to derive an analytical expression for the output ($I_D - V_{DS}$) characteristics, it may be calculated by using NEGF formalism discussed in Chap. 4, and is shown in Fig. 5.16. Clearly, there are two regions in the output characteristics; a linear region also called triode region and a saturation region above a drain bias written as $V_{DS,sat}$.

By using the output characteristics, one may calculate the output resistance as,

$$r_o = \frac{\partial V_{DS}}{\partial I_D} \tag{5.16}$$

Finally, the voltage gain is defined as follows,

$$A_v = \frac{\partial V_{DS}}{\partial V_{GS}} = \frac{\partial I_D}{\partial V_{GS}} \frac{\partial V_{DS}}{\partial I_D}$$

and given in terms of the transconductance (g_m) and output resistance (r_o) as follows,

$$A_v = g_m r_o \tag{5.17}$$

A gain value greater than unity is desirable for a transistor working as an amplifier. Operating a transistor in the saturation region of the output characteristics helps to achieve a higher output resistance, and hence a higher gain.

Nonideal Behavior

Consider the body factor again,

$$m = 1 + \frac{C_{\mathrm{dep}}}{C_{ox}}$$

Under ideal conditions, $C_{\mathrm{dep}} \ll C_{ox}$, and hence one may ignore the C_{dep}/C_{ox} ratio, which gives $m \approx 1$ and hence an inverse subthreshold slope of $S = 2.3\, k_{\mathrm{B}}T/q$ (about 60 mV/decade at room temperature). The gate voltage induced barrier in the channel region follows the gate voltage variation with one-to-one correspondence. However, for nanoscale transistors, the body factor may have a value much greater than unity, in which case, one has to include the body factor in (5.14) as follows,

$$S = 2.3 \frac{k_{\mathrm{B}}T}{q} \times m$$

It thus becomes difficult to turn a transistor OFF. In fact, $m = 2 - 3$ are not unheard of, which gives $S = 120 - 180$ mV/decade respectively. For a longer channel device, the influence of the source-drain voltage on the barrier height for electrons injected from the source contact is negligible as shown in Fig. 5.17a. However, with device scaling, a positive drain voltage may lead to a drop in the barrier height, known as drain induced barrier lowering (DIBL) as shown in Fig. 5.17b, where the blue dashed line schematically shows the location of the barrier close to the source contact in the absence of V_{DS}. With such a drain voltage the transfer characteristics show higher OFF current as shown in Fig. 5.17c.

With the channel length scaling, the barrier width may also become small enough that charge carriers may simply tunnel through the channel as shown in Fig. 5.18. This phenomenon is termed as source-drain tunneling leakage and should not be confused with the OFF current. With the barrier modification due to the drain bias, this source-drain leakage may become even more severe.

5.4 Resonant Tunneling Diode

Consider the multilayer device structure shown in Fig. 5.19, where a semiconductor film is sandwiched between two insulating films, probed by two semiconducting contacts. This unipolar device is called resonant tunneling diode (RTD). If we were to replace the right contact by a metal, the device operating principle still remains the

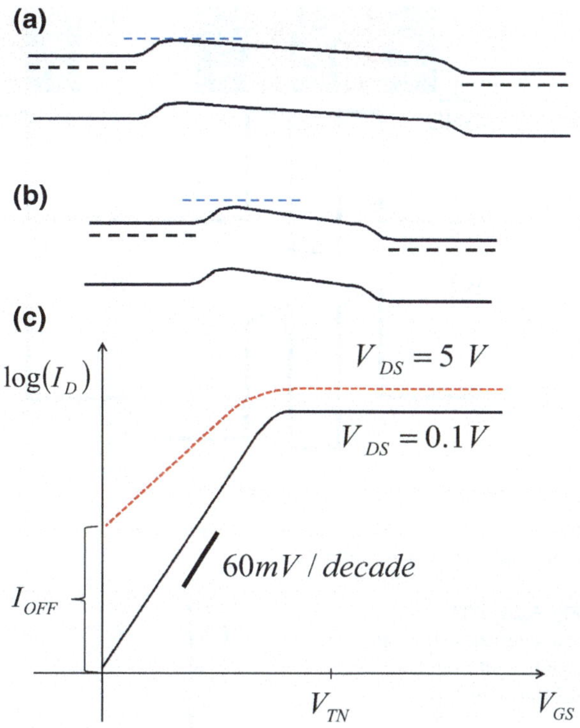

Fig. 5.17 nMOS DIBL. Band diagram for **a** long channel nMOS, and **b** short channel nMOS. **c** Effect of DIBL on the subthreshold characteristics

Fig. 5.18 nMOS. Source drain tunneling leakage

same. The key requirement is that one of the two contacts should be a semiconductor. Furthermore, the middle semiconducting region could be a semiconducting film, a quantum dot, or a molecule with quantized energy levels.

The equilibrium band diagram for this device is shown in Fig. 5.19a, where the blue dashed line represents the equilibrium chemical potential energy (μ_o) and the black solid lines show the conduction band profile in various regions. Due to the quantization in the transport direction, the energy levels are shown in the middle semiconducting region.

To begin with as shown in Fig. 5.19b, the energy levels are far from the chemical potential energy, i.e. in offresonant state, therefore the current is small as shown in Fig. 5.20. In this state, the current increases with increasing bias.

With a positive bias on the right contact, the energy levels move downward and at a certain bias as shown in Fig. 5.19c, the energy level becomes resonant with the chemical potential energy of the left contact, which results in increasing current

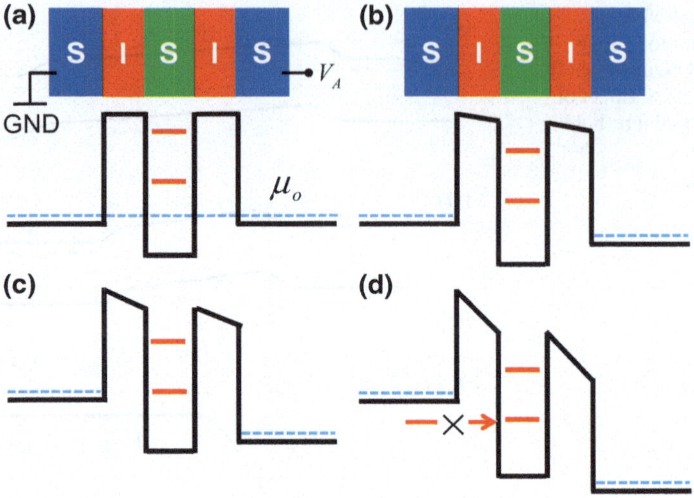

Fig. 5.19 RTD. Band diagram under **a** equilibrium, **b** increasing bias, **c** resonance, and **d** negative differential resistance condition

Fig. 5.20 RTD. IV characteristics showing negative differential resistance feature

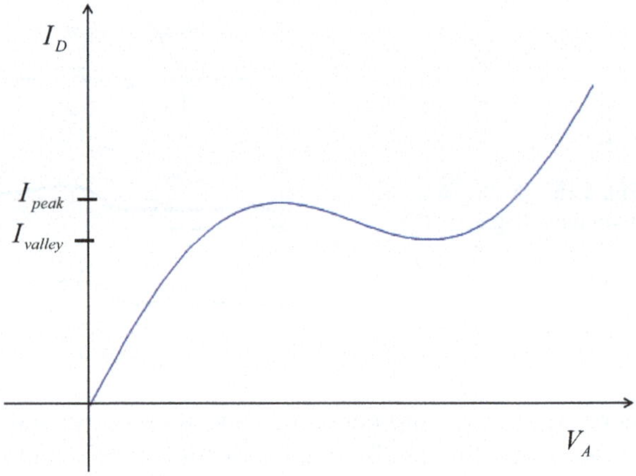

as shown in Fig. 5.20. With a further increase in the applied bias, the energy level moves downwards as shown in Fig. 5.19d, which results in zero coupling resulting in a reduced transmission. In this energy range, the current decreases with the increasing bias.

The maximum current and the minimum current in this region are called the peak current (I_{peak}) and the valley current (I_{valley}), respectively. Since $\mathrm{d}I_D/\mathrm{d}V_A$ or $\mathrm{d}V_A/\mathrm{d}I_D$ is negative in this range, this phenomenon is called negative differential conductance (NDC) or negative differential resistance (NDR). One should note that the conductance (I_D/V_A) and resistance (V_A/I_D) remain positive. It is the differential conductance ($\partial I_D/\partial V_A$) and differential resistance ($\partial V_A/\partial I_D$) that become negative. With a further increase in the applied bias, if additional energy levels enter the conduction window, the current starts to increase again as shown in Fig. 5.20. This bistable behavior may be used in oscillators, memories, and various other applications.

Problems

5.1 pn junction diode. Using single orbital basis set for a 1D channel and contacts with $t = 1$ eV, for a device with nine atoms, the approximate conduction band profile (to be added to the diagonal of the Hamiltonian) is shown in Fig. 5.21 and given as,
[0 0 0 0.25 0.50 0.75 1 1 1]
Calculate the IV characteristics for the conduction band (only) of this pn junction diode with bias applied to contact 2 (the p-region) with –1 V to 1 V range. Assume the voltage drop across the middle five atoms as below,
U = [0 0 linspace(0,–V,5) –V –V]
Take μ_o = –2 eV.

5.2 Resonant Tunneling Diode. Using single orbital basis set for a 1D channel and contacts with $t = 1$ eV, for a device with nine atoms, the approximate conduction band profile (to be added to the diagonal of the Hamiltonian) is shown in Fig. 5.22 and given as,
[0 1 1 0 0 0 1 1 0]

Fig. 5.21 Problem 5.1

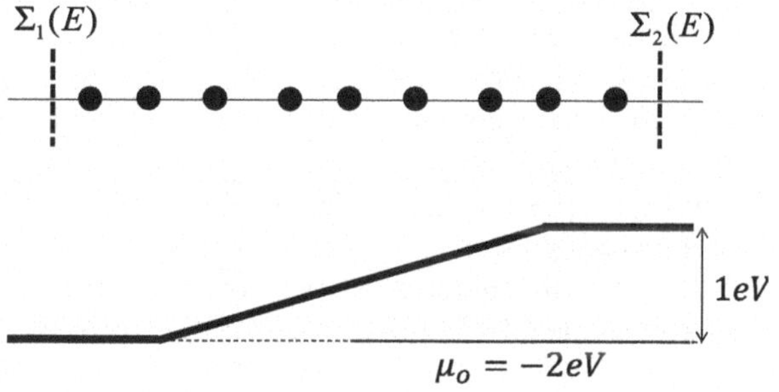

Fig. 5.22 Problem 5.2

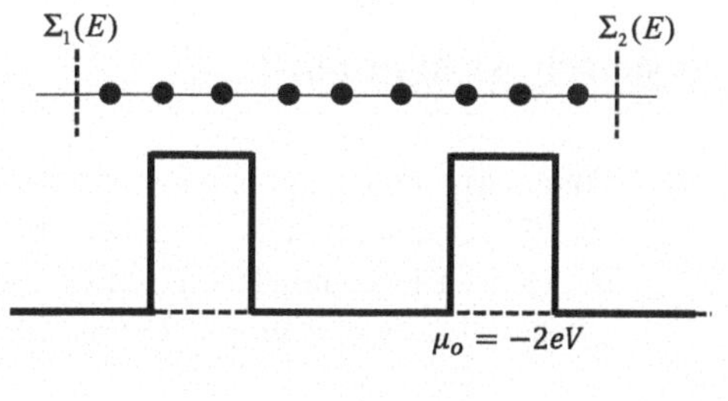

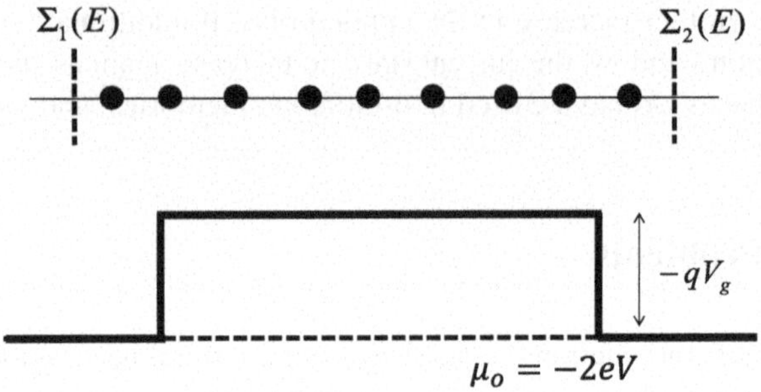

Fig. 5.23 Problem 5.3

Calculate the IV characteristics for the conduction band (only) of this resonant tunneling diode with bias applied to contact 2 with 0 V to 1 V range. Assume the voltage drop across the middle seven atoms as below,
U = [0 linspace(0,–V,7) –V]
Take $\mu_o = -2$ eV.

5.3 Field Effect Transistor. Using single orbital basis set for a 1D channel and contacts with $t = 1$ eV, for a device with 9 atoms shown in Fig. 5.23, calculate the output (I_d–V_{ds}) characteristics for the conduction band (only) of this channel with bias applied to contact 2 with 0 to 0.2 V range. Assume a linear voltage drop across the device due to drain voltage.
For the gate voltage, the conduction band profile (to be added to the diagonal of the Hamiltonian) is shown below and is given as,
[0 V_g V_g V_g V_g V_g V_g V_g V_g V_g 0]
Calculate the transfer (I_d–V_{ds}) characteristics for a drain bias of 0.1 V and gate voltage range of 0 V to −1 V.
Take $\mu_o = -2$ eV.

5.4 What is the value of the inverse subthreshold slope for an ideal FET at room temperature?

5.5 If $I_G = 1\mu A$ for $V_G = 0.1V$ and $I_G = 10\mu A$ for $V_G = 0.2V$, determine inverse subthreshold slope.

5.6 How does the subthreshold performance of the transistor in problem 5.5 compare with an ideal FET if one compares the respective inverse subthreshold slopes?

Research Assignments

R5.1 For an armchair graphene nanoribbon of atomic width 3, calculate the transfer (I_d–V_{gs}) characteristics for a drain bias of 0.1 V and gate voltage range of 0 V to −5 V. The channel consists of four unit cells. Source and drain contacts consist of armchair graphene nanoribbons of atomic width 5. $t = -3$ eV.

Chapter 6
Spin Based Devices

In the previous chapter, we discuss devices that control the flow of electrons or holes by making use of an electrostatic potential (intrinsic or extrinsic). Charge carriers (electrons and holes) have an additional degree of freedom in the form of spin, which may be thought of as a net magnetic moment pointing in two opposite directions, which are taken as either up-spin ($\uparrow$-spin) or down-spin ($\downarrow$-spin).[1] In this chapter, we discuss the spin dependent properties of materials and how one may use them in devices—an area of immense scientific and technological interest, known as Spintronics. We start the discussion with the magnetic, electronic, and transport properties of ferromagnetic (FM) materials. Next, we discuss GMR (giant magnetoresistance) and MTJ (magnetic tunnel junction) devices. Finally, devices based on spin transfer torque (STT) are discussed.

6.1 Ferromagnetic Materials

Ferromagnetic materials are the common magnets. With applied magnetic field intensity ($\boldsymbol{H}$) in a certain direction, a net magnetization ($\boldsymbol{M}$) is developed in these materials. Such a magnetization M–H curve is shown in Fig. 6.1. Even for the applied magnetic field ($\boldsymbol{H}$) approaching zero, a net remanent magnetization or remanence (M_r) is retained in the material. To reduce the magnetization to zero, additional magnetic field intensity is required, known as coercivity or coercive field (H_c) as shown in Fig. 6.1.

Generally, the net magnetization $\boldsymbol{M}$ may be written as a function of magnetization $\boldsymbol{m}$ of individual electrons as follows,

[1]There are two kinds of magnetizations associated with an electron, one due to its spin and the other one due to its orbital motion.

H. Raza, *Nanoelectronics Fundamentals*, NanoScience and Technology,
https://doi.org/10.1007/978-3-030-32573-2_6

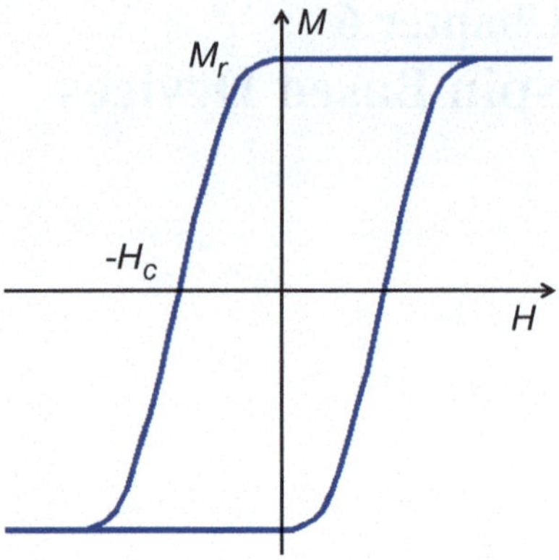

Fig. 6.1 Magnetization curve of a ferromagnet

$$\boldsymbol{M} = \sum_{i=1}^{N} \boldsymbol{m}_i$$

where N is the total number of electrons with individual magnetization $\boldsymbol{m}_i$. Furthermore, the magnetic flux density is given as,

$$\boldsymbol{B} = \mu_o (\boldsymbol{H} + \boldsymbol{M}) \tag{6.1}$$

where $\boldsymbol{M} = \chi_m \boldsymbol{H}$, which gives $\boldsymbol{B} = \mu_o (1 + \chi_m) \boldsymbol{H}$, where the magnetic susceptibility (χ_m) is given as,

$$\chi_m = \frac{\partial M}{\partial H}$$

Since spins may be associated with two distinct directions, the magnetization due to ↑-spin and ↓-spin electrons is respectively given as, $\boldsymbol{M}_\uparrow = N_\uparrow \boldsymbol{m}_s^\uparrow$ and $\boldsymbol{M}_\downarrow = N_\downarrow \boldsymbol{m}_s^\downarrow$; where $N_\uparrow$ and $N_\downarrow$ are the number of ↑-spin and ↓-spin electrons, respectively, and $\boldsymbol{m}_s^\uparrow$ and $\boldsymbol{m}_s^\downarrow$ are the magnetizations of the individual ↑-spin and ↓-spin electrons, respectively. The net magnetization is thus given as,

$$\boldsymbol{M} = \boldsymbol{M}_\uparrow + \boldsymbol{M}_\downarrow$$

The reference direction for spin magnetization is usually taken along z-axis, which gives the magnetization due to ↑-spin and ↓-spin electron as follows,

$$\boldsymbol{m}_s^\uparrow = +\frac{\hbar}{2} z$$

$$\boldsymbol{m}_s^\downarrow = -\frac{\hbar}{2} z$$

The factor of 1/2 is due to the Fermion nature of electrons. The net magnetization is thus given as,

$$\boldsymbol{M} = \left(N_{\uparrow} - N_{\downarrow}\right) \frac{\hbar}{2} z$$

which may be aligned along $+z$ for $\left(N_{\uparrow} > N_{\downarrow}\right)$ and vice versa. Here we introduce a new terminology about spin in addition to the ↑-spin and ↓-spin terms, namely majority spin and minority spin. The spin orientation that has the larger number of electrons is called the majority spin and the one that has lesser number of electrons is called the minority spin.

Thus for a material with $N_{\uparrow} > N_{\downarrow}$, $N_{\text{maj}} = N_{\uparrow}$ and $N_{\text{min}} = N_{\downarrow}$; whereas for $N_{\downarrow} > N_{\uparrow}$, $N_{\text{maj}} = N_{\downarrow}$ and $N_{\text{min}} = N_{\uparrow}$. The magnetic polarization (P) of a ferromagnetic material is given as,

$$P = \frac{N_{\text{maj}} - N_{\text{min}}}{N_{\text{maj}} + N_{\text{min}}} \tag{6.2}$$

which is always a positive quantity. For Fe, the polarization has a value of about 0.45.

The band structure of a nonmagnetic (NM) material has degenerate bands for ↑-spin and ↓-spin electrons, as discussed in Chap. 3, which allows one to perform calculations for only one spin and multiply the final result by two for various quantities of interest as discussed in Chap. 4. The case for ferromagnet (FM) materials is rather different. The band structure for ↑-spin and ↓-spin electrons may be written as follows,

$$E = E_{c\uparrow} + \frac{\hbar^2 k^{\uparrow 2}}{2m} \tag{6.3}$$

$$E = E_{c\downarrow} + \frac{\hbar^2 k^{\downarrow 2}}{2m} \tag{6.4}$$

where $E_{c\uparrow}$ and $E_{c\downarrow}$ are the conduction band edges for ↑-spin and ↓-spin bands, respectively.[2] The conduction band edge of the majority band is lower than that of the minority band as shown in Fig. 6.2. The difference in the band edges between the ↑-spin and ↓-spin bands is called Exchange Split, given as, $|E_{c\uparrow} - E_{c\downarrow}|$.

Within the 1D tight binding approximation, Hamiltonian matrix for the ↑-spin band with the conduction band edge $E_{c\uparrow}$, is given as,

$$[H]^{\uparrow} = \begin{bmatrix} E_c^{\uparrow} + 2t & -t & 0 & 0 & \cdots \\ -t & E_c^{\uparrow} + 2t & -t & 0 & \cdots \\ 0 & -t & E_c^{\uparrow} + 2t & -t & \cdots \\ 0 & 0 & -t & E_c^{\uparrow} + 2t & \cdots \\ \cdots & \cdots & \cdots & \cdots & \cdots \end{bmatrix} \tag{6.5}$$

[2] $E_{c\uparrow}$ and $E_{c\downarrow}$ are indeed equal to each other for a paramagnetic material in the absence of a magnetic field.

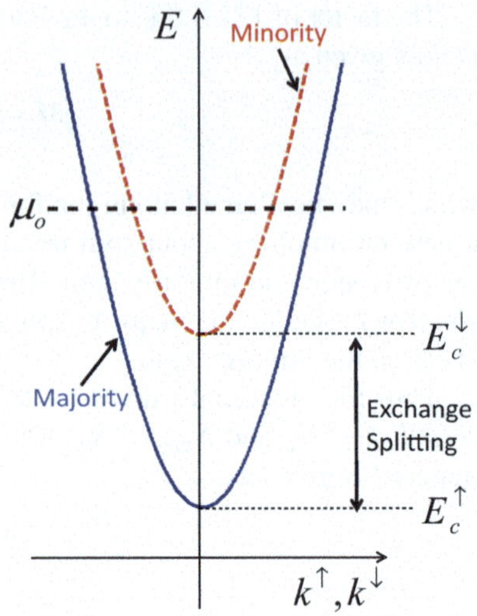

Fig. 6.2 Band structure of a ferromagnet

with the corresponding dispersion relation $E(k) = E_c^\uparrow + 2t\left[1 - \cos(k^\uparrow a)\right]$, where a is the lattice constant. Similarly, for the $\downarrow$-spin band, the Hamiltonian matrix with the conduction band edge $E_{c\downarrow}$, is given as,

$$[H]^\downarrow = \begin{bmatrix} E_c^\downarrow + 2t & -t & 0 & 0 & \cdots \\ -t & E_c^\downarrow + 2t & -t & 0 & \cdots \\ 0 & -t & E_c^\downarrow + 2t & -t & \cdots \\ 0 & 0 & -t & E_c^\downarrow + 2t & \cdots \\ \cdots & \cdots & \cdots & \cdots & \cdots \end{bmatrix} \tag{6.6}$$

with the corresponding dispersion relation $E(k) = E_c^\downarrow + 2t\left[1 - \cos(k^\downarrow a)\right]$.

It is interesting to note that the wavevector at the equilibrium chemical potential energy (μ_o) is larger for the majority spin band. Since, the group velocity for a parabolic band is given as, $v_g = dE/\hbar dk = \hbar k/m$, the group velocity for the majority band is higher than that of the minority band at the equilibrium chemical potential energy.

Fe (BCC) band structure for the majority and the minority spins is shown in Fig. 6.3. For each band, the band edge is lower for the majority spin. Additionally, each band has a dominant orbital symmetry. Δ_1 band has dominantly s-orbital symmetry. Δ_5 band dominantly has p-orbital symmetry, whereas Δ_2 and $\Delta_{2'}$ bands have mostly d-orbital symmetry. One may make use of these symmetry states in devices as discussed in Sect. 6.3.

It should also be noted that for Δ_1 band, one finds only majority spin states around the chemical potential energy, whereas for Δ_2 band, only the minority spin states are around the chemical potential energy. Such bands are called half metallic bands. If all the bands around μ_o exhibit half metallic behavior around the chemical potential

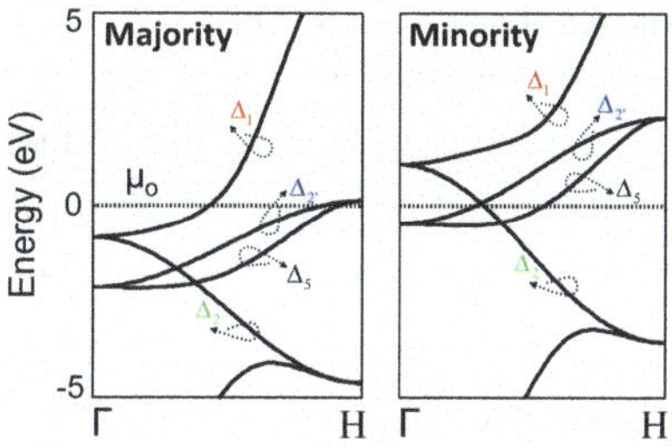

Fig. 6.3 Fe (BCC) band structure for majority and minority spin

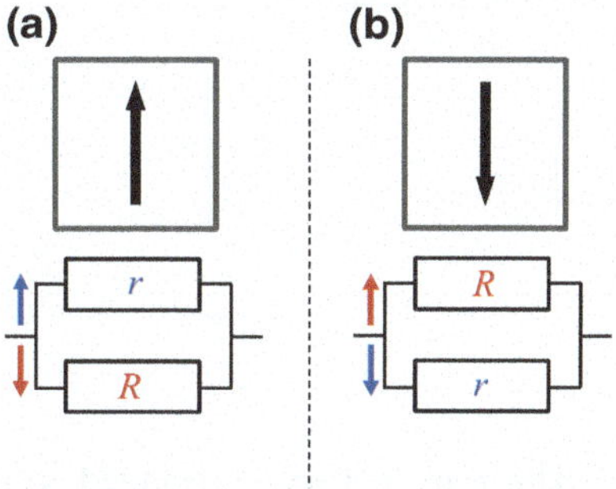

Fig. 6.4 Circuit model of a ferromagnet. **a** ↑-spin is the majority spin with a smaller resistance r, whereas ↓-spin is the minority spin with a larger R resistance. **b** ↓-spin is the majority spin with a smaller resistance r, whereas ↑-spin is the minority spin with a larger resistance R

energy, the material is called half metallic, which is desirable for optimum GMR and MTJ devices, as discussed later.

For the conduction properties, in the absence of any spin flip scattering, the transport through a ferromagnetic material may be thought of as current flowing through two independent spin channels. In Fig. 6.4a, a ferromagnet with ↑-spin majority is shown with a corresponding circuit model. Similarly, in Fig. 6.4b, a ferromagnet with ↓-spin majority is shown with the corresponding circuit model. In either case, the resistance for the majority spin may be written as, $R_{\text{maj}} = r$ and for the minority spin, $R_{\text{min}} = R$, where, $R_{\text{maj}} < R_{\text{min}}$ and hence $r < R$. The net resistance in this simple circuit model is the parallel combination of the two resistances and it is given as,

$$R_{\text{FM}} = \frac{rR}{r+R} \approx r$$

In the above two cases, it should be noted that $R_{\text{maj}} = r$ is for ↑-spin and ↓-spin in Fig. 6.4a, b, respectively.

6.2 Giant Magnetoresistance Devices

A GMR device is a trilayer structure of FM-NM-FM layers, where the FM (ferromagnetic) and NM (nonmagnetic, usually paramagnetic) layers are metallic. While trilayer is a basic requirement, the device could have more than three layers.

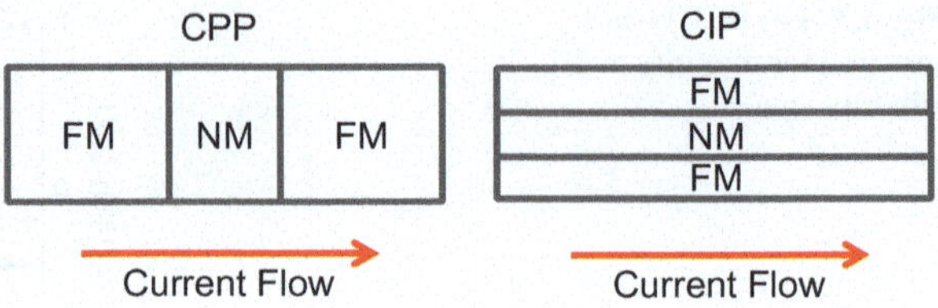

Fig. 6.5 GMR devices. CPP and CIP configurations

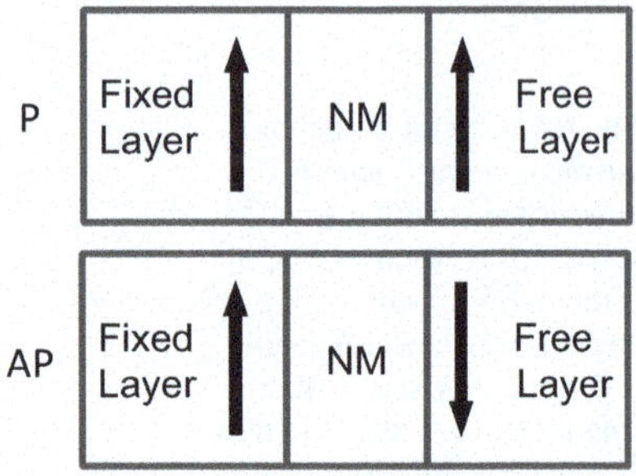

Fig. 6.6 GMR devices. Parallel (P) and Antiparallel (AP) configurations

The trialyers may be arranged in either CPP (current perpendicular to plane) or CIP (current in plane) configurations as shown in Fig. 6.5. It is usually CPP configuration, which gives a better performance and hence is the most commonly used. Therefore, we discuss CPP configuration in this section. The thickness of NM layer is usually a few nanometers to reduce spin flip scattering. Historically, Cu is used for the NM layer, whereas Fe, Co, and permalloy (an alloy of nickel and iron) are the preferred choices for the FM layers.

If the magnetization of the two FM layers is in the same direction, it is called Parallel (P) configuration, whereas the reverse direction configuration is called antiparallel (AP) as shown in Fig. 6.6. One of the layers has a fixed magnetization direction, called fixed layer or pinned layer, whereas the magnetization of the other layer may be switched by using an external magnetic field, which is called free layer. Also, terms like hard layer and soft layer are also used for fixed and free layers, respectively.

The P and AP configurations may be achieved by magnetization switching of the soft layer due to an external magnetic field by using a soft magnet (like permalloy) with a small coercivity for the free layer as shown in Fig. 6.7. A small magnetic field (larger than the coercivity of the soft magnetic) may easily switch the magnetization of the free layer without perturbing the magnetization of the hard layer. This switching asymmetry between the fixed and the free layers may also be achieved by using an additional antiferromagnetic (AFM) layer pinning the magnetization of the fixed layer due to exchange bias.

The circuit model for the P and AP configuration is shown in Fig. 6.8 without any spin flip scattering in the NM layer. The resistance in parallel configuration is given as,

$$R_{\mathrm{P}} = \frac{4rR}{2r + 2R} \approx 2r$$

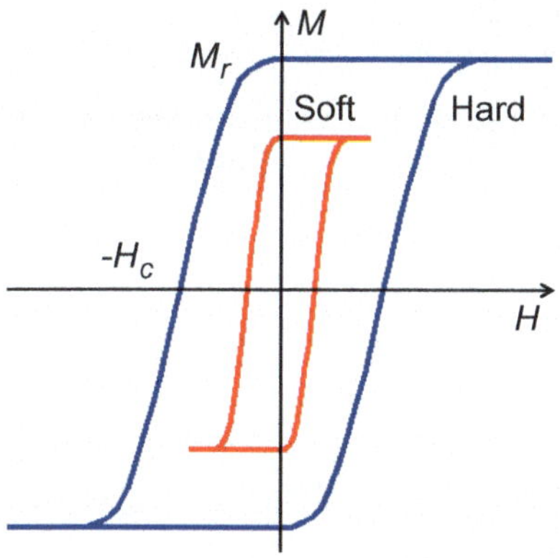

Fig. 6.7 Magnetization curves of a hard and a soft magnet

whereas the resistance in antiparallel configuration is as follows,

$$R_{AP} = \frac{r + R}{2} \approx \frac{R}{2}$$

IV characteristics for the P and AP configurations are shown in Fig. 6.9, which shows lesser current for the AP configuration due to higher resistance. The %GMR ratio in terms of resistance, conductance, and current is respectively defined as follows,

$$\begin{aligned} \%\text{GMR ratio} &= \left(\frac{R_{AP} - R_P}{R_P}\right) \times 100 = \left(\frac{R_{AP}}{R_P} - 1\right) \times 100 \\ &= \left(\frac{G_P}{G_{AP}} - 1\right) \times 100 = \left(\frac{I_P}{I_{AP}} - 1\right) \times 100 \end{aligned} \tag{6.7}$$

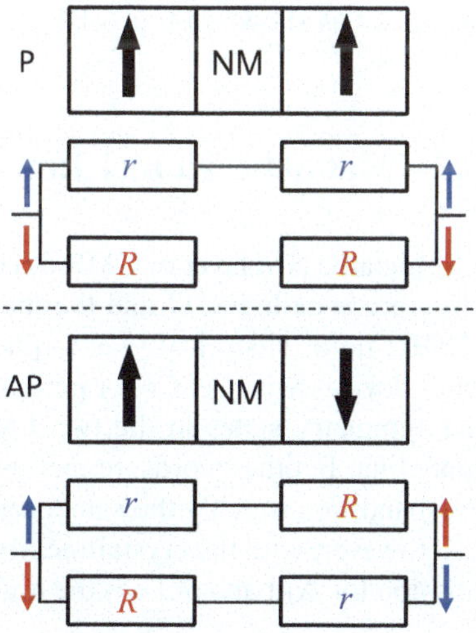

Fig. 6.8 Circuit model for Parallel (P) and Antiparallel (AP) configurations

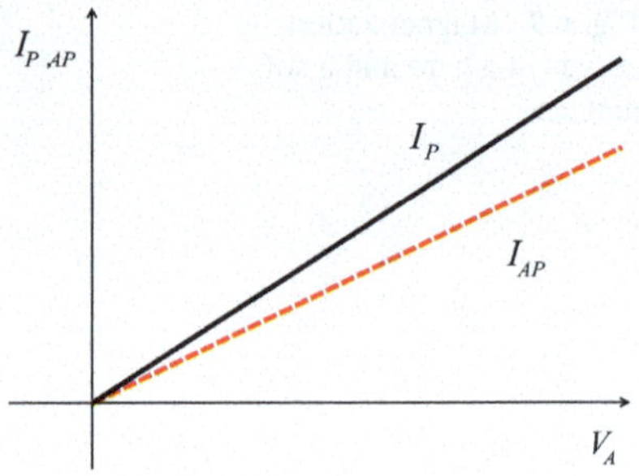

Fig. 6.9 GMR devices. IV characteristics

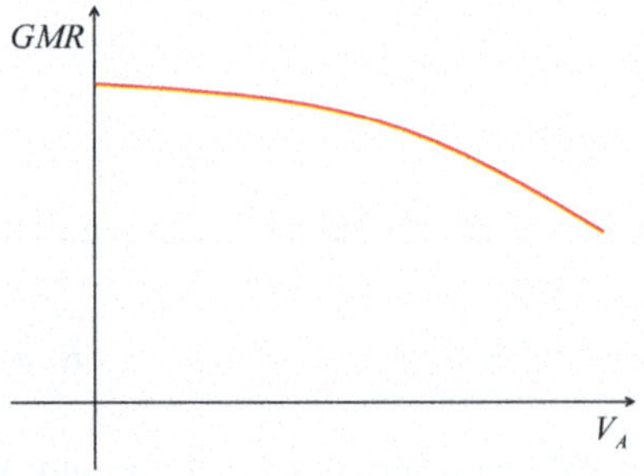

Fig. 6.10 GMR devices. Bias dependence of magnetoresistance ratio

For the above circuit model, the %GMR ratio is given as,

$$\%\text{GMR ratio} = \left(\frac{R}{4r} - 1\right) \times 100 \tag{6.8}$$

While in the circuit model, the resistances are not explicitly shown to be bias dependent, in reality these vary with the applied voltage, which makes the GMR ratio bias dependent as well. Usually with the increasing bias, the GMR ratio monotonically decreases as shown in Fig. 6.10.

6.3 Magnetic Tunnel Junction Devices

The metallic NM layer of GMR device may be replaced with an insulator. This device structure is called MTJ and the magnetoresistance (MR) ratio is called tunnel MR (TMR) ratio. Historically, amorphous aluminum oxide (Al_2O_3) has been used in MTJ devices with Fe, Co, or permalloy FM contacts. With the amorphous alumina, the symmetry states in the two FM contacts like Fe do not have any preferential tunneling. In other words, tunneling probability or transmission for Δ_1, Δ_5, Δ_2 and $\Delta_{2'}$ bands are roughly the same, schematically shown in Fig. 6.11a.

However, with the crystalline MgO insulator, the tunneling probability and transmission for Δ_1 band of Fe is the highest, followed by the Δ_5 band of Fe; whereas Δ_2

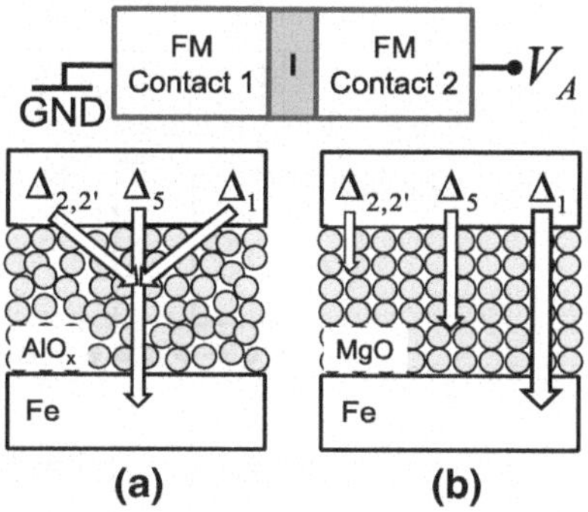

Fig. 6.11 Symmetry filtering in MTJ device. **a** Various bands tunnel through amorphous alumina with equal probability. **b** Δ_1 band in Fe (001) has a higher tunneling probability through crystalline MgO (001)

and $\Delta_{2'}$ bands of Fe have the least tunneling probability and transmission, schematically shown in Fig. 6.11b. This may be conceptually understood in a simple picture that the valence orbitals of MgO that constitute the conduction and the valence bands have *s*-orbital (due to Mg) and *p*-orbital (due to O) symmetry states, respectively. As shown in Fig. 6.3, Δ_1 band of Fe has predominantly *s*-orbital symmetry; Δ_5 band of Fe has predominantly *p*-orbital symmetry; whereas Δ_2 band and $\Delta_{2'}$ band of Fe have mostly *d*-orbital character. Due to this constitution, the *d*-orbitals of Fe are quenched during tunneling, where as Δ_1 band and Δ_5 band tunnel through due to *s*-orbital and *p*-orbital contribution in MgO. The preferential tunneling of the half metallic Δ_1 band due to symmetry filtering property of MgO leads to a high %TMR ratio, which is defined as,

$$\%\text{TMR ratio} = \left(\frac{I_\text{P} - I_\text{AP}}{I_\text{AP}}\right) \times 100 = \left(\frac{I_\text{P}}{I_\text{AP}} - 1\right) \times 100 \tag{6.9}$$

For calculating the transport properties, the band diagram is shown in Fig. 6.12 for P and AP configurations. For each configuration, one has to find current due to $\uparrow$-spin and $\downarrow$-spin channels. In the absence of spin flip scattering, one may treat these two spin channels independently. For the P configuration, the $\uparrow$-spin electrons remain majority spin in both contacts, whereas $\downarrow$-spin electrons remain minority spin. However, for the AP configuration, $\uparrow$-spin electrons are majority spin in the left contact and minority spin in the right contact; whereas, $\downarrow$-spin electrons are minority spin in the left contact and majority spin in the right contact. Thus, in each configuration, one has to pay attention to the book keeping of the correct band edge.[3]

For the $\uparrow$-spin channel in P configuration shown in Fig. 6.12, the device Hamiltonian for the nine atoms channel is given as,

[3] $E_c^\uparrow$ is 0.5 eV below μ_o, and $E_c^\downarrow$ is 0.25 eV below μ_o. These specific values are taken arbitrarily, and may vary for different materials.

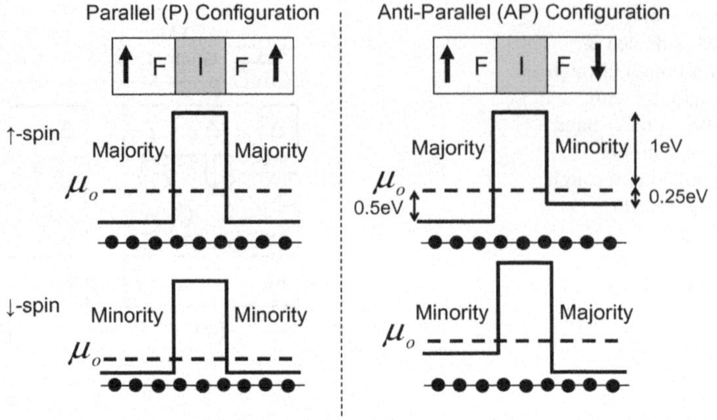

Fig. 6.12 MTJ device. Band diagram is shown in P and AP configuration. The equilibrium chemical potential energy (μ_o) is shown by a dashed line

$$[H_P]^{\uparrow} = \begin{bmatrix} -0.5+2t & -t & 0 & 0 & 0 & 0 & 0 & 0 & 0 \\ -t & -0.5+2t & -t & 0 & 0 & 0 & 0 & 0 & 0 \\ 0 & -t & -0.5+2t & -t & 0 & 0 & 0 & 0 & 0 \\ 0 & 0 & -t & 1+2t & -t & 0 & 0 & 0 & 0 \\ 0 & 0 & 0 & -t & 1+2t & -t & 0 & 0 & 0 \\ 0 & 0 & 0 & 0 & -t & 1+2t & -t & 0 & 0 \\ 0 & 0 & 0 & 0 & 0 & -t & -0.5+2t & -t & 0 \\ 0 & 0 & 0 & 0 & 0 & 0 & -t & -0.5+2t & -t \\ 0 & 0 & 0 & 0 & 0 & 0 & 0 & -t & -0.5+2t \end{bmatrix}$$

where the band edge for the majority spin electrons is taken at −0.5 eV and the barrier height in the insulator region is taken at 1 eV for $\mu_o = 0$. For the two contacts, the dispersion relations are given as,

$$E = U(1,1) - 0.5 + 2t\left[1 - \cos(k_1^{P\uparrow}a)\right]$$

$$E = U(9,9) - 0.5 + 2t\left[1 - \cos(k_2^{P\uparrow}a)\right]$$

where U is the potential energy due to an applied bias, and $k_{1,2}^{P\uparrow}a$ are the wavevectors for ↑-spin in the P configuration of the two contacts. By using the above equation, one may calculate the contact self energies, contact broadening functions, device Green's function, and eventually transmission $T_P^{\uparrow}(E)$ for ↑-spin electrons in the P configuration. The current due to ↑-spin channel in P configuration is thus given as,

$$I_P^{\uparrow} = \frac{q}{h}\int_{-\infty}^{+\infty} \mathrm{d}E\, T_P^{\uparrow}(E)\left[f_1(E) - f_2(E)\right]$$

Note that the factor 2 is missing due to spin nondegeneracy. For ↓-spin channel in P configuration shown in Fig. 6.12 with the chosen parameters, the device Hamiltonian for the nine atoms channel is given as,

$$[H_{\mathrm{P}}]^{\downarrow} = \begin{bmatrix} -0.25+2t & -t & 0 & 0 & 0 & 0 & 0 & 0 & 0 \\ -t & -0.25+2t & -t & 0 & 0 & 0 & 0 & 0 & 0 \\ 0 & -t & -0.25+2t & -t & 0 & 0 & 0 & 0 & 0 \\ 0 & 0 & -t & 1+2t & -t & 0 & 0 & 0 & 0 \\ 0 & 0 & 0 & -t & 1+2t & -t & 0 & 0 & 0 \\ 0 & 0 & 0 & 0 & -t & 1+2t & -t & 0 & 0 \\ 0 & 0 & 0 & 0 & 0 & -t & -0.25+2t & -t & 0 \\ 0 & 0 & 0 & 0 & 0 & 0 & -t & -0.25+2t & -t \\ 0 & 0 & 0 & 0 & 0 & 0 & 0 & -t & -0.25+2t \end{bmatrix}$$

where the band edge for the minority spin is taken as −0.25 eV.

For the two contacts, the dispersion relations are given as,

$$E = U(1,1) - 0.25 + 2t\left[1 - \cos(k_1^{\mathrm{P}\downarrow}a)\right]$$

$$E = U(9,9) - 0.25 + 2t\left[1 - \cos(k_2^{\mathrm{P}\downarrow}a)\right]$$

where $k_{1,2}^{\mathrm{P}\downarrow}a$ are the wavevectors for ↓-spin in the P configuration of the two contacts.

Given the transmission $T_{\mathrm{P}}^{\downarrow}(E)$, the current due to ↓-spin channel in P configuration is given as,

$$I_{\mathrm{P}}^{\downarrow} = \frac{q}{h}\int_{-\infty}^{+\infty} \mathrm{d}E\, T_{\mathrm{P}}^{\downarrow}(E)\left[f_1(E) - f_2(E)\right]$$

The total current in P configuration is thus given as,

$$I_{\mathrm{P}} = I_{\mathrm{P}}^{\uparrow} + I_{\mathrm{P}}^{\downarrow} \tag{6.10}$$

For ↑-spin channel in AP configuration shown in Fig. 6.12 with the chosen parameters, the device Hamiltonian for the nine atoms channel is given as,

$$[H_{\mathrm{AP}}]^{\uparrow} = \begin{bmatrix} -0.5+2t & -t & 0 & 0 & 0 & 0 & 0 & 0 & 0 \\ -t & -0.5+2t & -t & 0 & 0 & 0 & 0 & 0 & 0 \\ 0 & -t & -0.5+2t & -t & 0 & 0 & 0 & 0 & 0 \\ 0 & 0 & -t & 1+2t & -t & 0 & 0 & 0 & 0 \\ 0 & 0 & 0 & -t & 1+2t & -t & 0 & 0 & 0 \\ 0 & 0 & 0 & 0 & -t & 1+2t & -t & 0 & 0 \\ 0 & 0 & 0 & 0 & 0 & -t & -0.25+2t & -t & 0 \\ 0 & 0 & 0 & 0 & 0 & 0 & -t & -0.25+2t & -t \\ 0 & 0 & 0 & 0 & 0 & 0 & 0 & -t & -0.25+2t \end{bmatrix}$$

where band edges for the majority and minority spin electrons are at −0.5 and −0.25 eV respectively. The ↑-spin electrons are majority spin in the left contact and minority spin in the right contact. For the two contacts, the dispersion relations are given as,

$$E = U(1,1) - 0.5 + 2t\left[1 - \cos(k_1^{\mathrm{AP}\uparrow} a)\right]$$

$$E = U(9,9) - 0.25 + 2t\left[1 - \cos(k_2^{\mathrm{AP}\uparrow} a)\right]$$

where $k_{1,2}^{\mathrm{AP}\uparrow} a$ are the wavevectors for ↑-spin in the AP configuration of the two contacts.

Given the transmission $T_{\mathrm{AP}}^{\uparrow}(E)$, the current due to ↑-spin channel in AP configuration is given as,

$$I_{\mathrm{AP}}^{\uparrow} = \frac{q}{h}\int_{-\infty}^{+\infty} \mathrm{d}E\; T_{\mathrm{AP}}^{\uparrow}(E)\left[f_1(E) - f_2(E)\right]$$

For ↓-spin channel in AP configuration shown in Fig. 6.12 with the chosen parameters, the device Hamiltonian for the nine atoms channel is given as,

$$[H_{\mathrm{AP}}]^{\downarrow} = \begin{bmatrix} -0.25+2t & -t & 0 & 0 & 0 & 0 & 0 & 0 & 0 \\ -t & -0.25+2t & -t & 0 & 0 & 0 & 0 & 0 & 0 \\ 0 & -t & -0.25+2t & -t & 0 & 0 & 0 & 0 & 0 \\ 0 & 0 & -t & 1+2t & -t & 0 & 0 & 0 & 0 \\ 0 & 0 & 0 & -t & 1+2t & -t & 0 & 0 & 0 \\ 0 & 0 & 0 & 0 & -t & 1+2t & -t & 0 & 0 \\ 0 & 0 & 0 & 0 & 0 & -t & -0.5+2t & -t & 0 \\ 0 & 0 & 0 & 0 & 0 & 0 & -t & -0.5+2t & -t \\ 0 & 0 & 0 & 0 & 0 & 0 & 0 & -t & -0.5+2t \end{bmatrix}$$

where band edges for the majority and minority spin electrons are at −0.5 and −0.25 eV respectively. The ↓-spin electrons are minority spin in the left contact and majority spin in the right contact.

For the two contacts, the dispersion relations are given as,

$$E = U(1,1) - 0.25 + 2t\left[1 - \cos(k_1^{\mathrm{AP}\downarrow} a)\right]$$

$$E = U(9,9) - 0.5 + 2t\left[1 - \cos(k_2^{\mathrm{AP}\downarrow} a)\right]$$

where $k_{1,2}^{\mathrm{AP}\downarrow} a$ are the wavevectors for ↓-spin in the AP configuration of the two contacts.

Given the transmission $T_{\mathrm{AP}}^{\downarrow}(E)$ for ↓-spin electrons in the AP configuration. The current due to spin channel in AP configuration is given as,

$$I_{\mathrm{AP}}^{\downarrow} = \frac{q}{h}\int_{-\infty}^{+\infty} \mathrm{d}E\; T_{\mathrm{AP}}^{\downarrow}(E)\left[f_1(E) - f_2(E)\right]$$

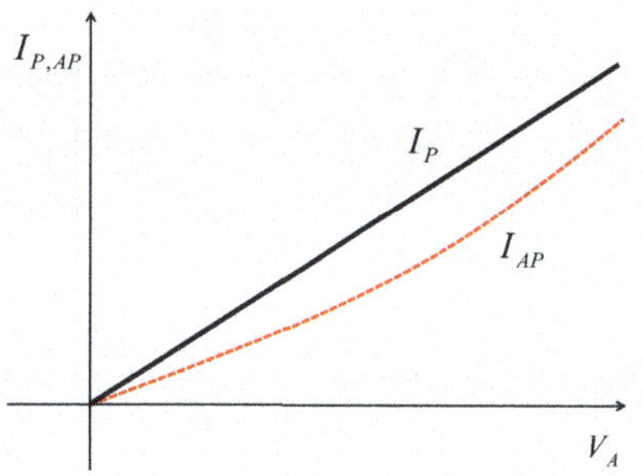

Fig. 6.13 MTJ device. IV characteristics

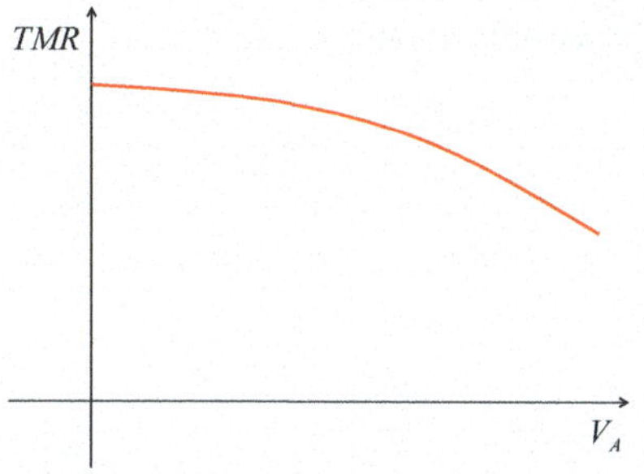

Fig. 6.14 MTJ device. TMR ratio trends

The total current in AP configuration is thus given as,

$$I_{\mathrm{AP}} = I_{\mathrm{AP}}^{\uparrow} + I_{\mathrm{AP}}^{\downarrow} \tag{6.11}$$

The P and AP IV characteristics are shown in Fig. 6.13, where the AP current is smaller than the P current—similar trend to that of GMR devices. TMR ratio is shown in Fig. 6.14, which shows a monotonically decreasing ratio with increasing bias—a trend usually observed in MTJ devices.

The current in AP configuration in MTJ devices is small due to nonoverlapping bands in the two contacts for either spin, as shown in Fig. 6.15a for zero bias. However, with applied bias, the bands of the same spin in the two contacts may start overlapping as shown in Fig. 6.15b. This may result in even a higher AP current than the P current, at high bias. Such a behavior leads to a negative TMR ratio as shown in Fig. 6.16—a trend also observed in experiments.

It should also be noted that the voltage drop in the device is only across the insulator, since electric field may not exist in an ideal metal. The Laplace's potential energy matrix with linear screening in P and AP configurations for both ↑-spin and ↓-spin channels is given as,

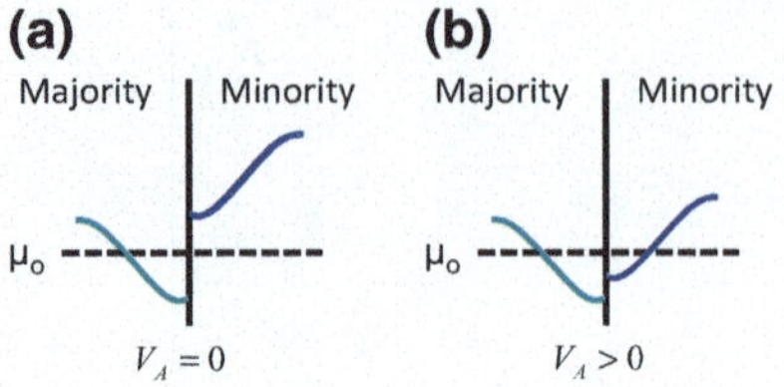

Fig. 6.15 MTJ device. **a** Equilibrium band diagram for a half metallic band with nonoverlapping bands, due to exchange bias, in AP configuration. **b** Bands may start overlapping with applied bias

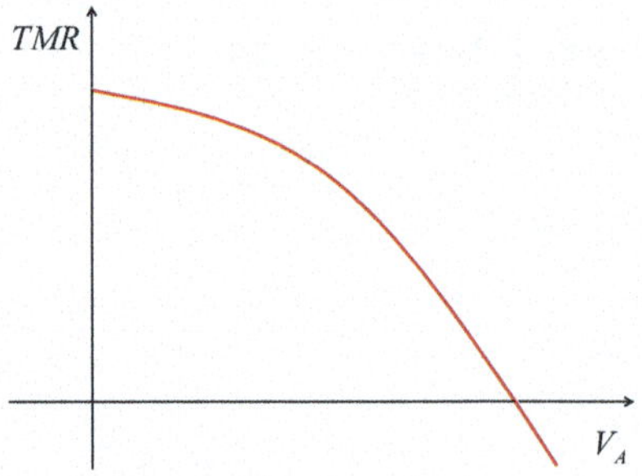

Fig. 6.16 MTJ device. Negative TMR ratio

$$[U_L] = \begin{bmatrix} 0&0&0&0&0&0&0&0&0 \\ 0&0&0&0&0&0&0&0&0 \\ 0&0&0&0&0&0&0&0&0 \\ 0&0&0&0&0&0&0&0&0 \\ 0&0&0&0&-0.5V_A&0&0&0&0 \\ 0&0&0&0&0&-V_A&0&0&0 \\ 0&0&0&0&0&0&-V_A&0&0 \\ 0&0&0&0&0&0&0&-V_A&0 \\ 0&0&0&0&0&0&0&0&-V_A \end{bmatrix}$$

where contact 1 is connected to ground and the V_A bias is applied at contact 2.

6.4 Spin Transfer Torque Devices

In GMR and MTJ devices, the magnetization of the free layer is switched by using an external magnetic field as shown in Fig. 6.17a. We discuss a novel class of devices based on Spin Transfer Torque (STT) in this section, where the magnetization of a free layer may be switched by using spin polarized current as shown in Fig. 6.17b.

Electrons have two types of magnetic moments, one due to the orbital motion and the other one due to the spin. The magnetic moment due to orbital motion is given as,

$$\boldsymbol{m}_l = \gamma \boldsymbol{L}$$

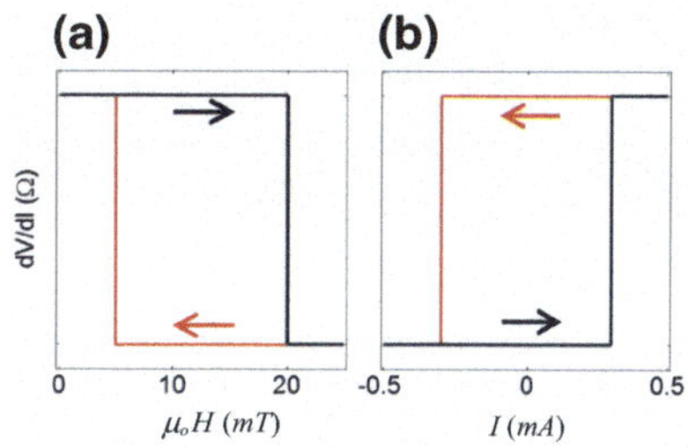

Fig. 6.17 Bistable switching, **a** due to an external magnetic field in GMR and MTJ devices, and **b** due to spin polarized current in spin transfer torque (STT) devices

where γ is the gyromagnetic ratio given as,

$$\gamma = -\frac{q}{2m_\mathrm{e}} \tag{6.12}$$

and m_e is the electron mass. The Bohr magneton, which gives the magnitude of the magnetic moment due to orbital motion of a single electron is given as,

$$\mu_\mathrm{B} = \frac{q\hbar}{2m_\mathrm{e}}$$

The torque due to the angular motion is given as the time rate of change of the angular momentum as follows,

$$\boldsymbol{N}_t = \frac{\partial \boldsymbol{L}}{\partial t} = \frac{1}{\gamma}\frac{\partial \boldsymbol{m}_l}{\partial t}$$

where the torque ($\boldsymbol{N}_t$) for an external applied field density $\boldsymbol{B}$ is defined as follows,

$$\boldsymbol{N}_t = \boldsymbol{m}_l \times \boldsymbol{B}$$

Combining the above two equations and writing in terms of a generic angular momentum ($\boldsymbol{m}$), one gets,

$$\frac{\partial \boldsymbol{m}}{\partial t} = \gamma\,(\boldsymbol{m} \times \boldsymbol{B})$$

For an applied magnetic field, $\boldsymbol{B} = B_o\boldsymbol{z}$, one obtains the following time dependent (x, y, z) components of the magnetic moment,

$$\begin{aligned} m_x(t) &= |m|\sin(\theta)\cos(\omega_\mathrm{L} t) \\ m_y(t) &= |m|\sin(\theta)\sin(\omega_\mathrm{L} t) \\ m_z(t) &= |m|\sin(\theta) \end{aligned}$$

The magnetic moment is oscillatory in (x, y) plane with the period of oscillation given by the Larmor frequency (ω_L) as follows,

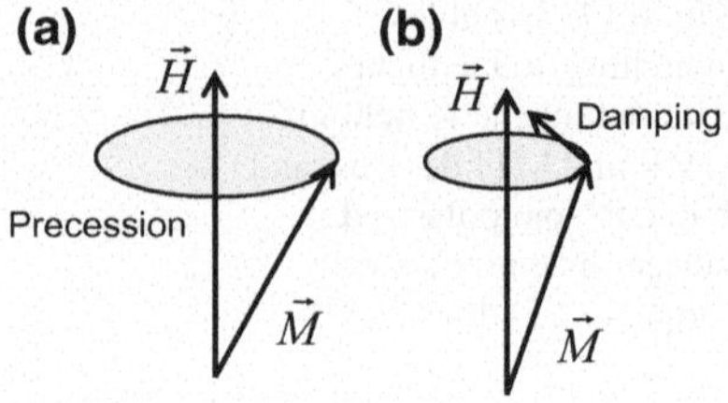

Fig. 6.18 Spin dynamics. **a** Precession. **b** Damping

$$\omega_{\mathrm{L}} = \gamma B_o = \frac{qB_o}{2m_{\mathrm{e}}} \tag{6.13}$$

This is called precession and is shown in Fig. 6.18a. By including a term for damping, one gets,

$$\frac{\partial \boldsymbol{m}}{\partial t} = \gamma\,[\boldsymbol{m} \times \boldsymbol{B} - R\,(\boldsymbol{m} - \boldsymbol{m}_o)]$$

where $\boldsymbol{m}_o$ is the initial magnetic moment and R is the relaxation matrix, giving rise to the damping as shown in Fig. 6.18b.

Similarly, in terms of the net magnetization $\boldsymbol{M}$, the dynamics equation is called the Bloch equation and is given as,

$$\frac{\partial \boldsymbol{M}}{\partial t} = \gamma\,(\boldsymbol{M} \times \boldsymbol{B} - R\,(\boldsymbol{M} - \boldsymbol{M}_o)) \tag{6.14}$$

Next, we consider representation of the complete spin polarized wavefunction. For the spin degree of freedom, the Pauli matrices are given as,

$$\begin{aligned} \hat{\sigma}_x &= \begin{bmatrix} 0 & 1 \\ 1 & 0 \end{bmatrix} \\ \hat{\sigma}_y &= \begin{bmatrix} 0 & -i \\ i & 0 \end{bmatrix} \\ \hat{\sigma}_z &= \begin{bmatrix} 1 & 0 \\ 0 & -1 \end{bmatrix} \end{aligned} \tag{6.15}$$

and the corresponding spin operators are given as,

$$\begin{aligned} \hat{S}_x &= \frac{\hbar}{2}\begin{bmatrix} 0 & 1 \\ 1 & 0 \end{bmatrix} \\ \hat{S}_y &= \frac{\hbar}{2}\begin{bmatrix} 0 & -i \\ i & 0 \end{bmatrix} \\ \hat{S}_z &= \frac{\hbar}{2}\begin{bmatrix} 1 & 0 \\ 0 & -1 \end{bmatrix} \end{aligned}$$

Since $\hat{S}_z$ operator is diagonal, the diagonal matrix elements give the Eigen values of spin magnetic moment as follows,

$$m_z = \pm\frac{\hbar}{2}$$

with the following spin polarized Eigen functions, also called Spinors,

$$+\frac{\hbar}{2} \Rightarrow \phi_{\uparrow} = \begin{Bmatrix} 1 \\ 0 \end{Bmatrix} \qquad -\frac{\hbar}{2} \Rightarrow \phi_{\downarrow} = \begin{Bmatrix} 0 \\ 1 \end{Bmatrix}$$

The total wavefunction in 1D including spin is thus given as,

$$\Psi = \frac{e^{ikx}}{\sqrt{L}}[a\phi_{\uparrow} + b\phi_{\downarrow}]$$

where a and b are the coefficients for the ↑-spin and ↓-spin, noting $a^2 + b^2 = 1$.

Substituting for the Spinors,

$$\Psi = \frac{e^{ikx}}{\sqrt{L}}\left[a\begin{Bmatrix} 1 \\ 0 \end{Bmatrix} + b\begin{Bmatrix} 0 \\ 1 \end{Bmatrix}\right] = \frac{e^{ikx}}{\sqrt{L}}\begin{Bmatrix} a \\ 0 \end{Bmatrix} + \begin{Bmatrix} 0 \\ b \end{Bmatrix}$$

Next we discuss the spin current. Starting with the charge current definition given in (4.1) as follows,

$$I = -q\frac{i\hbar}{2m}\left(\Psi\frac{\mathrm{d}\Psi^*}{\mathrm{d}x} - \Psi^*\frac{\mathrm{d}\Psi}{\mathrm{d}x}\right)$$

Since the electron current is a product of velocity (v) and the electron density (n), the spin current is also a product of velocity (v) and the spin density (S)[4] as follows $v \otimes S$, and is formally given as,

$$Q = -\frac{i\hbar}{2m}\left(\Psi S\frac{\mathrm{d}\Psi^*}{\mathrm{d}x} - \Psi^* S\frac{\mathrm{d}\Psi}{\mathrm{d}x}\right) \tag{6.16}$$

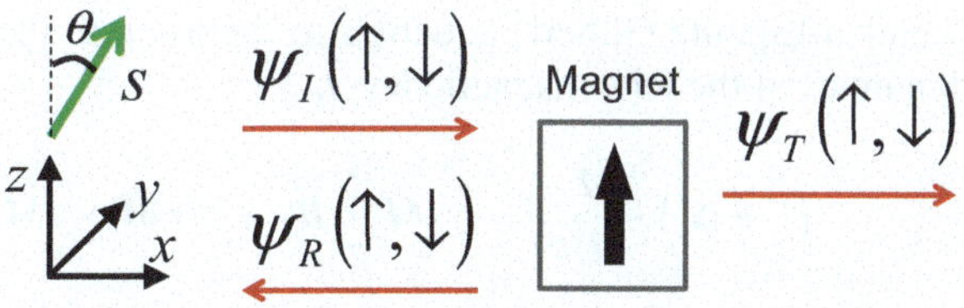

Fig. 6.19 Spin dependent tunneling and spin transfer torque

[4]Please make a note of the reuse of the symbol S for spin density, which has been used earlier for the cross sectional area.

For an electron incident on a magnet shown in Fig. 6.19, the spin polarized incident, transmitted and reflected wavefunctions are respectively given as,

$$\Psi_I = \frac{e^{ikx}}{\sqrt{L}}\left[\cos\left(\frac{\theta}{2}\right)\phi_\uparrow + \sin\left(\frac{\theta}{2}\right)\phi_\downarrow\right]$$
$$\Psi_T = \frac{e^{ikx}}{\sqrt{L}}\left[t_\uparrow \cos\left(\frac{\theta}{2}\right)\phi_\uparrow + t_\downarrow \sin\left(\frac{\theta}{2}\right)\phi_\downarrow\right]$$
$$\Psi_R = \frac{e^{ikx}}{\sqrt{L}}\left[r_\uparrow \cos\left(\frac{\theta}{2}\right)\phi_\uparrow + r_\downarrow \sin\left(\frac{\theta}{2}\right)\phi_\downarrow\right]$$

where the choice of $(\theta/2)$ ensures that for $\theta = 0$, one obtains $\uparrow$-spin with magnetization aligned along the positive z-axis and for $\theta = \pi$, one obtains $\downarrow$-spin with magnetization aligned along the negative z-axis, given as,

$$\boldsymbol{m}_s = \pm\frac{\hbar}{2}z$$

With this choice of the wavefunctions, the incident, transmitted and reflected spin current is respectively given as follows,[5]

$$\boldsymbol{Q}_I = \frac{\hbar^2 k}{2mL}\left[\boldsymbol{x}\sin(\theta) + \boldsymbol{z}\cos(\theta)\right]$$
$$\boldsymbol{Q}_T = \frac{\hbar^2 k}{2mL}\left[\boldsymbol{x}\sin(\theta)Re\left(t_\uparrow t_\downarrow^*\right) + \boldsymbol{y}\sin(\theta)Im\left(t_\uparrow t_\downarrow^*\right) + \boldsymbol{z}\left\{|t_\uparrow|^2\cos^2(\theta/2) - |t_\downarrow|^2\sin^2(\theta/2)\right\}\right]$$
$$\boldsymbol{Q}_R = \frac{\hbar^2 k}{2mL}\left[\boldsymbol{x}\sin(\theta)Re\left(r_\uparrow r_\downarrow^*\right) + \boldsymbol{y}\sin(\theta)Im\left(r_\uparrow r_\downarrow^*\right) + \boldsymbol{z}\left\{|r_\uparrow|^2\cos^2(\theta/2) - |r_\downarrow|^2\sin^2(\theta/2)\right\}\right]$$

The spin transfer torque is defined as, $\boldsymbol{N}_{\rm st} = \boldsymbol{Q}_I + \boldsymbol{Q}_T - \boldsymbol{Q}_R$ and is given as,

$$\boldsymbol{N}_{\rm st} = \frac{\hbar^2 k}{2mL}\sin(\theta)\left[\boldsymbol{x}\left\{1 - Re\left(t_\uparrow t_\downarrow^* + r_\uparrow r_\downarrow^*\right)\right\} + \boldsymbol{y}\left\{\sin(\theta) - Im\left(t_\uparrow t_\downarrow^* + r_\uparrow r_\downarrow^*\right)\right\}\right] \tag{6.17}$$

For $\theta = 0$ or $\theta = \pi$, the spin transfer torque is zero, i.e. $\boldsymbol{N}_{\rm st} = 0$. In addition, for $t_\uparrow = t_\downarrow$ and $r_\uparrow = r_\downarrow$, the spin transfer torque is also zero, i.e. $\boldsymbol{N}_{\rm st} = 0$, since $|t_\uparrow|^2 + |t_\downarrow|^2 = 1$ and $|r_\uparrow|^2 + |r_\downarrow|^2 = 1$.

One has to include this spin transfer torque in the phenomenological LLG (Landau-Lifshitz-Gilbert) equation to incorporate the effect of the magnetization dynamics of the ferromagnetic layer,

$$\left(1 + \alpha^2\right)\frac{\partial \boldsymbol{M}}{\partial t} = -\gamma\left(\boldsymbol{M} \times \boldsymbol{B}\right) - \gamma\alpha\hat{\boldsymbol{M}} \times \left(\boldsymbol{M} \times \boldsymbol{H}_{\rm eff}\right) + \boldsymbol{N}_{\rm st} \tag{6.18}$$

[5] *Re* stands for the real part of the complex number.

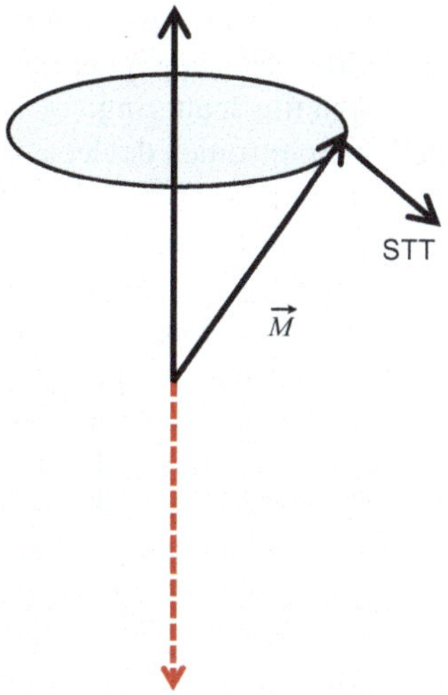

Fig. 6.20 STT. Magnetization switching due to STT

where α is the phenomenological Gilbert damping parameter whose value is usually taken as $0.001-0.1$. One has to additionally solve the LLG equation self consistently with the transport equations.

The effect of STT on the magnetization dynamics is shown in Fig. 6.20. With small amount of STT, the magnetization may still be damped. By increasing the STT, precession may be stabilized even in the presence of damping. This precession has been shown to generate electrical oscillations in GHz range—a feature that may be used in oscillators. With a further increase in the STT, the magnetization of the FM layer may be switched and subsequently stabilized due to damping in the opposite direction, schematically shown by the red dotted line, potentially switching the memory state from P to AP configurations. The magnetization may be switched back with the opposite direction of the spin polarized current. Thus, the spins of the conduction electrons may be used to flip the magnetization of the soft magnet, thus eliminating the need for an external applied magnetic field.

Problems

6.1 For Fe, determine the magnetic polarization by looking up the majority spin and the minority spin contribution to the electron density.

6.2 For Ni, determine the magnetic polarization by looking up the majority spin and the minority spin contribution to the electron density.

6.3 In a circuit model of a ferromagnet, the resistance of majority spin is given as 1 Ω and the resistance of minority spin is given as 10 Ω. Determine the net resistance of the ferromagnet.

6.4 If the ferromagnet of problem 6.3 is used in a GMR device, calculate the GMR ratio. You may ignore the resistance of the nonmagnetic spacer layer, as well as spin flip scattering.

6.5 In spintronics devices, why current in parallel configuration is usually higher than the antiparallel configuration? Can you conceive of a situation where the current in antiparallel configuration would be higher than the parallel configuration?

6.6 For an MTJ device with nine atoms (include the 3 metal atoms on the left/right of the insulator inside the device Hamiltonian), by using single orbital basis set for a 1D channel and contacts with $t = 1$ eV, calculate and plot the IV characteristics in P and AP configuration with bias applied to contact 2 with 0–0.2 V range. Take $\mu_o = 0$. The approximate conduction band profile (to be added to the diagonal of the Hamiltonian) is shown below and given as follows,
P-↑ spin: [−1 −1 −1 1 1 1 −1 −1 −1]
P-↓ spin: [−0.1 −0.1 −0.1 1 1 1 −0.1 −0.1 −0.1]
AP-↑ spin: [−1 −1 −1 1 1 1 −0.1 −0.1 −0.1]
AP-↓ spin: [−0.1 −0.1 −0.1 1 1 1 −1 −1 −1]
Assume the voltage drop across the middle 3-atoms as below,
U = [0 0 0 linspace(0, −V, 3) −V V −V];
(NOTE) Add E_c for up spin and down spin for both the contacts while calculating the product *ka* and the self energy.

6.7 Prove that for $\theta = 0$, the spin transfer torque is zero.

6.8 Prove that for $\theta = \pi$, the spin transfer torque is zero.

6.9 Calculate the value of θ for the maximum value of the spin transfer torque.

Research Assignment

R6.1 For a 1 nm MgO based tunnel junction, calculate spin polarized transport for Δ_1 and Δ_5 bands. Consult literature for band edges and effective masses. Furthermore, calculate the current in parallel configuration, the current in antiparallel configuration, and TMR ratio. Does TMR ratio become negative at high bias?

Part III
Circuits and Systems

Chapter 7
Memories

Memories are used extensively to store information in a binary manner. A memory circuit is a bistable circuit that may be in the high current (low resistance) state or the low current (high resistance) state. One should note that for the use of nanodevices in computing applications, e.g. micro and nanoprocessors, the ON/OFF current ratio has to be four to six orders of magnitude. However, for memory applications, the ON/OFF current ratio requirement is not as stringent. In fact, a ratio of only 10 or less (in most cases) is considered state of the art for the memory technology.

In the memory technology, volatile memory is the one in which the information is lost during power cycle, whereas the one in which the information is retained for an appreciable amount of time during a power cycle is called nonvolatile memory. From the perspective of functionality, memories may be clearly divided into two categories, read only memory (ROM) and random access memory (RAM). As the name suggests, in RAM each bit may be randomly accessed.[1] This feature makes RAMs impressively fast, but at the same time more expensive due to complex address circuitry. Furthermore, RAM is volatile, whereas ROM is nonvolatile. Due to their unique features, ROMs and RAMs have their niche applications. ROMs are mostly used in high volume data storage applications and RAMs are used where high speed data access to microprocessors is desired.[2]

One example of erasable and programmable nonvolatile memory for nonvolatile applications is the hard disk, which is based on the GMR device discussed in Chap. 6. The flash memory device discussed in this chapter is another example of nonvolatile memory, which is widely used in high volume data storage applications.

In this chapter, we discuss RAM devices based on CMOS (complementary MOS) technology. There are two types of RAM devices, namely static RAM (SRAM) and dynamic RAM (DRAM). SRAM has the best read/write characteristics with the least

[1]In fact, it is a group of bits, called a word, that are addressed as a whole. The size of the word depends on the size of the data bus as discussed later.

[2]ROMs may also be erasable and/or programmable.

H. Raza, *Nanoelectronics Fundamentals*, NanoScience and Technology,
https://doi.org/10.1007/978-3-030-32573-2_7

latency, but one needs at least six transistors (6T) for each memory bit, thus making this technology quite expensive, whereas DRAM has slower read/write speeds and need refreshing every few *ms*. One may implement DRAM with only one transistor and one capacitor (1T1C) making this technology extremely cost effective.

Since the inception of the charge based and the spin based devices, there has been a continued effort to invent and engineer memories with the best of RAM and ROM characteristics. Such an ideal memory device is called a universal memory and is the holly grail in the memory research.

7.1 CMOS Inverter

Consider the CMOS inverter with voltage transfer characteristics (VTC) shown in Fig. 7.1. An nMOS and a pMOS, labeled T1 and T2 respectively, are connected in series such that the drains of nMOS and pMOS are connected together. This terminal forms the output stage of the inverter. The gates of the two transistors are connected together, which form the input stage of the inverter. The source of nMOS is connected to the ground, hence $V_{GS}|_{nMOS} = V_{in}$ for the nMOS, whereas the source of pMOS is connected to the power supply voltage $V_{DD} = 5$ V, hence $V_{GS}|_{pMOS} = V_{in} - V_{DD}$ for the pMOS. One should note that the threshold voltage for nMOS (V_{TN}) is positive and that of pMOS (V_{TP}) is negative. Furthermore, $V_{GS} > V_{TN}$ for turning nMOS ON, and $V_{GS} < V_{TP}$ for turning pMOS ON.

With the input voltage $V_{in} = 0$, the nMOS transistor T1 is in OFF state since $V_{GS}|_{nMOS} = V_{in} < V_{TN}$, and pMOS is in the ON state, given $V_{GS}|_{pMOS} = V_{in} - V_{DD} < V_{TP}$. Since T1 is OFF, no current flows in the circuit despite T2 being ON. In this case, the supply voltage (V_{DD}) appears across T1, which makes the voltage at the output stage $V_{out} = V_{DD}$. Once the input voltage becomes greater than the threshold voltage for the nMOS transistor, T1 turns ON and the current starts to flow in the circuit, since T2 is in the ON state as well. $V_{TN} \leq V_{in} \leq V_{DD} + V_{TP}$ is the only range in which the current flows and hence power dissipates. Beyond V_{in} of $V_{DD} + V_{TP}$, the pMOS transistor T2 turns OFF since $V_{GS}|_{pMOS} = V_{in} - V_{DD} > V_{TP}$, whereas the

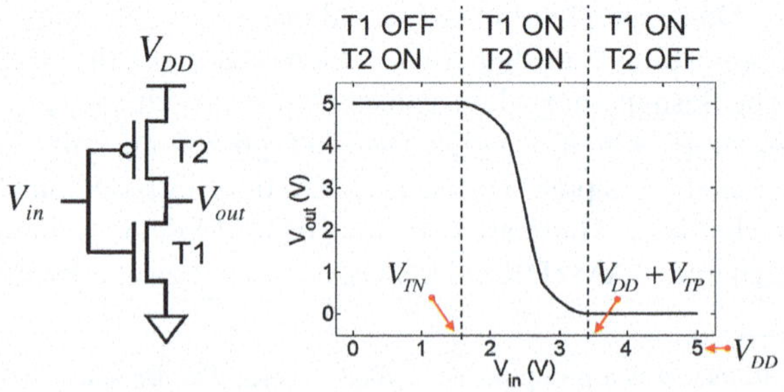

Fig. 7.1 CMOS inverter. Circuit and voltage transfer characteristics for a power supply voltage of $V_{DD} = 5$ V

nMOS transistor T1 is already ON, making the output voltage $V_{out} = 0\,V$.[3] Due to this peculiar VTC, the circuit is rightfully called an Inverter, i.e. it inverts the input voltage level. If the input voltage is LOW (0 V), the output voltage is HIGH (5 V)[4] and vice versa.

In a CMOS inverter, any variation in the input voltage between 0 to V_{TN} does not show up at the output voltage. Similarly any variation in the input voltage between $V_{DD} + V_{TP}$ to V_{DD} also get suppressed as shown in Fig. 7.1. These two input voltage ranges are defined as the noise margins for LOW input state and HIGH input state, and lead to noise immunity inherent in the digital circuits at the device level. The ability of digital circuits to correct for noise errors at the device level is the enabling technology behind digital revolution, that allows integration of billions of transistors without error accumulation. On the other hand, in an analog technology, the signal as well as noise is amplified.[5]

The voltage gain is defined as $|\partial V_{out}/\partial V_{in}|$. While the gain in the noise margin voltage range is zero, it does have a finite value during the transition from HIGH state to LOW state. A unity gain VTC is an inclined line (with intercepts of 5 V on V_{out} axis and 0 V on V_{in} axis) without any noise margins and hence lacks noise immunity. Therefore, the gain in the VTC should be greater than unity for digital circuits for noise suppression. While the concept of gain is a well established concept in analog electronics, it is equally important in digital electronics to maximize noise immunity and noise suppression at the device level.

7.2 SRAM

SRAM technology is implemented by using CMOS circuits. An SRAM cell consists of six MOS transistors, four for storing the logic information and two for selecting a memory cell in random access manner. Before discussing the CMOS 6T-SRAM circuit, let us consider a simpler four transistor (4T) circuit to elucidate the working principle.

Consider the four transistor circuit shown in Fig. 7.2a, which essentially consists of two CMOS inverters connected back to back, i.e. the output of the first CMOS inverter (with transistors T1 and T2) is connected to the input of the second CMOS inverter (with transistors T3 and T4).

---

[3]Here we assume the FETs as ideal gate operated switches, which have infinite resistance when OFF and zero resistance when ON. However, this is not the case in reality and hence LOW output voltage may be slightly higher than 0 V, and HIGH output voltage may be slightly lower than the supply voltage.

[4]Logic level LOW is defined as 0 V, and logic level HIGH is defined as 5 V. Yet another way to define the logic levels is the Boolean notation of logic 0 and logic 1. While, 0 V is usually defined as logic 0 (LOW) and 5 V as logic 1 (HIGH), but this choice is not universal. One could very well define 0 V as logic 1, and 5 V as logic 0.

[5]There are indeed some common mode noise elimination methodologies in analog circuits as well.

Fig. 7.2 SRAM working principle. **a** A bistable 4T-circuit with a feedback loop. **b** Voltage transfer characteristics

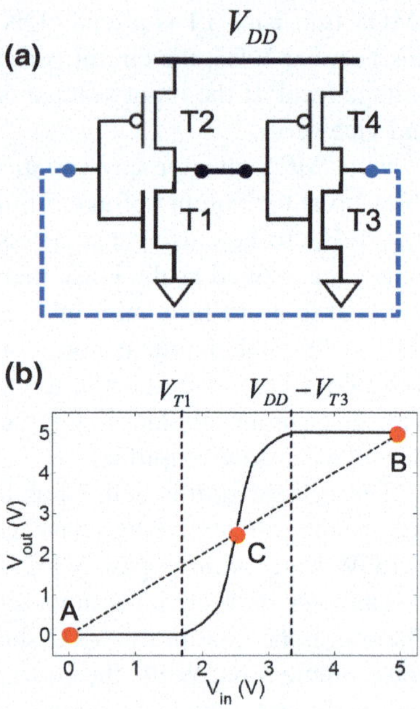

One may imagine connecting the output terminal and the input terminal by a feedback loop. In this case, the output and input terminals are simply connected by a dashed line, which depicts a linear relationship between the output and input voltages as follows, $V_{\text{in}} = V_{\text{out}}$. The resulting linear plot is shown in Fig. 7.2b with a dashed line superimposed on the VTC of the four transistor circuit without the feedback loop. The advantage of the feedback loop is that the circuit remembers its previous state even when the input is removed. Such a feedback loop becomes the primary feature of a physical circuit that may serve as a memory element. There are three operating points for this circuit now with the feedback loop which are labeled as A, B and C.

In order to understand the operation of the 4 T circuit, let us assume that to begin with the input voltage is LOW, i.e. $V_{\text{in}} = 0$. For $V_{\text{in}} < V_{T1}$ and $V_{\text{in}} - V_{\text{DD}} < V_{T2}$, the nMOS T1 is OFF and the pMOS T2 is ON, respectively, resulting in the voltage at the drain of T1 as HIGH, which equals the gate voltage for the nMOS T3 and is much greater than the threshold voltage for T3, driving T3 into the ON state. On the other hand, the pMOS T4 is OFF now, resulting in LOW voltage. Thus the output voltage for the LOW input voltage is also LOW, shown as point A in Fig. 7.2b. This result is consistent with the feedback loop, where $V_{\text{out}} = V_{\text{in}}$.

Similarly, let us assume that to begin with input voltage is HIGH. For $V_{\text{in}} = V_{\text{DD}}$, the nMOS T1 is ON and the pMOS T2 is OFF resulting in the voltage at the drain of T1 as LOW, which equals the gate voltage for the nMOS T3 and is less than the threshold voltage for T3, driving T3 into the OFF state. On the other hand, the pMOS T4 is ON now, resulting in HIGH voltage. Thus the output voltage for the HIGH input

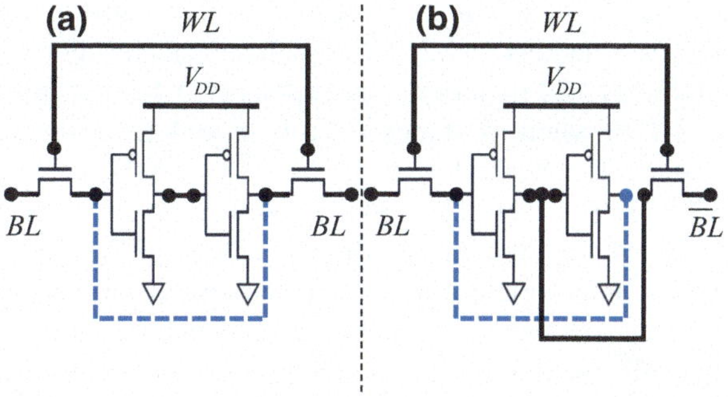

Fig. 7.3 CMOS 6T-SRAM cell. **a** WL and BL configuration. **b** WL and BL/$\overline{BL}$ configuration

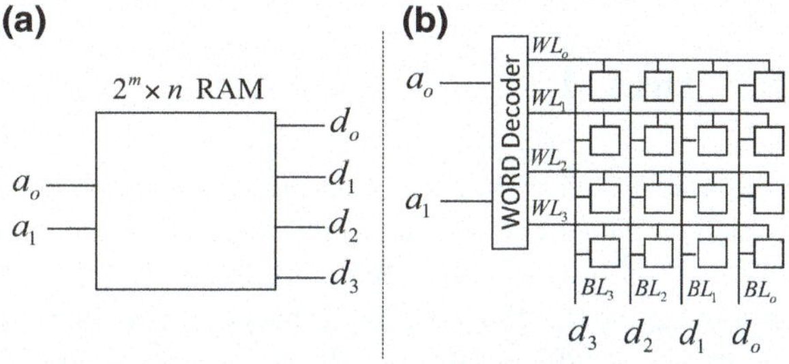

Fig. 7.4 RAM architecture. **a** Block diagram for $2^m \times n$ RAM showing $m = 2$ address bus and $n = 4$ data bus. **b** A word decoder drives WLs, which are complemented by the BLs to give a two dimensional array of memory cells

voltage is also HIGH, show as point B in Fig. 7.2b, and yet again is consistent with the feedback loop.

In summary, there are two useful operating points for this circuit, namely points A and B,[6] thus making it a bistable circuit which is used as a part of 6T-SRAM memory cell. Furthermore, since the four transistor circuit looses information in a power cycle, the memory cell is inherently volatile.

While the 4T-cell can store the logic state, one needs two additional transistors for reading/writing and selecting the memory cell. The CMOS 6T-SRAM cell is shown in Fig. 7.3a with two additional transistors used to select a memory element by making use of a word line (WL) for select and bit lines (BL) for input/output. Another CMOS 6T-SRAM cell circuit is shown in Fig. 7.3b, which makes use of WL for select and BL/$\overline{BL}$ for input/output. These memory cells are the basic storage unit in RAM architecture which we discuss next.

RAM architecture is shown in Fig. 7.4a as a block diagram with m-bits of address bus and n-bits of data bus. Due to the binary nature of addressing, for an m-bit address line and n-bit word size, the memory size has to be $2^m \times n$. For $m = 2$ address bus, one gets four words (and hence four WLs, labeled as WL3, WL2, WL1, and WL0), and with $n = 4$, each word can store 4-bits of data (and hence 4-bit data bus labeled as $d_3d_2d_1d_0$ accessible by using four BLs, labeled as BL3, BL2, BL1, and BL0, respectively), thereby making the total number of memory cells

[6]Point C is not used in the memory design due to its inherent instability and lack of noise margin. A memory state stored at point C would ultimately move to either point A or point B depending on the circuit noise.

as $2^2 \times 4 = 16$, which can store 16-bits of information. Similarly, for $m = 27$ and $n = 64$, one gets 134,217,728 words each storing 64-bits of information and hence 1,073,741,824 memory cells each storing 1-bit of information, conveniently written as 1 GB, where B stands for a byte and $1B = 8bits$. One should note that in the context of memory design, $1K = 2^{10} = 1,024$, $1M = 2^{20} = 1,048,576$, and $1G = 2^{30} = 1,073,741,824$. Finally, the memory cells are arranged in a two dimensional array, the architecture of which is shown in Fig. 7.4b. The rows form the WLs and the columns form the BLs. When a particular WL is selected, the intersection of the selected row and the columns selects specific set of memory cells. Once addressed properly, the read/write control circuitry is used to read/write data respectively.

7.3 DRAM

Dynamic RAM (DRAM) is based on the principle of charge storage in a MOS capacitor. The 1T1C memory cell consists of one nMOS and one storage capacitor placed right next to each other as shown in Fig. 7.5a, b. The capacitor's gate is used as a storage gate (SG) and the transistor's gate is used as a transfer control selected by the WL. At the input stage, the pn junction diode of the nMOS is connected to the BL, which sets the potential to introduce the charge state corresponding to the HIGH state or logic 1 state (absence of charge on the storage capacitor by connecting BL to +5 V), and LOW state or logic 0 state (presence of charge on the storage capacitor by connecting BL to 0 V).

Next, we discuss the detailed working principle of a DRAM cell. By connecting the BL to +5 V, the pn junction is made reverse biased and hence there is no injection of charge into the device. One should note that the p-type substrate is connected to the ground and SG is always tied to +5 V. By applying +5 V to the WL, charge transfer between the BL and SG is made possible. However, since there is no charge injection across the reverse biased pn junction, no charge is stored on the SG that corresponds to the logic 1 level as shown in Fig. 7.5a. After pulling the WL voltage to 0 V, the storage capacitor is isolated and hence the logic 0 level is retained.

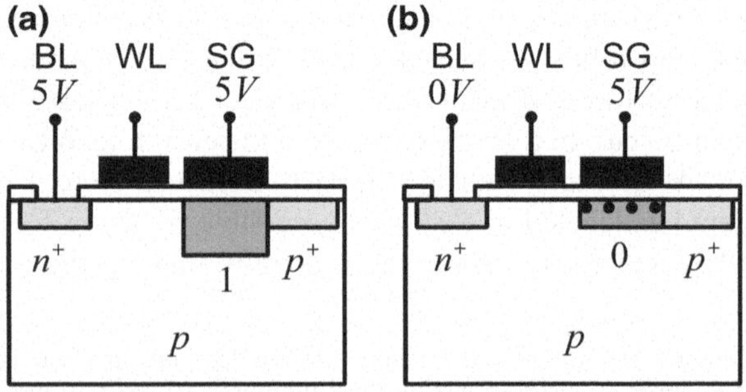

Fig. 7.5 1T1C DRAM cell. **a** HIGH (+5 V) state at BL is written to the cell. Storage gate (SG) always has +5 V applied with no charge stored for the logic 1 state. **b** LOW (0 V) state at BL is written to the cell. Negative charge is stored for the logic 0 state. p^+ region is for isolation

Next, consider the case where the BL is connected to 0 V. In this case, the pn junction at the input stage is forward biased if the WL is connected to +5 V, since SG is always connected to +5 V. This enables the negative charge injection from the BL into the transfer capacitor of WL, which gets stored in the storage capacitor as shown in Fig. 7.5b. Finally, when the voltage at the transfer capacitor of WL is pulled back to 0 V, which isolates the BL and the SG, the negative charge is retained in the storage gate, which corresponds to the logic 0 level. However, this stored charge may not be retained for a long time. Within few *ms*, the stored charge is leaked into the substrate or recombined with thermally generated holes. Therefore, in DRAM cells, refreshing the charge state every few *ms* is a necessary condition for proper functioning of the DRAM device. This brings an additional complexity of refresh circuitry, the details of which are beyond the scope of this book. However, the structural simplicity and the reduced cost of DRAM cells makes them a compelling choice for the RAM market. In fact, in the global market, DRAM holds about 90% of the RAM market share.

One key parameter for DRAM performance optimization is the charge stored in the SG. For a given voltage, the stored charge may be increased by enhancing the MOS capacitance. While the capacitance may be increased by decreasing the dielectric thickness and by using high-K dielectric materials, yet another parameter is the cross sectional area of the capacitor. Within the given area constraint of a memory cell, the cross sectional area of the capacitor may be increased by using the trench configuration as shown in Fig. 7.6. A trench may be etched into silicon and then a dielectric may be deposited followed by a metallic film as a gate by using techniques like Atomic Layer Deposition (discussed in Chap. 9) that are optimized for high aspect ratio material synthesis. Another avenue is the use of nanomaterials to enhance the surface area and hence the capacitance.

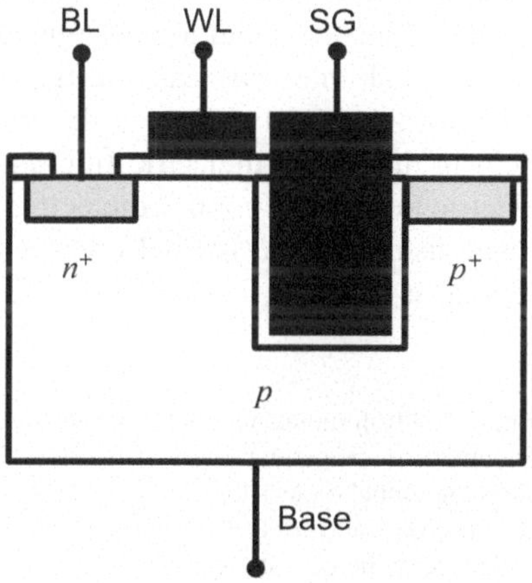

Fig. 7.6 Trench capacitor in DRAM cell for better charge storage

7.4 Flash Memory

A nonvolatile flash memory cell is shown in Fig. 7.7, where the key difference from the FET discussed in Sect. 5.4 is the design of the gate stack. In FET, while the electrical equivalent gate dielectric thickness is about 1/40th of the channel length, the gate dielectric in the flash memory is rather thick. In fact, it consists of three distinct regions, namely 3–5 nm tunnel dielectric for allowing a charge injection from the channel or the substrate, a storage node where the injected charge may be stored and 30–40 nm control dielectric that isolates the storage node from the gate contact.

Various multi dimensional materials have been used in the storage node. For example, the initial flash memories had two dimensional semiconducting films as the storage node. Since then, one dimensional nanowires and nanotubes have been incorporated. Zero dimensional nanostructures, like quantum dots, nano dots, nano crystals, organic molecules, etc., have also been used. One of the advantages of using nanostructures, particularly zero dimensional nanomaterials, is the improvement in reliability. Consider a pinhole defect in the tunnel dielectric. If the storage node consists of continuous medium, presence of such a defect may cause complete loss of the stored charge. However, if the storage node consists of discrete entities like nanomaterials, a defect leads to charge loss from only a few discrete storage locations without changing the memory state.

The material content of the storage node (in 0D, 1D, 2D) could be metallic (e.g. Au, Pt, etc.), semiconducting (Si, carbon nanostructures, etc.), and even traps in some insulators (e.g. traps in silicon nitride, etc.). Use of these materials improves the device performance of the flash memories. For example, the use of metal nanocrystal results in higher electric field due to the stored charge. This helps in electrostatically modulating the channel transport and enables reading the charge state of the storage node with better signal to noise ratio.

The goal for electrostatic design optimization of the gate stack is to optimize the effect of the stored charge on the channel and not on the gate contact. The electric field in the tunnel dielectric (region 1) is designed to be high as compared to the electric field in the control dielectric (region 2) as shown in Fig. 7.8. The continuity relationship for the normal components of the electric flux densities for the two regions is given as,

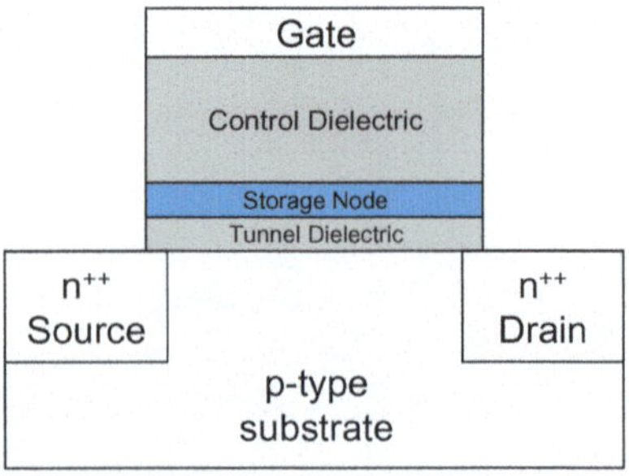

Fig. 7.7 Flash memory. Schematic device structure showing tunnel dielectric, storage node, and control dielectric in the gate stack

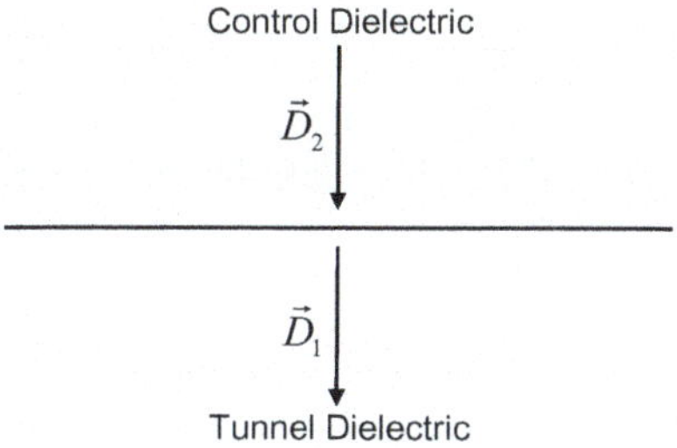

Fig. 7.8 Flash memory. Electrostatic design with high-K dielectric for the control dielectric

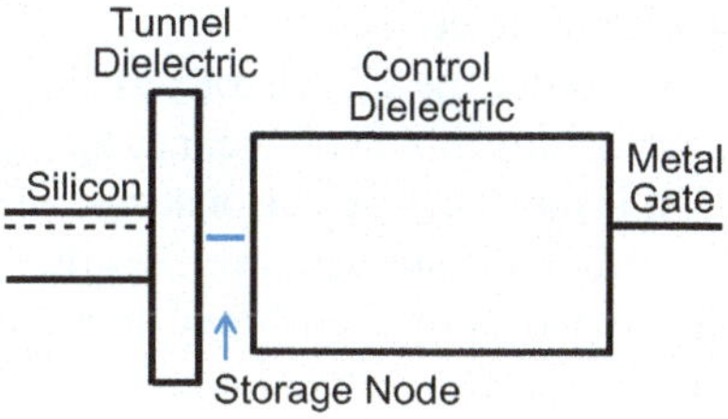

Fig. 7.9 Flash memory. Equilibrium band diagram

$$\mathcal{D}_{1n} - \mathcal{D}_{2n} = \rho_s \tag{7.1}$$

where ρ_s is the surface charge density.

Assuming, zero charge stored at the interface of the two dielectrics for simplification, i.e. $\rho_s = 0$, one obtains,

$$\mathcal{D}_{1n} = \mathcal{D}_{2n} \tag{7.2}$$

Writing the normal components of the electric flux densities ($\mathcal{D}_{1n}$ and $\mathcal{D}_{2n}$) in terms of the normal components of the electric field intensities ($\mathcal{E}_{1n}$ and $\mathcal{E}_{2n}$) and relative dielectric constants (K_c and K_t) of the tunnel dielectric and the control dielectric respectively,[7] one obtains,

$$\frac{\mathcal{E}_{1n}}{\mathcal{E}_{2n}} = \frac{K_c}{K_t} \tag{7.3}$$

Hence, in order to enhance the electric field intensity in the tunnel dielectric ($\mathcal{E}_{1n}$), high-K dielectric for the control dielectric and low-K dielectric for the tunnel dielectric are optimum for better electrostatic design.

It is well known that high-K dielectrics have smaller bandgap. Thus the bandgap of the control dielectric is usually smaller than that of the tunnel dielectric as shown in Fig. 7.9 for the equilibrium band diagram. With a smaller bandgap for the control dielectric, the barrier from the storage node to the gate is smaller, which requires a thicker control dielectric. The change in the threshold voltage due to the stored charge is given as,

[7] $\mathcal{D} = K\varepsilon_o\mathcal{E}$.

$$\Delta V_{\mathrm{TN}} = -\frac{\Delta Q_n}{C_c} \tag{7.4}$$

where ΔQ_n is the stored charge and C_c is the capacitance of the control dielectric given as,

$$C_c = S\frac{K_c \varepsilon_o}{t_c} \tag{7.5}$$

where t_c is the control dielectric thickness. For a higher shift in the threshold voltage ΔV_{TN}, the control dielectric capacitance should be small, which may be achieved by a thicker dielectric.

Consider the CV characteristics in Fig. 7.10. By applying a gate voltage greater than the threshold voltage, i.e. $V_{\mathrm{GS}}|_{\mathrm{nMOS}} > V_{\mathrm{TN}}$, results in forming an electron inversion layer. These electrons may tunnel through the tunnel gate dielectric and get trapped in the storage layer, resulting in *Write* or *Program* process. Similarly a large negative bias may push any stored electrons away from the storage node, resulting in *Erase* or *Reset* operation.

Without any charge stored (the logic 0 level), the CV curve is shown in Fig. 7.10 with gate voltage sweep from the negative (Erase) to the positive (Write or Program) direction. If one applies a write pulse of appropriate duration, negative charge is stored in the storage node resulting in the logic 1 level. Due to this negative charge, the threshold voltage is shifted towards positive voltage as shown in Fig. 7.10. While CV characterization is a convenient method to characterize the threshold voltage shift characteristics, it is usually the IV characteristics that are used in integrated circuits as shown in Fig. 7.11 for the logic 0 level and the logic 1 level.

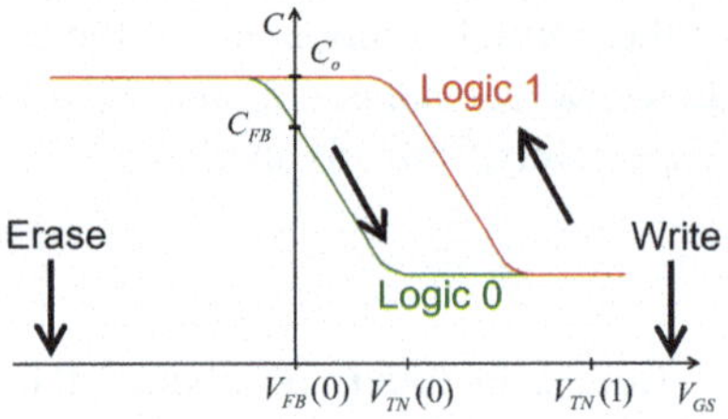

Fig. 7.10 Flash memory. CV characteristics

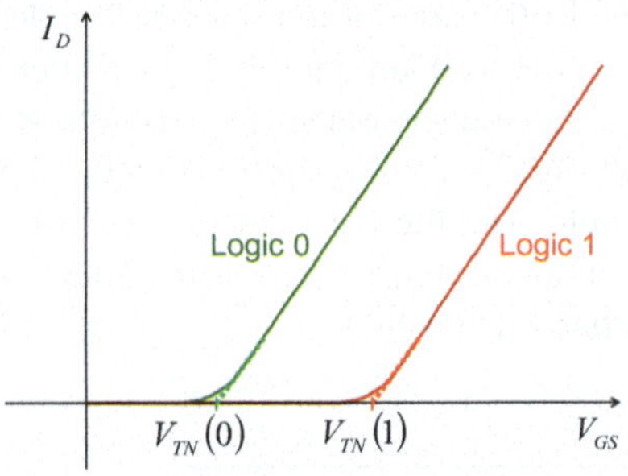

Fig. 7.11 Flash memory. Transfer characteristics for the logic 0 level and the logic 1 level

Flash memories are also designed to retain charge for up to 10 years. A characteristic retention curve is shown in Fig. 7.12, where the threshold voltage window[8] decreases over time due to charge leakage in the logic 1 level or even charge trapping due to defect generation in the logic 0 level.

The amount of charge stored during Write or Program process also depends on the program pulse width (τ_p) which results in the the upward shift in V_{TN} for the logic 1 level as shown in Fig. 7.13. Similarly, during erase cycle, the threshold voltage shift depends on the erase pulse width resulting in downward shift in V_{TN} for the logic 0 level. Given current through the tunnel dielectric (I_t) and the control (I_c) dielectric, the stored charge ΔQ_n is given as,

$$\Delta Q_n = (I_t - I_c)\,\tau_p \tag{7.6}$$

Here, I_t and I_c may be calculated by using NEGF with an appropriate description of self energy for the storage node.

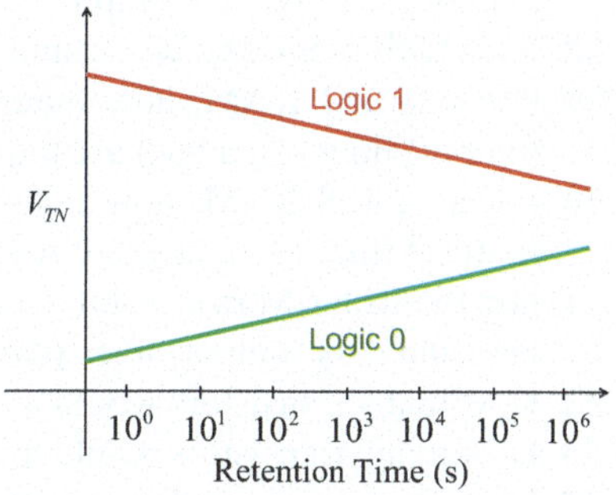

Fig. 7.12 Flash memory. Retention characteristics for the logic 0 level and the logic 1 level

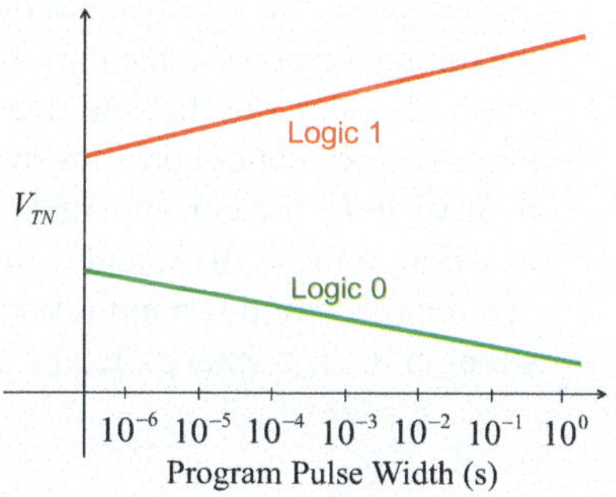

Fig. 7.13 Flash memory. Write/erase characteristics for the logic 0 level and the logic 1 level

[8]Difference between the threshold voltages for logic 1 and logic 0 states.

Problems

7.1 In the context of CMOS inverter, what are the two main advantages of using CMOS for digital applications?

7.2 NMOS. $V_{TN} = 2.5\,V$. Would the transistor be ON/OFF if:
(a) $V_G = 0\,V$ and $V_S = 0\,V$ (b) $V_G = 5\,V$ and $V_S = 0\,V$ (b) $V_G = 0\,V$ and $V_S = 5\,V$ (b) $V_G = 5\,V$ and $V_S = 5\,V$

7.3 PMOS. $V_{TP} = -2.5\,V$. Would the transistor be ON/OFF if:
(a) $V_G = 0\,V$ and $V_S = 0\,V$ (b) $V_G = 5\,V$ and $V_S = 0\,V$ (b) $V_G = 0\,V$ and $V_S = 5\,V$ (b) $V_G = 5\,V$ and $V_S = 5\,V$

7.4 For a CMOS inverter with $V_{TN} = 1.5\,V$, and $V_{TP} = -2.5\,V$, calculate the noise margin for HIGH (5 V) input.

7.5 For a CMOS inverter with $V_{TN} = 1.5\,V$, and $V_{TP} = -2.5\,V$, calculate the noise margin for LOW (0 V) input.

7.6 For a CMOS inverter with $V_{TN} = 2\,V$, and $V_{TP} = -1\,V$, calculate the noise margin for HIGH (5 V) input.

7.7 For a CMOS inverter with $V_{TN} = 2\,V$, and $V_{TP} = -1\,V$, calculate the noise margin for LOW (0 V) input.

7.8 For a CMOS inverter with a unity gain, draw the voltage transfer characteristics.

7.9 For a 16×1 RAM, how many address (input) lines (address bus) and data (output) lines (data bus) are required? What is the word size?

7.10 For a 32×8 RAM, how many address (input) lines (address bus) and data (output) lines (data bus) are required? What is the word size?

7.11 For 16 GB RAM in a 64-bit computer, how many address (input) lines (address bus) and data (output) lines (data bus) are required? What is the word size?

7.12 How many transistors would a 16 GB 6T-SRAM chip have?

7.13 How many transistors would a 16 GB 1T1C-DRAM chip have?

7.14 For a flash memory device, the change in threshold voltage is 1 V. Design the values of the thicknesses and dielectric constants for the control dielectric and the tunnel dielectric. One may assume an appropriate value of the stored charge.

7.15 Flash Memory. For the gate stack with metal nanocrystal shown in the Fig. 7.14, plot 2D electric field profiles (using COMSOL or equivalent software) with the relative dielectric constant of 4 for the tunnel dielectric and the relative dielectric constant of (a) 4, (b) 8, and (c) 12 for the control dielectric. The diameter of the metal nanocrystal is 5 nm (choose any material). Comment on the electric field strength in the tunnel dielectric for various values of control dielectric constant.

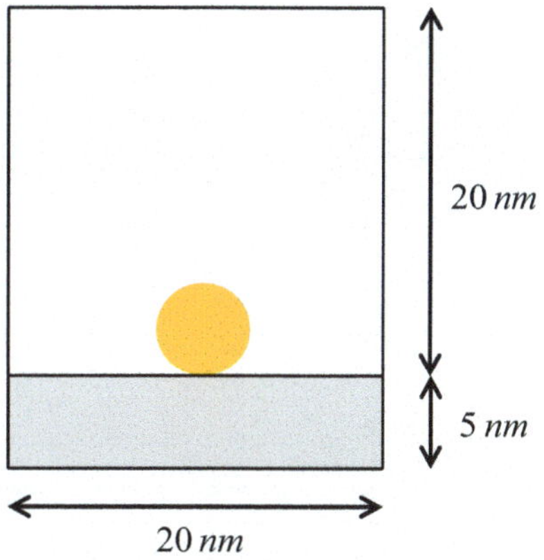

Fig. 7.14 Problem 7.15

Research Assignment

R.1 For a flash memory device, the gate stack may have nanoparticles made of either metal, semiconductor or organics. Write a one page summary of which of the nanoparticles (metal, semiconductor, or organics) would affect the channel the most.

Chapter 8
Circuits and Systems

In this chapter, we discuss various circuits and systems based on the CMOS technology. While, it is simply not possible to cover the broad area of the circuits and systems related to nanoelectronics in a single chapter, we do discuss a few representative circuits and systems, in order to emphasize the key concepts.

In the big picture of the relationship between the system level design to the circuit level design through various design approaches at the algorithmic level, the RT (register[1] transfer) level, the logic level, and the circuit level, the Gajski Kuhn chart (better known as Y chart, as shown in Fig. 8.1) is very much applicable. At the system level, the architecture (memory,[2] for example), the system specifications, and the chip as a whole relate to structure, behavioral, and physical/geometrical domains, respectively. Furthermore, at the circuit level, the description of transistor's function, transistor itself, and the transistor layout that is applicable in behavioral, structure, and physical/geometrical domains, respectively. In between the system level and the circuit level description, one may have to be considerate of the algorithmic, register transfer, and logic level description. While we disucss the logic level description in the context of the system design in this chapter, algorithmic and RT level discussion is beyond the scope of this book.

nMOS and pMOS are the building blocks for the CMOS technology. As discussed in Sect. 7.1, the CMOS inverter provides a low power switching mechanism, where the current flows and hence the power dissipates only during the switching process. A CMOS inverter with voltage gain may also enable noise suppression at the device level, which is the key reason behind digital revolution over the past few decades. The design level above the circuit level is the logic level, where the logic gates and

[1]A register is a physical circuit that may store multiple bits. e.g., an 8-bit register has the capacity to store 8-bits of data.

[2]System level design of memory is already discussed in Chap. 7.

H. Raza, *Nanoelectronics Fundamentals*, NanoScience and Technology,
https://doi.org/10.1007/978-3-030-32573-2_8

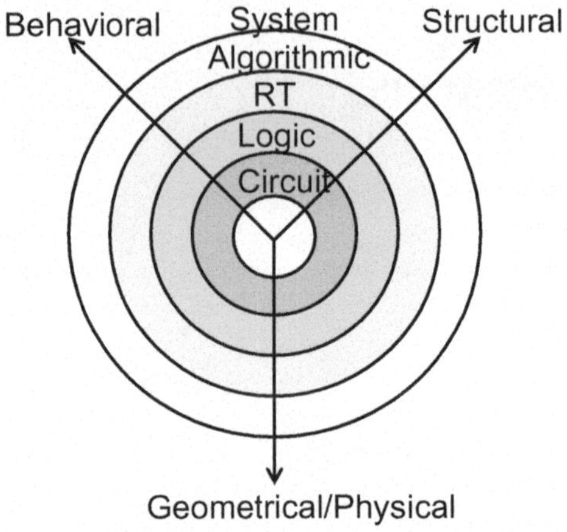

Y-Chart	Structural	Behavioral	Physical
System Level	CPU, Memory	Specifications	Chip
Algorithmic Level	Data Paths	Algorithms	Block
RT Level	ALU	Register Transfer	Floor Plan
Logic Level	Gates	Boolean Logic	Standard Cell
Circuit Level	Transistor	Transistor	Transistor Layout

Fig. 8.1 Y chart describing the structural domain, the behavioral domain, and the geometrical/physical domain

the Boolean logic are the tools for the structural domain and the behavioral domain, respectively. We discuss the combinational circuits used for implementing Boolean logic in the next section.

8.1 Combinational Circuits

A combinational circuit is a digital circuit, where the output depends only on the present inputs.

NOT Gate

The CMOS inverter shown in Fig. 8.2 provides the function of a NOT gate and hence NOT logic operation. The circuit symbol for the NOT gate is shown as well. The truth table[3] for the NOT gate is given in Table 8.1. By designating 0 V as the logic 0

[3]A table of logic outputs from all possible combinations of the logic inputs.

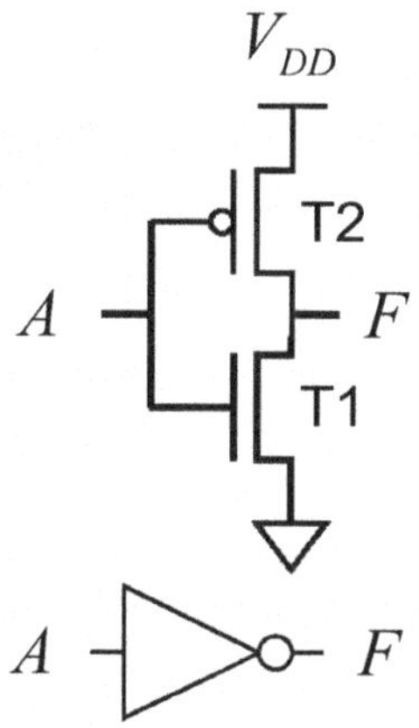

Fig. 8.2 NOT gate. CMOS inverter circuit and the circuit symbol for the NOT gate

Table 8.1 NOT gate. Truth table

Input A	Output F
0	1
1	0

(LOW) level and +5 V as the logic 1 (HIGH) level,[4] one may deduce that the circuit corresponding to the NOT gate inverts the input signal A. Given logic level 0 at the input A, one gets logic level 1 at the output F, and vice versa.[5] However, the NOT gate is not a universal gate and hence the NOT Boolean operation is not a universal Boolean operation.[6] The universal gates are indeed the NAND and the NOR gates, which we discuss next.

NAND Gate

The CMOS circuit for the two input NAND gate is shown in Fig. 8.3. The circuit symbol is shown as well, which is represented either by AND-NOT or NOT-OR combination. The corresponding truth table is given in Table 8.2. To get the logic level 0 output, one has to make sure that both T1 and T2 nMOS are ON. This may be enabled by applying logic level 1 at both the inputs A and B. For any other combination, either T1 or T2 or both are OFF, and one of the pMOS T3 or T4, or both are ON, leading to a logic level 1 at the output.

---

[4]This assignment is not universal, and one may very well assign 0 V to the logic 1 level and +5 V to the logic 0 level.

[5]One should note that there are two additional logic levels in digital circuits. Logic level Z represents high impedance state, whereas logic level X represents indeterminate state.

[6]To qualify as a universal gate, one should be able to implement all other logic gates by using one particular logic gate.

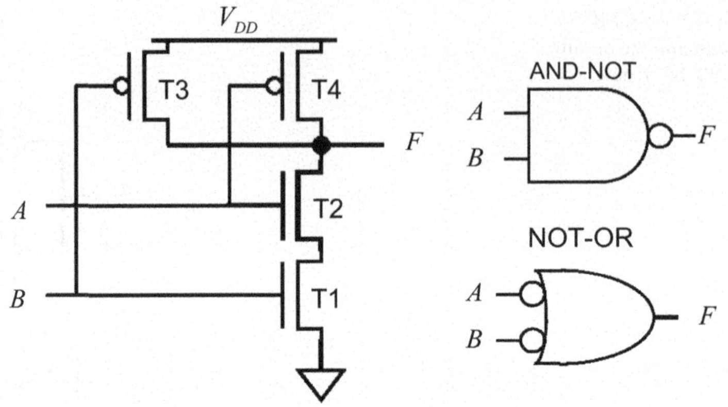

Fig. 8.3 Two input NAND gate with associated circuit symbol

Table 8.2 Two input NAND gate. Truth table

Input A	Input B	Output F
0	0	1
0	1	1
1	0	1
1	1	0

One may further show that all the logic functions may be implemented by using NAND gates only, thereby making it a universal gate. For example, a NOT gate may simply be formed by connecting the two inputs A and B with each other.

NOR Gate

The CMOS circuit for the NOR gate is shown in Fig. 8.4. The circuit symbol is shown as well, which is represented either by OR-NOT or NOT-AND combination. The corresponding truth table is given in Table 8.3. To get the logic level 1 output, one has to make sure that both T1 and T2 nMOS are OFF. This may be enabled by applying logic level 0 at both the inputs A and B. For any other combination, either T1 or T2 or both are ON, and one of the pMOS T3 or T4, or both are OFF, leading to a logic level 0 at the output.

One may further show that all the logic functions may be implemented by using NOR gates only, thereby making it a universal gate. For example, a NOT gate again may simply be formed by connecting the two inputs A and B with each other.

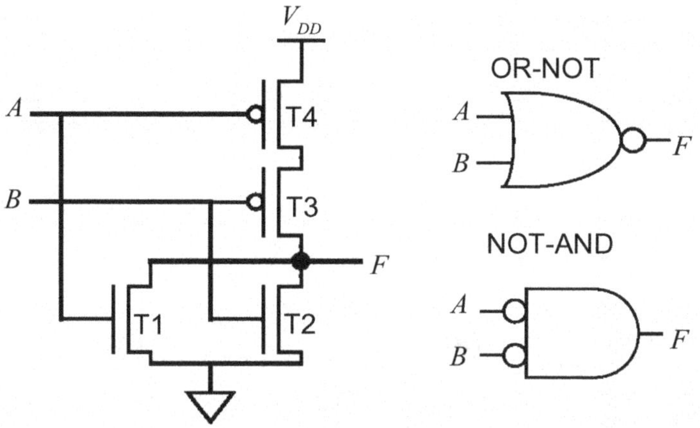

Fig. 8.4 Two input NOR gate with associated circuit symbol. CMOS circuit with two inputs and one output

Table 8.3 Two input NOR gate. Truth table

Input A	Input B	Output F
0	0	1
0	1	0
1	0	0
1	1	0

Transmission Gate

Most digital circuits have a bottleneck that is illustrated by the pass transistor. Consider nMOS in Fig. 8.5 with the input A at the source and the output F at the drain. For a gate logic level (called Select signal, labeled as S) of logic 1, the nMOS is ON. With an input logic level 0 at the source, the output level at the drain is a strong logic 0. However, with the input logic level 1 at the source, the logic level 1 output at the drain is degraded due to a finite channel resistance in the ON state. The pMOS, also shown in Fig. 8.5, exhibits the opposite trend.

One may combine the above mentioned complementary strengths and weaknesses of the nMOS and pMOS in the form of a transmission gate shown in Fig. 8.6, which

Fig. 8.5 Pass transistor. Signal may be degraded for HIGH and LOW inputs for nMOS and pMOS by applying $S = 1$ and $\overline{S} = 0$, respectively

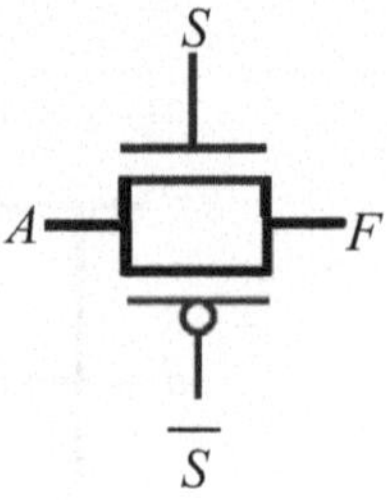

Fig. 8.6 Transmission gate. No signal degradation due to complementary features of nMOS and pMOS

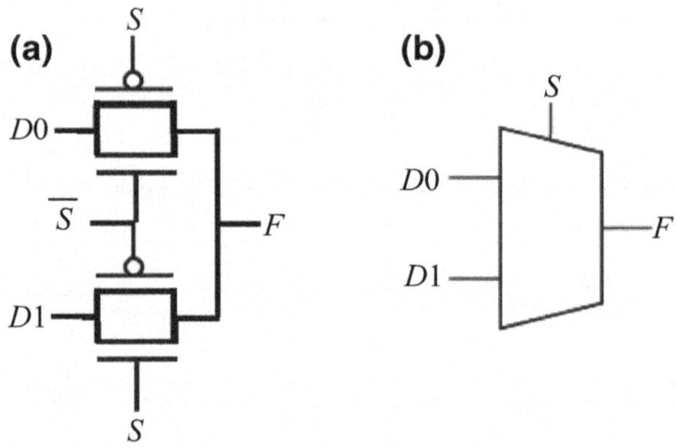

Fig. 8.7 Duplexer. **a** Two input duplexer circuit with one select signal S. **b** Circuit symbol with the inputs $D0$ and $D1$, one select signal S, and the output F

is basically a parallel combination of the nMOS and pMOS with complementary gate inputs of S and $\overline{S}$ respectively. By using the transmission gate, a strong logic level 0 or logic level 1 may always be observed irrespective of the gate select signal $(S/\overline{S})$.

Duplexer

By using two transmission gates (consisting of two nMOS and two pMOS in total), one obtains a duplexer circuit[7] as shown in Fig. 8.7a. The associated circuit symbol is shown in Fig. 8.7b, whereas the truth table is reported in Table 8.4.

The logic level 0 Select (S) enables the top transmission gate and hence the output F equals the input $D0$, irrespective of the logic level at the input $D1$ (represented by the logic level X in the truth Table 8.4). Similarly, the Select logic level 1 enables the bottom transmission gate, which leads to the output F being equal to the input $D1$, irrespective of the $D0$ input. Hence, by making use of the Select (S) signal, one may let either of the two inputs pass through, making this circuit a two input duplexer circuit.

[7] A more general circuit with the same functionality with 2^n inputs and n select signals is called a multiplexer. A duplexer is simply a two-input multiplexer.

Table 8.4 Truth table for the two input duplexer circuit

Select $S/\overline{S}$	Input $D1$	Input $D0$	Output F
0/1	X	0	0
0/1	X	1	1
1/0	0	X	0
1/0	1	X	1

8.2 Sequential Circuits

So far, we have only discussed the combinational circuits, where the output at any given time depends on the concurrent inputs. Such circuits have no memory of their own and the output at any instance simply depends on the inputs. Next, we discuss circuits where the output not only depends on the input at a given instance of time, but also the previous inputs, i.e. output gets affected by the history of the circuit. In addition, the output may be synchronized by the clock signal, thus making the circuit synchronous sequential circuit.

An asynchronous sequential D-latch circuit diagram is shown in Fig. 8.8a and the circuit symbol is shown in Fig. 8.8b. The latch consists of a duplexer and three inverters. The two transmission gates making up the duplexer are explicitly shown as well, where the top transmission gate is in the feedback loop. For the clock (CLK) logic 1 signal, the bottom transmission gate is ON, where the top one inside the feedback loop is OFF, thereby making the output Q equal to the input D, i.e. for the logic level 1 D input, one gets logic level 1 Q output and so on. For the clock

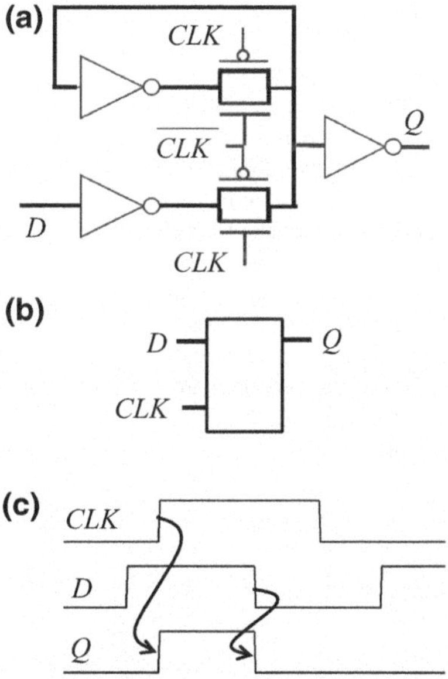

Fig. 8.8 D latch. **a** Circuit diagram. **b** Circuit symbol. **c** Timing diagram

logic level 0 signal, the bottom transmission gate is OFF, whereas the top gate in the feedback loop is ON, thus isolating the output *Q* from the input, where the latch remembers its state that was written under the clock logic level 1 due to the feedback loop. Thus, the asynchronous latch starts changing its output at the positive edge trigger, and keeps on following the input during the clock logic level 1 as shown in the timing diagram of Fig. 8.8c. At the negative edge trigger, the output stops following the input and stays in the previous state. Since the latch's output *Q* follows the input *D* as far as clock (CLK) signal is logic 1, latch inherently is asynchronous.

A synchronous D-flip flop serves a different purpose than that of an asynchronous latch. The flip flop consists of two latches connected in series as shown in Fig. 8.9a—the two latches are labeled as master latch and slave latch respectively. The circuit symbol for the flip flop is shown in Fig. 8.9b. For the clock (CLK) logic 0 signal, the master latch is enabled, thereby storing the incoming *D* logic value, whereas the slave latch is disabled. At the positive edge triggered CLK transition, when the CLK signal turns logic 1, the master latch is disabled and the slave latch is not enabled, thereby transferring the previously stored logic value in the master latch to the slave latch and hence to the output of the flip flop. Thus, the synchronous flip flop changes its output only at the positive edge trigger, and does not follow the input afterwards as shown in the timing diagram of Fig. 8.9c.

While we discuss the positive edge triggered flip flop here, it is indeed feasible to implement a negative edge triggered flip flop, as well as double edge triggered flip flop. Such synchronous circuit elements have the unique advantage that the outputs change at a given instance of time controlled by the clock signal. In a digital circuit, since various signals may arrive from different circuit elements with different latency, by using synchronous logic, one ensures that the right inputs are used for further signal processing and computation in order to avoid a runaway condition.

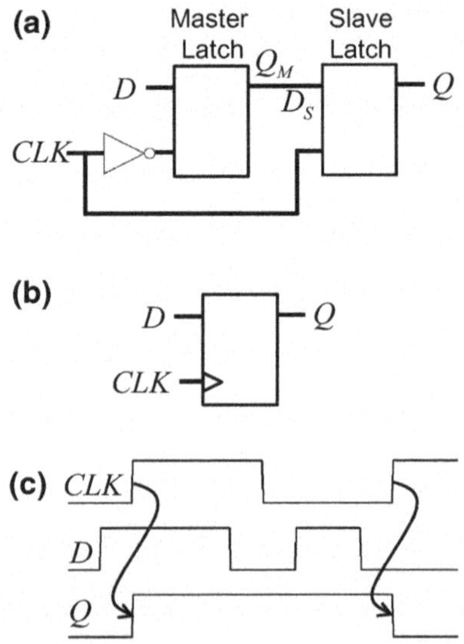

Fig. 8.9 D flip flop. **a** Circuit diagram. Output and input for the master latch and the slave latch are labeled as Q_M and D_S, respectively. **b** Circuit symbol. The triangle at the clock input represents synchronous operation and distinguishes the flip flop symbol form the latch one. **c** Timing diagram depicting positive edge triggered operation

8.3 Charge Coupled Devices

In this section, we discuss a system level design of charge coupled devices (CCD), which not only provides an interesting example, but also ties up the discussion in the context of the design in behavioral domain, the structural domain, and the physical/geometric domain. As shown in the Y chart (Fig. 8.1), the system specification for the CCD usually includes the total number of pixels, pixel array design, etc. The structural domain of CCD is concerned with the detailed architecture, i.e. whether a complete frame is transferred, etc. Finally, the physical/geometrical domain is what we focus on in this section.

We first discuss the working principle of CCD. Si has a bandgap of 1.18 eV at room temperature. Incident photons with an energy equal to the bandgap and above may be absorbed,[8] which create electron-hole pairs. The electrons in the conduction band and the holes in the valence band thereby increase the conduction in silicon. A simple pn junction diode may be used as a photodiode, where the light is detected by the additional current flow due to the photogenerated electron-hole pairs. [9]

With the incident photons on the silicon surface, the number of electron-hole pairs generated are directly proportional to the photons absorbed, which in turn are directly related to the incident photon intensity. The idea of CCD is based on the conversion of the photoabsorbed electron-hole pairs into an easily measurable electric current, and transferring the charge by using a simple circuitry, but complex timing. One may be able to perform signal processing to convert these electric signals into useful 2D (two dimensional) data.

CCD have attracted tremendous attraction due to widespread applications in imaging devices. Such applications have not only led to more popular usage in photographic digital cameras, but also have propelled scientific discovery. Modern microscopes and telescopes are equipped with CCD based high resolution digital cameras. Progress in biology, metrology, telescopy, and sensing depends greatly on the advances in CCD technology.

A charge coupled device may be thought of as a 2D array of MOS capacitors divided into rows and columns as shown in Fig. 8.10. With the properly engineered output stages and timing circuitry, the charge may be transferred from one row of MOS capacitors to the next in a parallel fashion. At the output stage, the charge transfer is serial as depicted in the readout node. Furthermore, the CCD array under discussion is a three phase (Φ_1, Φ_2, Φ_3) one, where each pixel consists of three MOS capacitors, the operation of which we discuss next.

The device structure of one pixel out of the two dimensional array of MOS capacitors in CCD configuration is shown in Fig. 8.11a. The photogenerated electron-hole pairs are usually lost at the Si/SiO_2 interface. Therefore, a buried n^{++} channel is usually fabricated by using ion implantation (a technique that is briefly discussed in

[8]Photons with energy less than the bandgap may also be absorbed leading to exciton peaks in the absorption spectrum.

[9]In fact, silicon is such a good sensor for photons and humidity, that one has to encapsulate silicon devices in dark and hermetically sealed packaging to ensure proper operation.

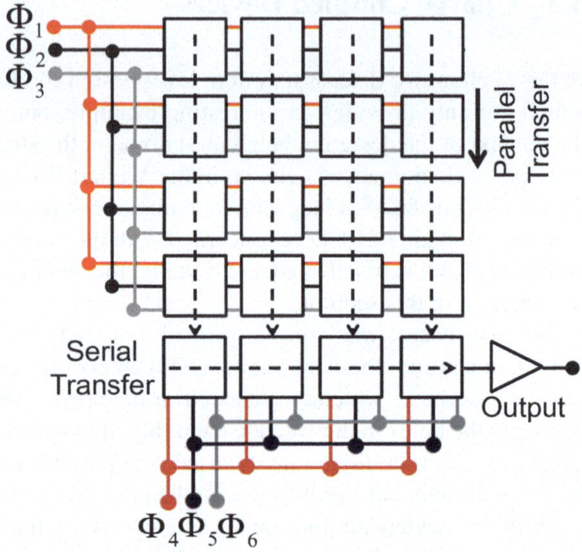

Fig. 8.10 CCD array. Charge in the three phases (Φ_1, Φ_2, Φ_3) is transferred in a parallel mode, and finally read out serially at the output stage

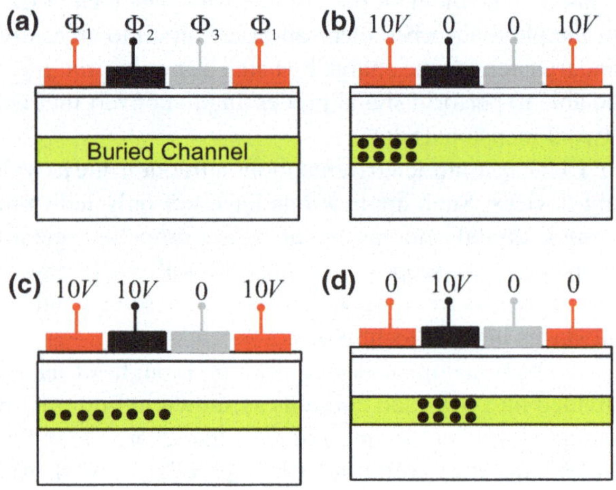

Fig. 8.11 Three phase CCD. **a** Device structure. **b** Charge storage in phase Φ_1. **c** Charge transfer from phase Φ_1 to phase Φ_2. **d** Charge storage in phase Φ_2

the next chapter). With the light input, electrons are collected in the buried channel by making use of the appropriate gate voltage of +10 V for phase Φ_1 as shown in Fig. 8.11b. The basic problem in transferring the charge state of each one of these capacitors is solved by properly timing the gate voltage of the MOS capacitor of phase Φ_1 and phase Φ_2, so that the charge from phase Φ_1 capacitor is transferred to the phase Φ_2 capacitor as shown in Fig. 8.11c, d. In this fashion, the charge is transferred between various phases within a pixel, as well as between pixels. This parallel operation, although slow in speed, results in highly cost effective solution.

One may characterize the CCD based on the quantum efficiency (QE), which is defined as follows,

$$QE = \frac{Generated\ electron\ hole\ pairs}{Number\ of\ incident\ photons} \tag{8.1}$$

Charge loss is an important parameter, which not only depends on the inherent lifetime of the photogenerated electrons, but also the Si crystal quality. Dark current is another important parameter that is present even in the absence of light. The dark current leads to the noise in CCD, and is instrumental in defining the SNR (signal to noise ratio). Moreover, the charge transfer efficiency (CTE) is defined as follows,

$$CTE = \frac{Photogenerated\ electrons\ transferred\ to\ the\ next\ phase}{Total\ number\ of\ the\ photogenerated\ electrons} \tag{8.2}$$

As a side note, we have presented the material in this chapter and the previous one in a rather nonchronological order. In fact, it was CCD and the associated working principles, which led to various other applications, including memories, shift register, signal processor, filter, etc.

8.4 VLSI

The development of the IC (integrated circuit) chips has been so fast paced that it is hard to compare with any other technological development in the recorded history. The *good* and *bad* of the IC technology has been summed up very well in the following quote:

> If the automobile had followed the same development cycle as the computer, a Rolls-Royce would today cost $100, get a million miles per gallon, and explode once a year, killing everyone inside.
>
> — Robert X. Cringely

VLSI (very large scale integration) is the technique of integrating circuits containing billions of transistors in a chip.[10] To enable this integration, Y chart introduced in Fig. 8.1 is again a very useful tool. We present the same information in the form of a flow diagram in Fig. 8.12. The design process starts with the system specifications, followed by the architectural design, which leads to the functional and logic design. Circuit design and the physical design are the next steps in the process. In particular, physical design consists of partitioning, floor planning, placement, clock tree synthesis, signal routing, and timing closure.

[10] There are various definitions of VLSI in terms of the number of transistors. Our focus here is only on the state of the art and not the historical development.

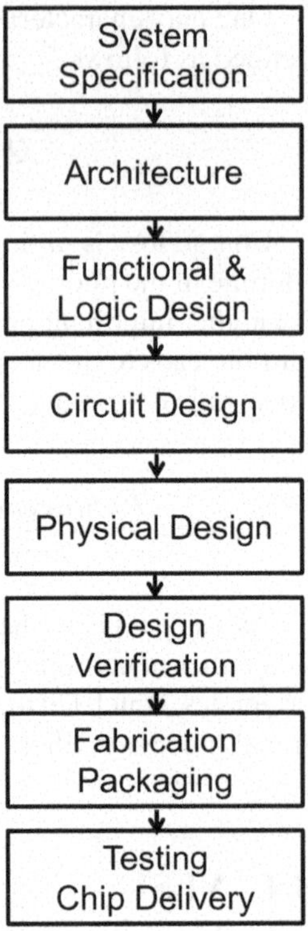

Fig. 8.12 VLSI design process

After a successful physical design, physical verification and sign off is carried out. At this point, the design is taped out for the chip fabrication, which is discussed in the next chapter. The chip design in the form of mask layouts are sent to the chip manufacturer electronically by using GDS (global distribution system) file format. Academic and smaller industry users may be interested in using MOSIS, EUROPRACTICE, and VDEC, in USA, Europe, and Japan, respectively for the chip fabrication. The bigger companies usually send out their designs to TSMC, UMC, Global Foundries, etc. Since chip manufacturing has become quite expensive over the past few decades, there are only a handful of companies left with their own dedicated fabs, like Intel and Samsung, etc.

Once the IC chips are fabricated and packaged, these are sent out for testing. After successful tests, the chip is ready for use. While most of the chip design and fabrication is silicon based. There are various semiconductor materials for niche applications, like GaAs (Gallium Arsenide), GaN (Gallium Nitride), SiC (Silicon Carbide), CdTe (Cadmium Tellurides), etc. However, one should note that even in silicon chip fabrication, there are more than 40 other materials being used.

One should note that chips are designed mostly for digital and analog applications. Each type of chip has its own design rules. Hence, one should look into digital as well as analog design rules for each application. There are various high level hardware description languages (HDL) that may be used in CAD (computer aided design), to aid in the design process by describing the analog and digital circuits. VHDL and Verilog HDL are the two most popular HDLs. In addition, there are various versions of VHDL and Verilog HDL for specific applications. Moreover, Verilog may not only be used for designing new chips, but may also be used for reconfiguring FPGA (field programmable gate array) chips for ASIC (application specific IC) applications. It is instructive to note that Verilog HDL syntax is very similar to that of the C language.

Being a high level language, the design in Verilog HDL is straight forward. Consider the following Verilog code for implementing a NAND gate out of an AND gate and a NOT gate,

```
module nandgate(in1, in2, out1)
// module and file name with inputs and output

input in1;
input in2;
// defining input ports

output out1;
// defining output port

assign out1 = ~(in1 & in2);
// & is the AND operation, and ~ is the NOT operation

endmodule
```

8.5 Power Dissipation

There are three kinds of power dissipation mechanisms in a circuit from the design and the operation perspectives. The first one is the static power that is present even in the absence of any switching event. Static power is usually due to a leakage current, e.g. a gate leakage or source to drain leakage current in a MOSFET, and therefore is also called leakage power. Given the leakage current (I_L) and the power supply voltage (V_{DD}), the static power (P_S) is given as,

$$P_S = V_{DD} I_L \tag{8.3}$$

The second one is the dynamic power, which is due to the switching events, which involve charging and discharging of capacitors in the circuit. One finds that the energy required to charge a capacitor having a capacitance C is $CV_{DD}^2/2$, whereas during the discharging event, $CV_{DD}^2/2$ is dissipated as well. The total energy dissipation per switching event involving charging and discharging a capacitor thus becomes CV_{DD}^2. Given the frequency of operation f, usually the clock frequency which is also the switching frequency, the dynamic power (P_D) is given as,

$$P_D = CV_{DD}^2 f \tag{8.4}$$

There is yet another kind of dissipated power that is attributed to the inherent device operation and is based on the entropy and the uncertainty of the switching devices. To describe a computing device[11] with N states, one needs $2^H = N$ bits for a binary system, where H is the Shannon entropy. Rearranging the above equation by assuming that the states are equally probable with the probability $p = 1/N$ for a specific state, one obtains,

$$H = \log_2 N = -\log_2(p) \tag{8.5}$$

On the other hand, the Boltzmann-Gibbs entropy for a specific state of the device under consideration is given as,

$$S = -k_B \ln(p) = -k_B \ln(2) \log_2(p) = k_B \ln(2) H \tag{8.6}$$

where k_B is the Boltzmann's constant. The heat difference is then given as,

$$\Delta Q = T\Delta S = k_B T \ln(2) \Delta H \tag{8.7}$$

where T is the temperature in K. This heat difference is the energy dissipated and is provided by the power supply. For ΔH of unity, i.e. a single binary switching event, one may show that the minimum energy dissipation per switching event (E_b) is given as,

$$E_b = k_B T \ln(2) \tag{8.8}$$

which is indeed the Shannon-von Neumann-Landauer (SNL) limit per switching event. At room temperature of 300 K, the SNL limit is about 17.88 meV. One has to multiply E_b with the switching frequency f to calculate the power associated with the device switching operations.

Problems

8.1 1-bit number. (a) What are the four logic operations one may implement? (b) Which of the two operations are reversible? (c) Which of the two operations are irreversible?

8.2 How many logic operations may be implemented on a 2-bit number?

8.3 How many logic operations may be implemented on a 4-bit number?

[11] Please make a note of the reuse of the symbols H, N, P, and p, which have been used earlier for Hamiltonian, total number of electrons, total number of holes, and hole concentration.

8.4 How many logic operations may be implemented on an 8-bit number?
8.5 Draw the CMOS circuit for a three input NAND gate.
8.6 Draw the CMOS circuit for a three input NOR gate.
8.7 Draw the CMOS circuit for a four input NAND gate.
8.8 Draw the CMOS circuit for a four input NOR gate.
8.9 The WL transistor used in 6T-SRAM cell is not a pass transistor. Comment on the use of pass transistors in SRAM cell.
8.10 How many select signals would a 4-to-1 multiplexer have?
8.11 How many select signals would a 8-to-1 multiplexer have?
8.12 How many select signals would a 16-to-1 multiplexer have?
8.13 How many bits can a latch store?
8.14 How many bits can a flip flop store?
8.15 What is the difference between a latch and a flip flop?
8.16 Asynchronous and Synchronous Circuits. Complete the timing diagram for D latch without clock signal, D latch with clock signal, and positive edge triggered D flip flop (Fig. 8.13).
8.17 Calculate the SNL limit per switching event at room temperature.
8.18 For the clock frequency of 1 GHz, calculate the minimum dissipated power according to SNL limit.

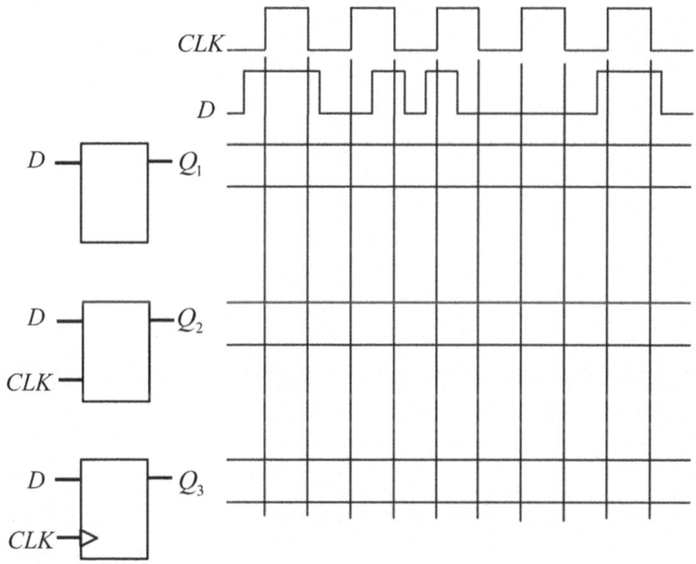

Fig. 8.13 Problem 8.16

Research Assignments

R8.1 Write a one page summary on the state of the art CCD for high end astronomical imaging.

R8.2 Follow the VLSI design process (Fig. 8.12) for a 3-phase 8 × 8 CCD chip and tape out the design to MOSIS.

Part IV
Fabrication, Characterization and Metrology

Chapter 9
Nanofabrication

In this chapter, we discuss device fabrication at the nanoscale. There are two approaches to nanofabrication, namely the conventional top down approach and the more novel bottom up approach. It is the bottom up approach that is most fascinating due to its emphasis on the use of atoms and molecules as the building blocks during the synthesis and fabrication, whereas in the top down approach, one follows the conventional approach of device scaling. We discuss both approaches with their pros and cons in this chapter.

To give a perspective, a fabricated silicon wafer is shown in Fig. 9.1, with the scientist/engineer wearing the clean room attire. The varying shades within a die[1] on this particular wafer depict various materials and varying length scales fabricated on each chip, where the nanoscale features may be connected with microscale interconnects. We also discuss microfabrication techniques in this chapter since these methods complement the fabrication at the nanoscale.

Irrespective of the nanofabrication approach and technique, a contamination free atmosphere is essential for a successful process run. Imagine the devastation caused by a microscale particle sitting on an integrated circuit containing nanoscale devices. Therefore, it is no surprise that nanofabrication is performed in clean room environments. In industrial production facilities, robotic automation is embedded to the level that manual handling of wafers is eliminated altogether. In a research environment, users handle wafers manually often complemented by some automation. Various classes of clean rooms represent decreasing amount of contaminants per cubic feet as shown in Fig. 9.2. For a certain class p, the amount of contaminants with a diameter d in μm should not exceed, $p\,(0.5/d)^{2.2}$. It is not unusual to find a class 100 or class 1000 clean room in a university campus these days, which respectively have less than 100 or 1000 particles (of 0.5 μm diameter) per cubic feet. This is achieved

[1] A silicon wafer is cut into various pieces, each containing a complete copy of a circuit or a system, e.g. processor chip or memory chip, etc. Each one of these pieces is called a die.

H. Raza, *Nanoelectronics Fundamentals*, NanoScience and Technology,
https://doi.org/10.1007/978-3-030-32573-2_9

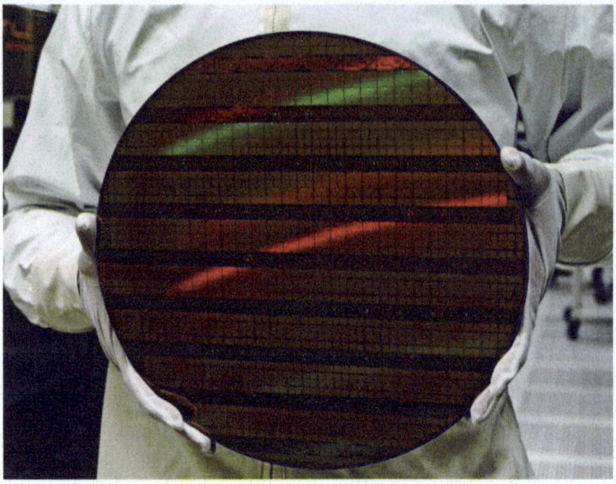

Fig. 9.1 Fabricated silicon wafer in a clean room environment. (Courtesy of Intel)

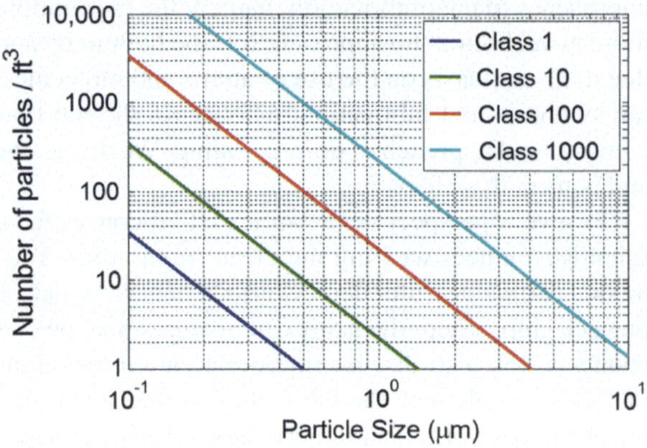

Fig. 9.2 Clean room classification

by using high performance HEPA (high efficiency particulate air) filters. In fact these filters are so effective that in a pollen season, spending a day in a clean room is quite relaxing for someone with the allergies.

Each process run starts with an appropriate substrate. Silicon is the material of choice for nearly 80% of IC fabrication. GaAs (gallium arsenide) is used in opto-electronic applications. Mercury cadmium telluride is quite popular with night vision applications. The wide bandgap semiconductors like GaN (gallium nitride) and SiC (silicon carbide) are vital for power electronics devices. Some substrates may be quite unorthodox and may not come to mind as the starting point for device fabrication. Consider the use of glass as a starting substrate for LCDs (liquid crystal displays), where size could easily be in several feet. Flexible substrates are yet another example and are an emerging research area.

9.1 Chemical Safety and Environmental Protection

Chemical safety and environmental protection are important components of the device fabrication. It is vital to understand the chemical toxicity to ensure one's well being as well as the environmental sustainability. Some of the chemicals used in silicon fabrication industry are well known mutagens and carcinogens.

Depending on the nature of the chemical, appropriate personal protection equipment (PPE) should be used. The occupational safety and health is regulated by various agencies in different countries, which include OSHA (Occupational Safety and Health Administration) in USA, EU-OSHA in European Union, CCOHS (Canadian Centre for Occupational Health and Safety), JNIOSH (National Institute of Occupational Safety and Health, Japan), and COSHA (China Occupational Health and Safety Association). On the other hand, environmental health and sustainability is ensured by Environmental Protection Agencies (EPA) of the respective countries.

In some cases, there is international cooperation to standardize the system of occupational safety and health, as well as environmental protection. For example, the material safety data sheet (MSDS) system is being replaced with the UN (United Nations) mandated Globally Harmonized System (GHS) of the classification and labeling of chemicals.

Kyoto protocol is an example of an international treaty that binds state parties to control green house gas emission, which very much applies to the manmade fluorinated greenhouse gases widely used in the silicon processing. Center for Disease Control (CDC) is yet another important player in the context of the use of nanomaterials and biological species during chip fabrication, and their potential hazards.

Manufacturers are required to provide information about the potentially hazardous properties of the chemicals under use. However, it is the user's responsibility to understand the consequences of the exposure and exercise proper care. Another useful standard for chemical labeling is the NFPA (National Fire Protection Association) 704 diamond as shown in Fig. 9.3a. The scale ranges from 0 (no hazard) to 4 (acute hazard). The blue part of the diamond is related to personal health, the red part to the flammability of the chemical, yellow part to the instability or explosiveness of the chemical, and finally the white part provides any special instructions while handling the chemical (e.g. reactivity with water, oxidizer, corrosive, biological hazard, poisonous, radiological, cryogenic, etc). One should note that the mixture of various chemicals may have a very different diamond diagram than those of the individual chemicals. Therefore, care should be exercised when mixing chemicals and/or storing various chemicals in the same area, e.g. a chemical cabinet. Furthermore, the diamond diagrams may be replaced by a more recent GHS pictograms with different hazard categories as shown in Fig. 9.3b.

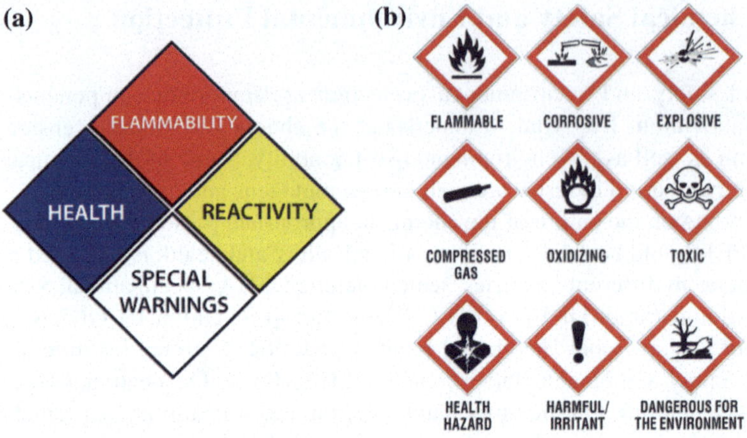

Fig. 9.3 Hazard communication standard. **a** NFPA 704 diamond. **b** GHS pictogram

9.2 Substrate

Single crystal silicon is grown by using two methods, Czochralski method and Float Zone method. Czochralski method is favored for producing large wafers, where a single crystal seed with proper orientation of silicon is lowered into a molten polycrystalline silicon melt in a controlled atmosphere. The seed crystal is gently pulled (sometimes over 24 hours or more) and rotated to make single crystal ingot. The ingot is sliced to make wafers. Either one side or both sides are polished, called single side polished (SSP) or double side polished (DSP) wafers, respectively. One should note that for various materials, the growth of single crystal substrate may require different methods, all of which are challenging to cover here and hence left for the interested reader to explore as an exercise.

For IC chip fabrication, increasing wafer size decreases the cost. Therefore, the wafer size has been continuously increasing. Currently,[2] state of the art commercial fabs (fabrication factories) process 450 mm wafers. The orientation of semiconducting single crystalline wafers is also an important parameter for defect density control as well as interaction and deposition of novel materials. Most popular surfaces have either (100) or (111) orientation and depend on the seed crystal during growth. Finally, the doping is an important consideration that determines the material resistivity as shown in Fig. 9.4 for n-type and p-type silicon doping at room temperature.

Once a substrate is chosen, the starting and perhaps the most important step is the initial cleaning with various chemicals. One should always wear appropriate PPE for working with solvents and/or acids. Since the processing is done in a clean room environment, the PPE should also be clean room compatible. The PPE should include bouffant hairnet and hoods, goggles, face mask (particularly working with acids),

[2]As of 2019.

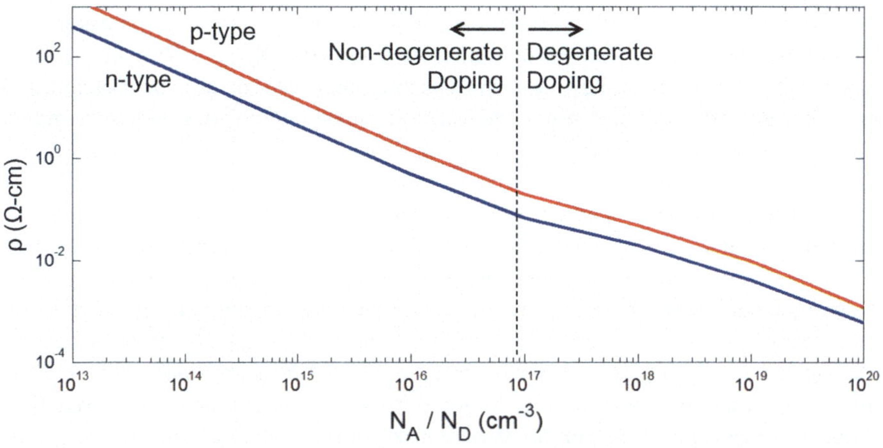

Fig. 9.4 The resistivity dependence on the substrate doping for silicon at room temperature. N_A/N_D are acceptor/donor atom concentrations for p-type and n-type dopings

chemical resistant and/or flame resistant lab gown (with additional acid gown cover), gloves (sometime multiple gloves or long sleeved acid gloves when applicable), shoe covers, and chemical and/or flame resistant boots. Additionally the supporting materials being carried in the clean room should also be clean room compatible, e.g. lab notebooks, writing pens, etc. A good sense of clean room ethics and protocols is vital for successful fabrication processes. In fact, the human user is perhaps the most *unclean* intruder in a clean room. Efforts are made to reduce the sources of contamination as much as possible. Apart from the clean room compatible PPE, incorporating air showers is yet another feature. The clean air flow may be laminar or turbulent in a clean room. However, the flow patterns are always designed in a way to take the contaminants away from the user into the exhaust, and are optimized to reduce the transfer of contaminants from the user to the sample area.

Various industries have standard cleaning procedures, consider RCA (Radio Corporation of America) clean for example. However, for the university research, the following procedure suffices for silicon wafers.[3] The standard time for each step is 10 min, unless otherwise specified. One starts with solvent cleaning, that involves tolune (optional) immersion, acetone dip, methanol soak and finally DI (deionized) water rinse followed by nitrogen drying (manually or in an automatic dryer). One should note that no liquid including water should be allowed to dry on the wafer surface. The use of DI water is essential, since undesired ions are detrimental to proper device function. Consider the use of stored charge in the gate stack of a flash memory that shifts the threshold voltage. With ions inside the gate stack, the same electrostatic potential arbitrarily changes the transistor threshold voltage, making the whole chip inoperable. In fact, this is precisely one of the major reasons behind the delay in implementation of the field effect transistor after its prediction. The researchers

[3]Reader is encouraged to check the procedure for a specialty substrate.

were not able to get rid of and eliminate the ions during sample preparation. One should note that even our hand touch introduces ions like K^+ and Na^+, etc.

After the solvent cleaning to remove any organic contaminants, acid cleaning is performed to further decontaminate the wafer from any organic or inorganic contaminants. The starting point is Piranha clean for 20 min, followed by DI water rinse, and nitrogen drying. Piranha is a name give to sulfuric acid and hydrogen peroxide mixture. Hydrogen peroxide is indeed thermodynamically unstable at room temperature and gives off atomic oxygen that reacts with the elemental carbon and hence gets rid of the carbon residues. The bubbling in the hot solution helps agitate the solution to clean the sample in an acidic environment. Due to hydrogen peroxide instability, the Piranha solution has a very short shelf life. Usually it is advised to prepare fresh solution and use right away.[4] A reminder about the AAA (always add acid) rule, i.e. sulfuric acid should be added to hydrogen peroxide. Being an exothermic reaction, adding hydrogen peroxide (or for that matter simple water) to sulfuric acid may cause significant splash. Finally, dipping silicon wafers in a dilute hydrofloric (HF) acid[5] solution for a few seconds removes the native oxide. This step prepares a fresh silicon surface that is truly contamination free. The HF acid drip is followed by the standard DI water rise and nitrogen drying.

9.3 Oxidation and Annealing

Oxidation of silicon wafers is the process of growing an amorphous silicon dioxide film by a thermal process inside a tube furnace as shown in Fig. 9.5. Since highest quality oxide films are desired for device fabrication, the tube is made of highest quality quartz (a metal free form of silicon dioxide). Additionally, ultra high purity (UHP) gases are used with up to 99.9999% purity, which are considerably expensive, however inevitable for growing the highest quality films. The wafers are loaded at low temperature (mostly at room temperature) in an inert gas ambiance like nitrogen to eliminate low temperature oxide growth by using ambient oxygen.

Historically, the oxidation tube furnaces have been placed horizontally as shown in Fig. 9.5. However, it was realized later that placing the tube furnaces in the vertical position reduces contamination. Although the gases of highest available purity are used, it is still possible to have some contaminants. By using the horizontal furnaces, all the wafers are exposed to such contaminants due to the laminar flow of gases inside the tube. By placing the furnace vertically, the first wafer towards the gas inlet is exposed to the contaminants, whereas the rest of the wafers are protected. The first and the last wafers are thus treated as *dummy* wafers and are not used in subsequent process run due to their exposure to contaminations and variations in the film thicknesses owing to the abruptness in gas flow.

[4]Pirahna substitutes with longer shelf life are commercially available.

[5]To check HF acid's toxicity, please consult the MSDS or GHS.

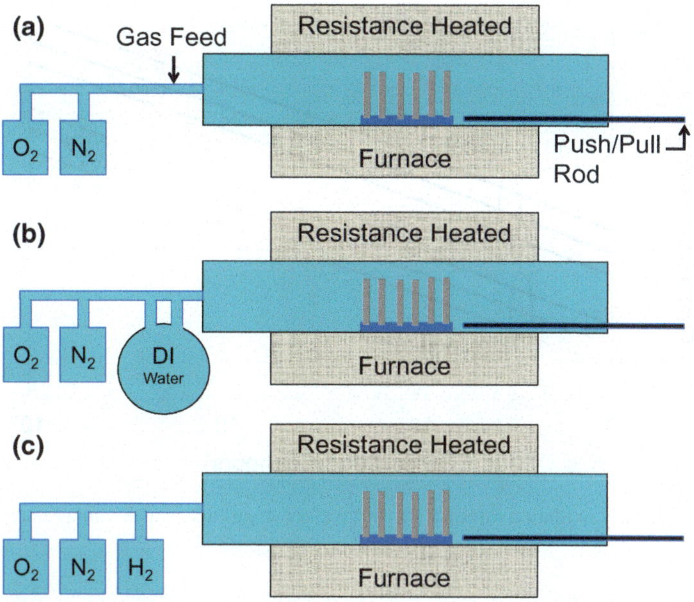

Fig. 9.5 Various types of oxidation processes. **a** Dry oxidation. **b** Wet oxidation. **c** Pyrogenic oxidation

Oxidation is divided into three types, depending on the type of oxidizing agent. In dry oxidation shown in Fig. 9.5a, one uses molecular oxygen to convert Si into SiO_2 at 800–1200 °C as follows,

$$Si + O_2 \rightarrow SiO_2$$

In wet oxidation shown in Fig. 9.5b, the feed consists of water vapors, whereas hydrogen is the byproduct.

$$Si + 2H_2O \rightarrow SiO_2 + 2H_2$$

In pyrogenic oxidation shown in Fig. 9.5c, the feed consists of oxygen and hydrogen gases that react to form water vapors, which is finally used for oxidation in much the same chemical reaction as wet oxidation. However, the advantage of pyrogenic oxidation is the purity of the feed gases. Since it is quite difficult to purify liquids to the same level as that of gases, it is always preferable to use gases for ultra high purity applications than liquids.

While dry oxidation gives the highest quality films, it has the least growth rate at a given temperature. However, for thin films, this is not a bottleneck and the quality of film takes precedence over the oxidation time. Therefore, dry oxidation is mostly used for ultrathin (less than 100 nm or so) films of highest quality. The oxide thickness as a function of oxidation time is reported in Fig. 9.6 for Si(100) surface. For wet

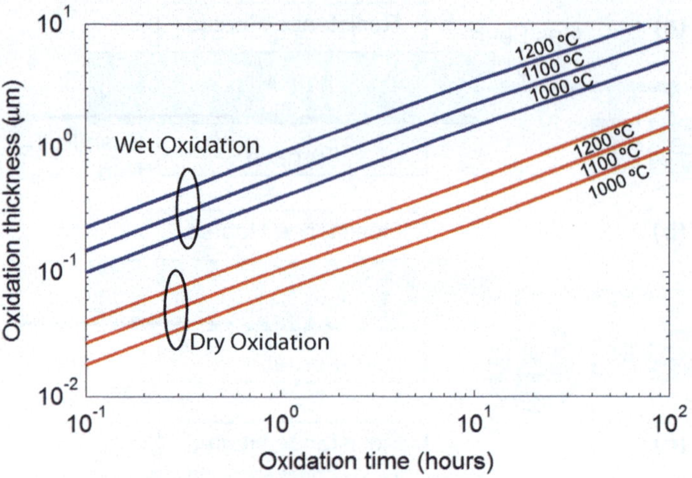

Fig. 9.6 Wet and dry oxidation rates for Si(100) wafers at various growth temperatures

oxidation, the oxidation rates are much higher as compared to dry oxidation. This becomes the primary advantage of the wet oxidation and the pyrogenic oxidation, which are routinely used to grow oxides up to few microns in thickness. Furthermore, the oxidation rates are also higher for (111) orientation as compared to the (100) surface due to higher atomic concentration per unit surface area.

Oxidation is perhaps the only process in micro and nanofabrication, where the substrate is consumed in the growth process. After oxidation, 45% of the oxide film is subsurface for the otherwise silicon wafer. In addition, the initial growth is chemical reaction limited. But once an oxide film is grown, further oxidation is limited by the diffusion of oxygen through the grown oxide film. While the initial growth is relatively fast, the additional growth of film takes much longer time. Gases like chlorine may also be added to oxygen to continually eliminate mobile charges like Na^+ and K^+ contaminating the Si wafers and also the walls of the furnace tube.

Due to the amorphous nature of silicon oxide and also the bond length mismatch between silicon and silicon oxide, defects get accumulated on the Si/SiO_2 interface, which are also called dangling bonds (i.e. unbonded Si atoms), despite the fact that this interface is one of the best known interfaces to humans and very reliable. In fact, the defect density is lower for (100) surface as compared to the (111) surface due to smaller atomic coverage per unit area, nonetheless is high enough to degrade the device quality. This defect density or surface state density is well characterized and increases with decreasing oxidation temperatures. One solution is to anneal the Si wafer in a forming gas mixture of nitrogen and hydrogen (of up to 10% hydrogen). The hydrogen helps to passivate the dangling bonds, which leads to a reduction in the surface state density. Nitrogen not only provides an inert atmosphere but also helps to stop the spontaneous burning and explosion of hydrogen by reducing its concentration. The net result is a steep drop of the surface state density as shown

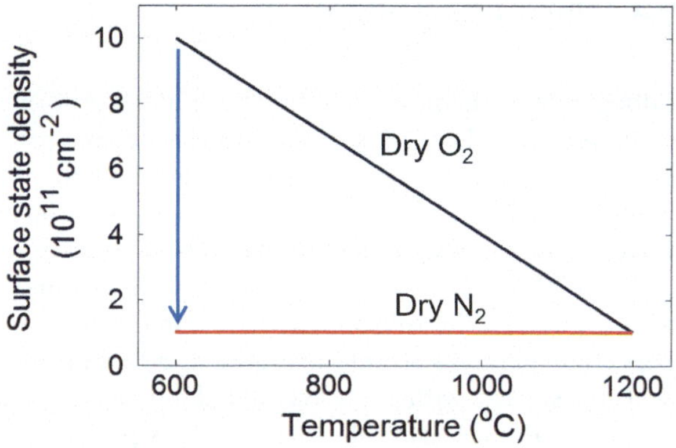

Fig. 9.7 Deal's triangle. Annealing reduces defect density

in Fig. 9.7. Due to the shape of this curve, the viewgraph is called *Deal's triangle*. This annealing step is also required after the final metal deposition for contacts. The post metal forming gas annealing step is performed to get a better interface between the metallic contacts and the channel surface. In addition, another annealing process is Rapid Thermal Annealing (RTA), also called Rapid Thermal Processing (RTP), where the sample is heated to elevated temperatures of up to 600 °C within a few seconds and kept at this high temperature for relatively short time period. The advantage here is that the dopant atoms do not get redistributed, whereas in a typical annealing process, dopant profile redistribution is inevitable. Such rapid heating is achieved by radiative means with the use of arc or halogen lamps. RTP is also used for oxidation and deposition.

Next step is to characterize the oxide film. It may be characterized crudely by the color chart. Varying film thicknesses exhibit different shades due to the interference between the reflections from the top surface and the Si/SiO_2 interface. Another method for precisely characterizing oxide films is by using Ellipsometry, which is routinely used for this purpose. One may also fabricate metal oxide semiconductor capacitors as discussed in Sect. 5.3 to electrically characterize the oxide film. Such characterization may include capacitance-voltage, leakage, and/or reliability studies. A useful technique for reliability characterization is the time dependent dielectric breakdown, where the oxide film is stressed with a constant voltage near the dielectric breakdown at room or elevated temperatures, and the leakage current is monitored as a function of time. At a certain time, the oxide film breaks down resulting in a high leakage current. The integral over the leakage current and the time-to-breakdown gives the charge-to-breakdown, a useful number in analyzing the reliability of the grown film.

9.4 Photolithography

Lithography is the process of transferring patterns on a substrate. Photolithography is the process where light is used to transfer the pattern by using a light sensitive photoresist film. Photolithography is also sometimes incorrectly termed as optical lithography. While optical wavelength is in 400–700 nm range, in fact much smaller wavelength of UV light is used in photolithography.

Photolithography process starts with spin coating a photon sensitive resist (photoresist) film on the substrate as shown in Fig. 9.8a, where the substrate is Si/SiO_2. For certain photoresist and substrates, a surface priming step may be required, which coats the surface with a self assembled monolayer of an appropriate chemical. Hexamethyldisilazane (HMDS) is a common primer used for oxide surfaces. A pre-exposure bake (also called soft bake) is performed to get rid of the solvent in the photoresist film. This bake may be performed in an oven or on a hot plate. The latter is preferred due to the solvent evaporation from the film.

A UV source is used to selectively expose the photoresist to UV light by using a photo mask as shown in Fig. 9.8b. In the case of the positive photoresist, the exposed parts of the photoresist film are converted into more acidic form, which may be dissolved in a developer solution of alkaline nature for a certain time, followed by DI water rise and nitrogen dry. The exposed photoresist film is thus removed, thereby creating an exact replica of the photo mask as shown in Fig. 9.8c. In case of the negative photoresist, the unexposed part of the photoresist film is dissolved in the developer, creating a negative of the mask, hence called negative photoresist. While the positive photoresist may give smaller feature size and better step coverage, the negative photoresist is usually less expensive and has excellent chemical resistance. The negative photoresists however have usually more health hazards and should be dealt with extra care.

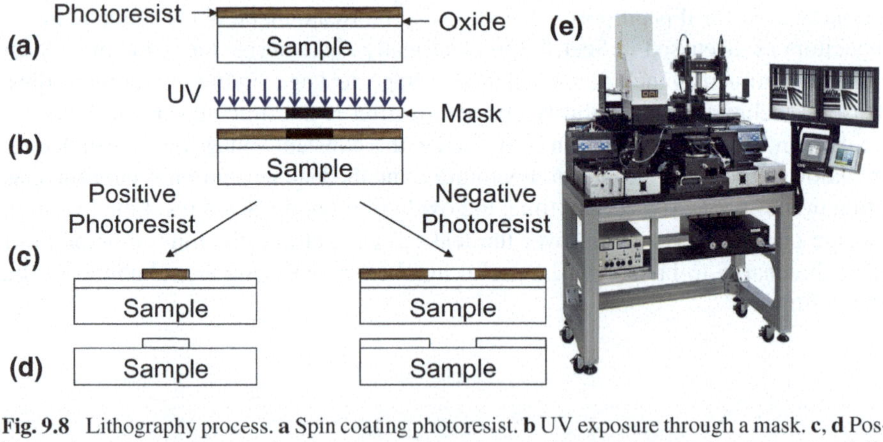

Fig. 9.8 Lithography process. **a** Spin coating photoresist. **b** UV exposure through a mask. **c, d** Positive and negative photoresist development, oxide etch, and photoresist removal. **e** A commercially available mask aligner. (Courtesy of OAI)

After the UV exposure, the photoresist is baked again in a process called post exposure bake or hard bake (due to a higher temperature than the pre exposure soft bake) to remove the developer solution from the film and to get a better edge profile and sidewall slope. One should note that with certain positive photoresists, it is possible to do image reversal by using ammonia, which is advantageous in the lift off process during metal deposition. The sidewall slope of a positive or negative photoresist is such that the evaporated metal film uniformly covers the photoresist. With the image reversal, the sidewall gives a shadowing effect with an undercut without metal coverage. Here the lift off solution may penetrate with ease to remove the photoresist and hence the deposited metal. The negative photoresist also has the advantage of the above mentioned undercut in the sidewall profile. In the final step, oxide may be selectively etched and photoresist stripped off from the substrate as shown in Fig. 9.8d. The resulting oxide feature on the substrate is an exact replica for a positive photoresist, and a negative of the mask for a negative photoresist. A commercially available mask aligner is also shown in Fig. 9.8e.

For any photolithography process, there are five important parameters to consider, resolution, alignment, yield, throughput and cost. Resolution is defined as the minimum feature size that may be transferred. This is a parameter that needs to be optimized since it depends on various factors like equipment optics quality, light wavelength, photoresist process, substrate, etc. Alignment is defined as the ability to overlay features from one layer of photolithograhpy process to another over a given surface area. One may achieve excellent alignment and overly over a small area, but fail to do so over a larger area. Yield is the ratio of dies without defects to the total number of dies manufactured on a wafer. Throughput is the quantity of wafers processed per unit time, which adds to the running cost of a fabrication run. The running cost is also affected by the supplies, chemicals, and maintenance. The fixed acquisition cost of a system is another important parameter.

The tools to transfer the features to the photoresist films are of two types, mask aligners and steppers. All these equipments have a light source, the output of which is collimated. The light source could be a simple UV. Further extensions include DUV (deep UV) and EUV (extreme UV) light sources. When mercury arc lamps are used in older equipments (still being used in universities world wide), the three characteristic emission peaks, with which the equipment is classified, are at the wavelengths 0.435, 0.405 and 0.365 μm, and are called G-line, H-line and I-line, respectively. Furthermore, 0.248, 0.193 and 0.157 μm wavelength peaks may be achieved with KrF, ArF, and F_2 excimer LASER sources, respectively. Decreasing the wavelength increases the resolution.

In the case of a mask aligner, the whole wafer is exposed at the same time. The photo mask may be either in contact with the wafer or in close proximity, also called contact lithography and proximity lithography, respectively. The advantages here are better resolution, high throughput and low cost due to its simplicity. Since exposure of the whole wafer is done at the same time, there is no opportunity to correct for local misalignments. In addition, since the mask is in contact with the wafer, it may get scratched or material from the wafer (e.g. photoresist chunks, etc) may get transferred to the mask. To overcome this bottleneck, a gap between mask and wafer is allowed, which should not be more than a few tens of microns, since the minimum feature size is given as $\sqrt{k\lambda\Delta}$, where k is the resist parameter, λ is the wavelength and Δ is the gap between the mask and the wafer. Operating the mask aligner in noncontact mode increases the mask life and reduces the defect density hence increases the yield, but at the cost of the resolution.

For steppers, projection lithography is used as shown in Fig. 9.9a. There is considerable gap between the reticle (note that it is not called mask here) and the wafer. The reticle is placed in between the condenser and the objective lens. Clearly the mask is not touching the reticle and hence the defect density is expected to be small leading to excellent yield. However, the projection lithography is achieved by the use of more complex optics in steppers, which not only increases the fixed cost at the acquisition time, but also increases the running cost due to expensive maintenance. Furthermore, during the projection step, the image on the wafer may be reduced by virtue of lens focusing. This may easily lead to five, ten or even fifteen times reduction in the feature size trend on the wafer surface. Such steppers are indeed classified as 5X, 10X, and 15X, respectively. One die is exposed at a time in steppers, thereby reducing the throughput. On the other hand, exposing a smaller area allows for better overlay, which is not possible with mask aligners. However, steppers are far more complex than mask aligners and hence are more expensive. A commercially available stepper is shown in Fig. 9.9b.

For the projection systems, the numerical aperture is defined as, $NA = n \sin(\alpha)$, where α is the angle between the center point of the wafer and the edge of the projector lens, n is the refractive index. Using Rayleigh's criteria, the minimum feature size, i.e. the resolution is given as,

$$W = k_1 \frac{\lambda}{NA}$$

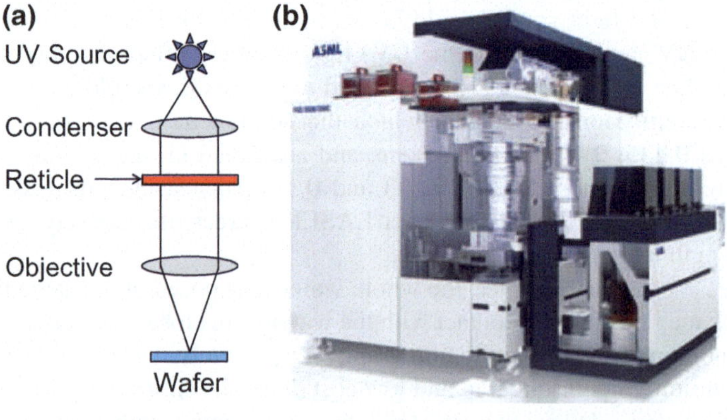

Fig. 9.9 Stepper. **a** Projection lithography system exposes one die at a time with feature reduction. **b** A commercially available stepper. (Courtesy of ASML)

where k_1 is an equipment parameter. The minimum feature size trend over the past few decades is shown in Fig. 9.10. The corresponding wavelength for the respective technology nodes is also shown emphasizing the subwavelength lithography since 1997. One may reduce the minimum feature size (i.e. increase the resolution) by decreasing k_1, decreasing λ, or by increasing the numerical aperture, which may be achieved by increasing the refractive index n, and the angle α. The historical trends for the numerical aperture and the k_1 parameters are shown in Figs. 9.11 and 9.12, respectively.

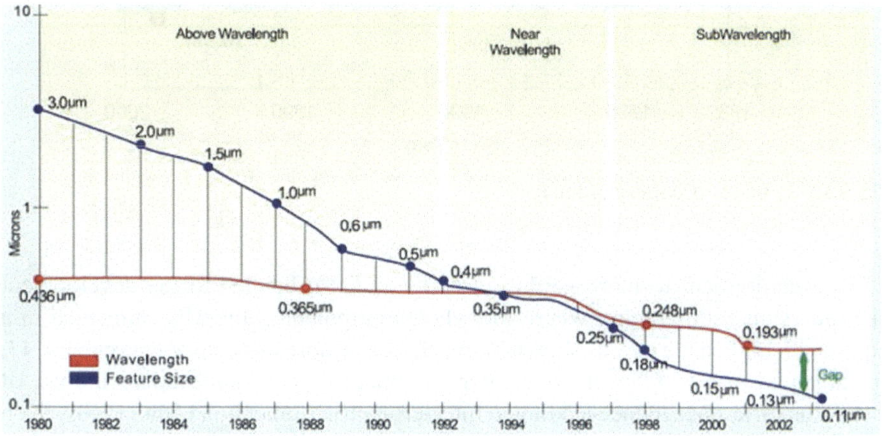

Fig. 9.10 Lithography scaling and the associated wavelengths. (Courtesy of UMC)

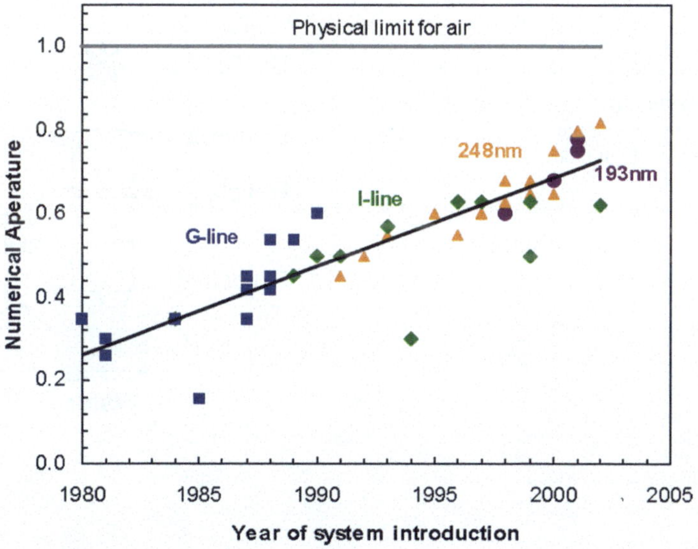

Fig. 9.11 Numerical aperture trend. (Courtesy of Intel)

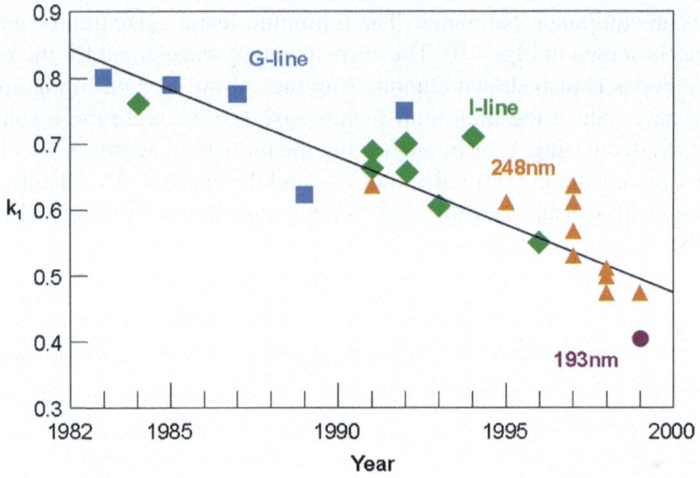

Fig. 9.12 k_1 trend. (Courtesy of Intel)

The requirement to increase the refractive index n has led to the development of Immersion Lithography, where the whole equipment is literally immersed in a liquid with high value of n as shown in Fig. 9.13. e.g. aqueous solution has $n = 1.43$, as compared to $n = 1$ for air. A greater alignment may be achieved by the use of interferometric stage, hence leading to Interferometric Immersion Lithography. Such interferometric stages are very useful in electron beam lithography as well, which is discussed in the next section.

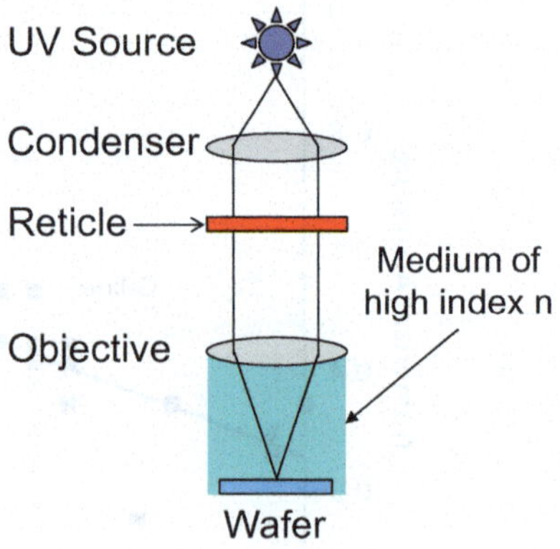

Fig. 9.13 Immersion lithography. A higher numerical aperture is obtained by immersing the mask aligner in a high index liquid

9.5 Electron Beam Lithography

As shown in Fig. 9.14, the electron optics consists of electric and magnetic lenses. By applying appropriate electric and magnetic fields, the electron beam shape and width is controlled. However, since the electron beam is quite sensitive to the stray charges, vacuum inside the optics column is an essential requirement. This ensures that the gas molecules do not get charged and affect the beam in an adverse manner.

Photons used in photolithography have two limitations, first is the diffraction limited resolution based on the wavelength, whereas the second is the diffraction limited depth of focus, which is defined as $k_2\lambda/(NA)^2$ in the far field, where k_2 is an equipment parameter. With the increasing numerical aperture, the depth of focus degrades severely. One solution is to use electrons instead of photons, where the energy and hence the wavelength is controlled by the acceleration voltage. An electron accelerated with a 10 KV voltage indeed possesses 10 KeV energy, and would behave as a wave with 0.04 nm wavelength. For this sub-nm wavelength, the diffraction limit does not pose any bottlenecks at the nanoscale. Historically, EBL has widely been used in mask fabrication.

The electron beam lithography thus has distinct advantage of nanometer scale resolution with excellent depth of focus. The writing process is Direct Write Lithography, where the electron beam scans the sample in a raster fashion. This leads to

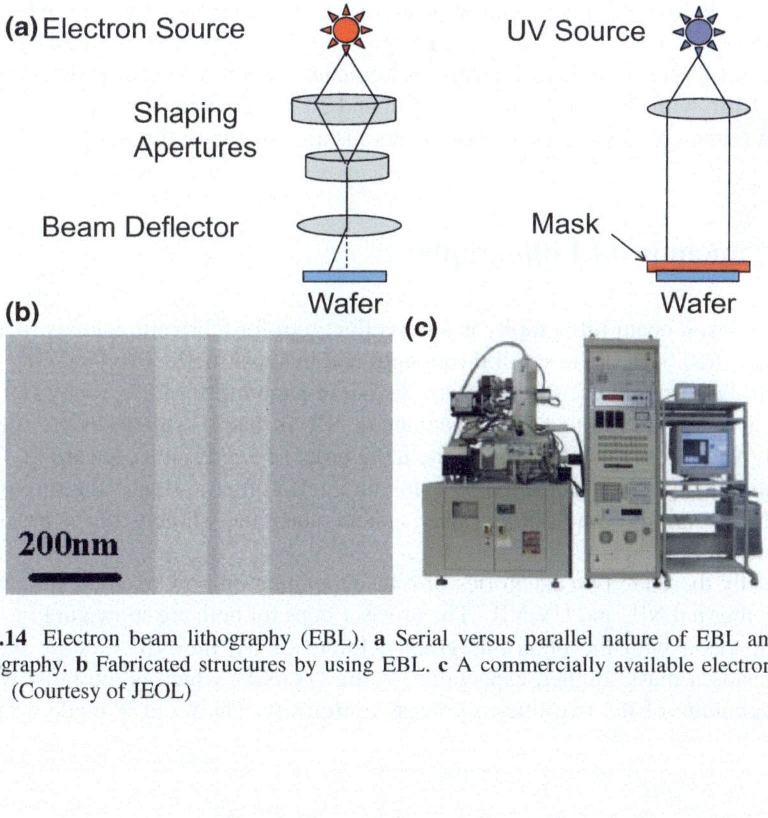

Fig. 9.14 Electron beam lithography (EBL). **a** Serial versus parallel nature of EBL and photolithography. **b** Fabricated structures by using EBL. **c** A commercially available electron beam writer. (Courtesy of JEOL)

a serial mode of lithography compared to the parallel mode in photolithography resulting in very small throughput. Multi beam writers may overcome the small throughput bottleneck. By using interferometric stage, one may obtain excellent alignment and overlay stitching capabilities resulting in high yield.

Electron beam resists are perhaps the most important aspect of the lithography process. Usually a bilayer process recipe is used and extensive characterization is performed to optimize the writing conditions. EBL may also be complemented with photolithography, where the critical dimension features are written by using EBL, and the microscale features are written by using the photolithography.

The optics for a typical EBL system is shown in Fig. 9.14a in comparison with the projection lithography. The column in EBL is similar to that of a scanning electron microscope (discussed in greater detail in the next chapter). However, the beam deflection coils have added circuitry to deflect the beam in a more controlled manner in order to obtain nanometer resolution.

A typical EBL process run starts with a schematics drawing for the pattern to be transferred to the wafer. Then wafer is cleaned and primed. The photoresist is spin coated followed by a prebake step. The critical step of electron beam direct write is performed, followed by the resist development. One should note that the electrons deposit energy on a wider region than the beam size. There is always a proximity affect that one has to account for in the design process and should be addressed by engineering the exposure pattern. Secondly, the chemical change in the electron beam resist is reversible in contrast to that in photolithography. Therefore, it is desirable to develop an electron beam resist right after the exposure. Afterward a post development bake is performed to get rid of any solvents, followed by either etching or lift off to complete the steps. A fabricated structure by using EBL is shown in Fig. 9.14b with about 130 nm line width and 20 nm gap between the two metallic lines. A commercially available EBL system is also shown in Fig. 9.14c.

9.6 Nanoimprint Lithography

While electron beam lithography is an excellent tool for fabricating nanostructures, the major drawback is the small throughput and the cost ineffectiveness. One alternative to EBL to address these drawbacks is the nanoimprint lithography (NIL) as shown in Fig. 9.15. Another key advantage of NIL is that 3D structures may be fabricated. Apart from this, the resolution on the order of 5 nm is not unheard of. Fairly good alignment may be achieved by using an interferometric stage. The throughput could be as high as a photolithography system, since the whole wafer is processed at a time.

Broadly there are two categories of NIL depending on how one cures the resist, namely thermal NIL and UV NIL. The process steps for both are shown in Fig. 9.15a in comparison with the photolithography. Interestingly, the NIL module may be fitted inside a mask aligner, especially for the UV-NIL, which is the most widely used technique of the two due to process uniformity. The mold is made by using

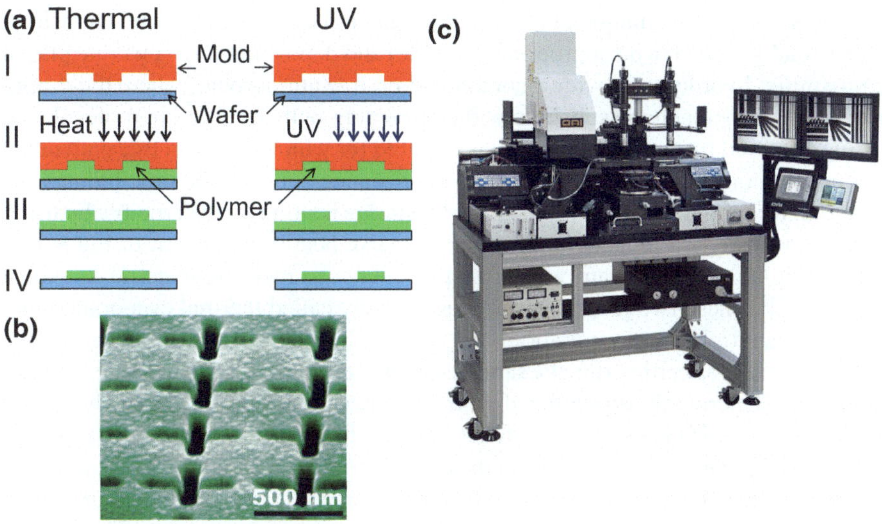

Fig. 9.15 Nanoimprint lithography. **a** Comparison of thermal and UV nanoimprint lithography. **b** Fabricated structures. **c** Commercially available system. (Courtesy of OAI)

EBL and consists of a hard material, like SiO_2. The process starts with dispensing or spin coating the liquid resist on the substrate. In case of thermal NIL, the resist is essentially a thermoplastic polymer, which may be cured by heat, whereas the UV resist is a light sensitive polymer which is cured by applying UV light.

Once alignment is achieved, the mold is pressed against the resist. This process transfers the negative of the pattern on the mold as shown in Fig. 9.15a. The wafer and mold should be horizontally aligned in addition to the lateral alignment usually used for uniform pattern transfer. For the thermal NIL, the resist is heated beyond the activation temperature of the polymer, which solidifies on cooling. In case of UV-NIL, the UV light hardens the resist. After removing the mold, the transferred features on the substrate are negative of the features on the mold. If there is still some residual layer, it is removed by descum or reactive ion etch as discussed later in this chapter. The fabricated patterns are also shown in Fig. 9.15b with about 50 nm resolution. A commercially available NIL system is shown in Fig. 9.15c.

9.7 Physical Vapor Deposition

There are various methods of material deposition on substrates, which may broadly be categorized based on the physical or the chemical means, and hence are called physical vapor deposition (PVD) and chemical vapor deposition (CVD), respectively. In this section, we discuss the PVD techniques of evaporation and sputtering.

Evaporation is a technique of heating up a material close to or beyond its melting point (usually placed in an appropriate crucible) and depositing on a substrate placed in proximity. In order to provide a contamination free atmosphere, where the evaporated atoms may reach the substrate without reacting with the ambient molecules or getting scattered from them, a vacuum is required for which the mean free path of the evaporated atoms should be greater than the distance between the crucible and the substrate. Heating may be achieved either by simply heating up the crucible by using thermal means e.g. a tungsten filament or resistive heating as shown in Fig. 9.16a, or by using an electron beam with sufficient energy to melt the material as shown in Fig. 9.16b. These two techniques are respectively called thermal evaporation and electron beam (e-beam) evaporation.

In a typical evaporation run, the samples are placed in the sample holder with shutter closed. The crucible is populated with the material to be evaporated. The quartz crystal oscillator is checked for its functioning, which monitors the film thickness. The material parameters are loaded in the crystal monitor apparatus. The evaporation chamber is then closed and vacuum is established over the next few hours, first with a roughing pump and then with either a diffusion or a cryogenic pump. The latter is preferred due its cleanliness, a diffusion pump may deposit oil residues in the evaporation chamber. Once the material evaporation rate is established, shutter is opened and subsequently closed once the desired film thickness has been deposited. It should be noted that some metals react very well with oxygen, e.g. chromium, titanium, etc. When evaporating such metals, one notices a sudden drop in the chamber pressure at the start of the evaporation step due to the reaction between the evaporated metal atoms and the oxygen in gas phase, thereby forming respective oxides in gas phase.

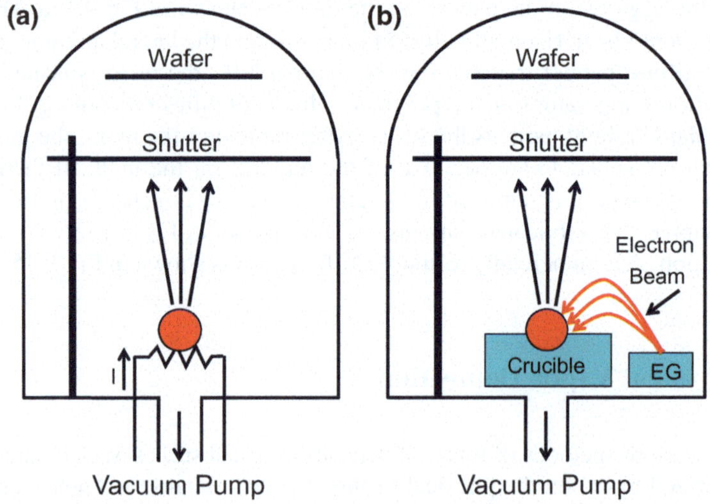

Fig. 9.16 Evaporation. **a** Thermal evaporation for metallic films and nanomaterials. **b** Electron beam evaporation for finer control of film morphology

These oxides can easily be removed by the vacuum pumps. For such reactive metals, it is desirable to wait for a few minutes before opening the shutter to make sure that the metal itself gets deposited, and not its oxide.

While evaporating oxides, e.g. SiO_2, the oxygen may be lost during the evaporation process and care must be taken as the composition of the deposited film may be SiO_x, where $x = 1-2$. After evaporation, appropriate amount of time is given for the wafer and the crucible to cool down. The chamber is then purged after isolating the pumps from the chamber. The samples are sometimes cooled or heated during the evaporation to obtain required film morphology. Sample rotation during evaporation is highly recommended as it improves film uniformity.

Both techniques have their strengths and weaknesses. Thermal evaporation is mostly used for certain metals like chromium, which may otherwise be oxidized during the heating/cooling cycle in the electron beam evaporator's crucible due to residual oxygen. Thermal evaporation is particularly advantageous for evaporating nanomaterials, like C_{60} molecules, etc. E-beam evaporation is desirable for a higher quality metallic, semiconducting, or dielectric films with better morphology. However, metal alloys, compound semiconductors, and compound dielectrics are tricky to evaporate due to different melting temperatures of the constituent elements.

With electron beam evaporation, refractory metals are difficult to evaporate since they tend to reflect off the electron wave. With magnetic materials, like Ni, Fe, Co, etc, the intrinsic magnetization of the crucible material affects the beam shape, pattern and location since an electromagnetic lens is used to focus the electron beam to heat the crucible. However, once the material is heated beyond the Neil temperature, the ferromagnetic properties are lost and the beam reverts back to the parameters for an otherwise nonmagnetic material. Furthermore, one should be cautious that the evaporation inherently damages the substrate, in particular if there is a soft layer like photoresist or polymer, etc.

Such considerations and corrections should be kept in mind while working with various materials. The evaporated films usually do not have a good step coverage, which is disadvantageous for most reasons but could in fact be beneficial for the lift off process. The step coverage may further be improved by rotating the substrate during evaporation.

It is interesting to note that the first 1D nanostructures were synthesized by using evaporation. Consider the shadowing effect inside a trench during the evaporation. By evaporating a thin film, a triangular shaped nanowire may emerge.

For metal alloys, compound semiconducting, and insulating films, sputtering provides an alternative as shown in Fig. 9.17. One could sputter refractory metals as well as magnetic materials with ease. The chamber consists of a parallel plate arrangement with the target material placed at the cathode and the sample placed at the anode, which is usually connected to ground. The anode may also be biased, which leads to a good step coverage. The chamber is first evacuated followed by Ar gas refill to appropriate pressure levels. A plasma is then generated either by DC electric field or RF field. RF is more advantageous with refractory metals. One should note that higher amount of energy is needed to generate plasma than to sustain it. Therefore, the plasma is usually generated with small amount of gas pressure and once sus-

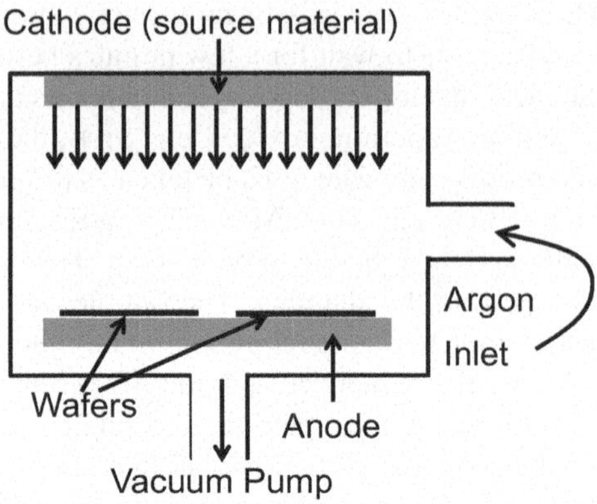

Fig. 9.17 Sputtering for metallic, insulating, and semiconducting film deposition

tained the gas pressure is increased to get a dense plasma, along with an increase in DC or RF energy to sustain the plasma. The Ar^+ ions in the plasma are attracted towards cathode, which is essentially a disk of the target material. Once the energetic Ar^+ ions hit the cathode, they knock off the target atoms, which get deposited on the wafer placed on the anode plate. Sample is sometimes heated to facilitate the diffusion of deposited atoms to improve film morphology and uniformity. The ideal sputtered film has small grain size, which may be obtained by optimizing the pressure of plasma gas, energy delivered to the plasma, and temperature.

In magnetron sputtering, instead of electric field between the two parallel plates, one may apply magnetic field in orthogonal direction, which causes the electrons to follow circular paths. This increases the probability of collision and hence a denser plasma is formed. Yet another method is reactive sputtering, where a metal is sputtered in the presence of a reactive gas, e.g. Ti (titanium) sputtering in the presence of N_2 (nitrogen) gives TiN (titanium nitride) film and cosputtering of Ti and Dy (dysprosium) in the presence of O_2 leads to TiDyO (titanium dysprosium oxide) film growth.

The process of sputtering may cause damage to the substrate, however due to a lower temperature, the damage is usually less than that due to evaporation. However, it is inevitable that some of the gaseous molecules get incorporated in the sputtered film, which may be removed by post sputtering annealing at the right temperatures.

9.8 Chemical Vapor Deposition

Chemical Vapor Deposition (CVD) is the process of material deposition by chemical reactions. CVD may be achieved with or without a catalyst. The one without catalysis involves various feed gases reacting into residual compounds which are then deposited on various substrates. Consider the following chemical reactions,

$$SiH_4 \rightarrow Si + 2H_2$$
$$SiH_4 + O_2 \rightarrow SiO_2 + 2H_2$$
$$3SiH_4 + 4NH_3 \rightarrow Si_3N_4 + 12H_2$$

where under the appropriate conditions, the silane (SiH_4) decomposition leads to silicon film, silane reacts with oxygen to form silicon dioxide film, and reaction with ammonia leads to silicon nitride film. One should note that silane and ammonia are dangerous gases and should be dealt with care. CVD by catalysis is an enabling deposition technique for nanomaterials like graphene and carbon nanotubes etc, which we discuss later.

A typical CVD setup may either have a tube furnace configuration or a chamber configuration as shown in Fig. 9.18a, b, respectively. The tube furnace has filament heating elements, whereas the reactor has quartz lamps as heating elements. In either case, the CVD is carried out at atmospheric pressure (AP) or low pressure (LP), respectively called APCVD or LPCVD. APCVD is a simple technique, but the deposition rates may be very high. By introducing low pressure in LPCVD, the deposition rates and hence film morphology and uniformity may be optimized. Another technique is plasma enhanced CVD (PECVD), where a plasma is used to facilitate the chemical reaction by creating ions which are more reactive than the bare atoms or molecules. The sample may be cleaned in the same chamber by using plasma before deposition, resulting in a cleaner surface and hence better film deposition.

In either tube furnace or chamber configuration, the process starts with loading the samples in an inert atmosphere. In case of LPCVD and PECVD, the chamber is evacuated to the base pressure, and the temperature is ramped to the process temperature. For LPCVD, it is filled with the reaction gas till the required gas pressure is achieved. For PECVD, the plasma is established in an additional process step. The feed gases are then introduced into the tube or the chamber to start the CVD process. Once the deposition is complete, the sample is cooled down and the system is purged for unloading the sample. By using either APCVD, LPCVD, or PECVD, one

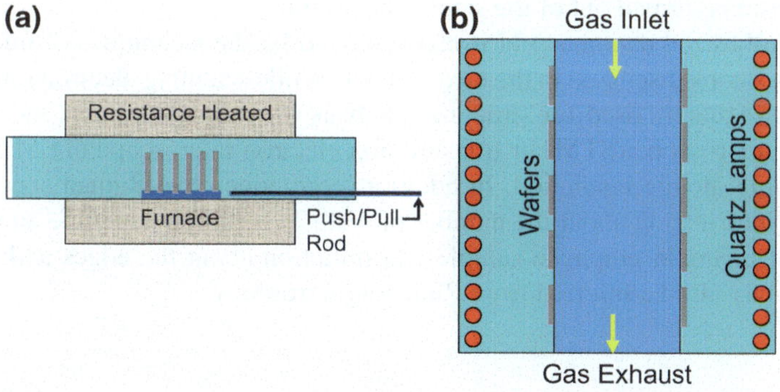

Fig. 9.18 Chemical vapor deposition of insulating and semiconducting films in **a** tube furnace, and **b** chamber configuration. Graphene and carbon nanotube may also be grown by furnace CVD

may synthesize polysilicon, silicon oxide, and/or silicon nitride films by selectively using silane, oxygen and ammonia as the feed gases. Optical and/or electromagnetic characterization is performed to determine the thickness, dielectric constant, and the index of refraction, etc.

In recent years, tube furnace CVD shown in Fig. 9.18a, has also widely been used for nanomaterial growth and synthesis. Two notable examples are that of graphene and carbon nanotube synthesis. However, for nanomaterials growth, an appropriate catalyst is essential. Consider graphene synthesis, where a uniform two dimensional metallic layer of copper gives an ideal platform for monolayer graphene growth, and the use of nickel layer leads to bilayer growth. In addition, the thickness of the metal layer should be greater than 500 nm for low defect growth. For carbon nanotube synthesis, the catalyst consists of nanoparticles or nanocrystals of Fe containing salts.

For nanomaterial growth in APCVD, the sample is loaded in the tube furnace in an inert argon ambience at room temperature. The temperature is ramped to the required annealing temperature, which is usually done under argon and hydrogen flow. It is further ramped to the growth temperature and the process gases are introduced with the right flow rates, followed by removing the inert argon. For graphene and carbon nanotube, the process gas is methane (CH_4). The methane flow rate is quite small for graphene synthesis (usually in *sccm*, standard centimeter cube per minute), whereas for carbon nanotube, the methane flow rate is quite high (on the order of *slm*, standard liter per minute[6]). If ethylene is used for carbon nanotube growth, the flow rates are in *sccm* range. After the growth step is complete, the inert gas is reintroduced and the process gas is turned off. The temperature is ramped down to room temperature and the sample is pulled out. One exception to this recipe is the ultrafast cooling, where the sample or the whole tube may be pulled out of the hot zone of the furnace in a matter of seconds, drastically reducing the sample temperature.

While the growth mechanism is a topic of considerable scientific interest and beyond the scope of this book, a heuristic explanation is discussed next. In a simple picture, the catalyst absorbs the carbon and when it gets saturated, it gives off the extra carbon. Under the right conditions of temperature, pressure and flow rate, the nanomaterial is formed out of the saturated carbon.

The synthesized nanomaterial is characterized by the techniques of microscopy and spectroscopy discussed in the next chapter. While scanning electron microscope (SEM) is frequently used for structural analysis of carbon nanotube, the scanning tunneling microscope (STM) or transmission electron microscope (TEM) are used for obtaining atomic resolution. In addition, techniques like Raman spectroscopy may give information about the number of walls in a carbon nanotube and also the number of layers in graphene samples. Information about the edges and the edge structure may also be inferred from Raman spectroscopy.

---

[6]where a liter equals 1000 cc.

9.9 Molecular Beam Epitaxy

Molecular beam epitaxy (MBE) is an epitaxial growth technique which falls under the bottom up approach to nanoscale film growth and quantum dot synthesis as shown in Fig. 9.19. MBE is usually carried out for III–V compound semiconductors like GaAs (gallium arsenide), AlGaAs (aluminum gallium arsenide), etc, or II–VI semiconductors like CdTe (cadmium telluride), etc. While III–V semiconductors find applications in optoelectronics like LEDs (light emitting diodes) and LASER (light amplification by stimulated emission of radiation) devices, etc, II–VI semiconductors are used for photovoltaic applications. HgCdTe (mercury cadmium telluride) is another popular material used for night vision applications. MBE is used for growing heterostructures of various compound semiconductors as a stack of various films. MBE equipment for a certain material set are kept isolated from other materials to eliminate cross contamination. Therefore, in a research and development environment, one finds dedicated III–V MBE reactor or II–VI MBE reactor, which are not intermixed.

The MBE consists of a UHV (ultrahigh vacuum) chamber. The effusion cells contain the feed materials with shutters closed when not in use. The walls of the MBE chamber are cooled with liquid nitrogen. This is done to capture any undesirable atoms and molecules, and to keep the chamber ultra clean. The sample is loaded through a load lock system, which helps to keep the UHV system intact. The growth process is monitored atomic layer by atomic layer by using RHEED (reflective high resolution electron energy diffraction) technique. The sample is rotated and heated as well for uniform growth and morphology by facilitating atomic diffusion. Once the film growth is complete, the sample is cooled down and transferred out of the chamber through the load lock system.

One of the most important components of MBE reactors is the UHV system. The quality of grown films and quantum dots is directly related to the vacuum quality. Once the integrity of this vacuum is compromised, one has to bake the whole system under vacuum to recover the UHV conditions.

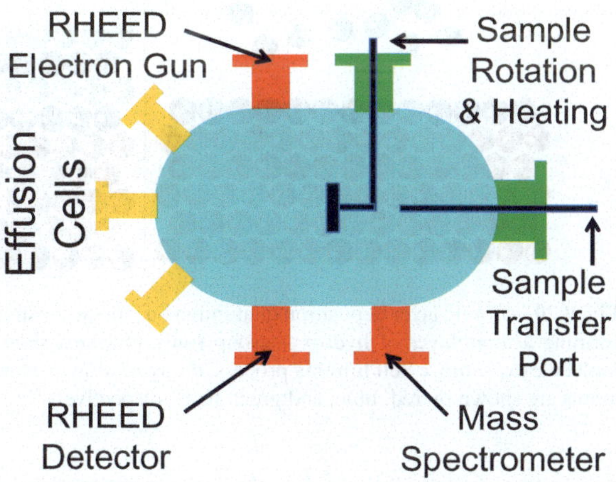

Fig. 9.19 Molecular beam epitaxy chamber for heterostructures growth

9.10 Atomic Layer Deposition

Atomic layer deposition (ALD) is a truly self limiting bottom up growth technique, based on the principle of molecular self assembly. From a humble beginning, this technique has matured now to the state of the art gate dielectric material deposition in the integrated circuits.

The growth process is shown in Fig. 9.20a, where the silicon atoms are exposed to the DI water vapors (only oxygen atoms are shown). The reaction between the oxygen atoms and the silicon atoms results in a silicon substrate terminated with a monolayer of hydroxyl (–OH) group as shown in Fig. 9.20b. The chemical reaction is given as,

$$2\text{Si} + 2\text{H}_2\text{O} \rightarrow 2\text{Si} - \text{OH} + \text{H}_2$$

The growth of a monolayer of hydroxl group is a self-limiting process according to the growth conditions, since additional oxygen atoms may not reach the silicon surface. In the next step, after removing the water vapors, TMA (trimethyl alu-

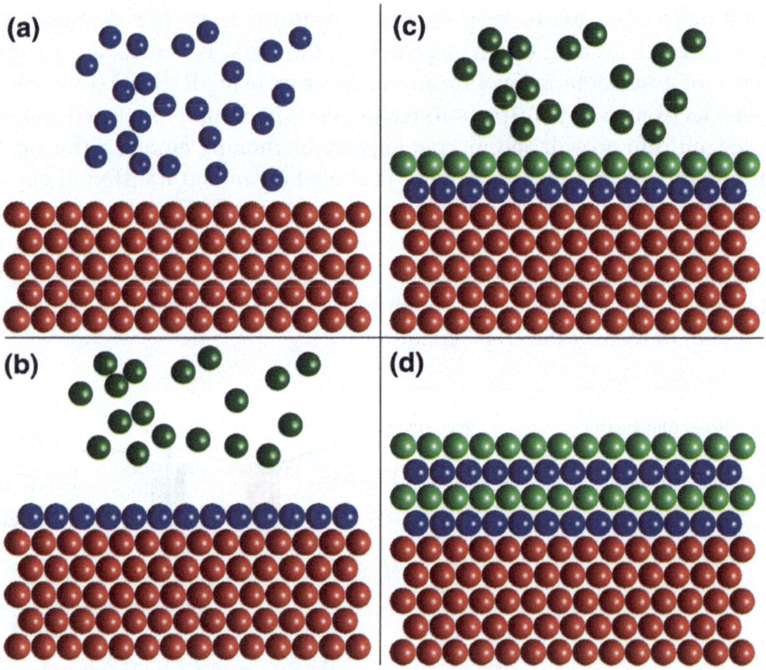

Fig. 9.20 Atomic layer deposition of alumina on silicon substrate. **a** DI (deionized) water exposure forming a monolayer of hydroxyl group (only O atoms shown). **b** TMA (trimethylaluminium) molecule exposure. **c** Self limiting process. **d** Layer by layer growth. Silicon, oxygen, and aluminum atoms are shown by red, blue, and green balls, respectively

mina) molecules are introduced in the vapor phase. The aluminum atom in the TMA molecule reacts with the oxygen atom on the surface resulting in Al–O bond as shown in Fig. 9.20c, with the methane molecule as the byproduct (not shown). The process is given by the following chemical equation,

$$Al(CH_3)_3 + Al - O - H \rightarrow Al - O - Al(CH_3)_2 + CH_4$$

TMA molecules thus form a monolayer, in a self limiting process, i.e. extra TMA molecules do not contribute to the growth beyond the first monolayer. At this stage, TMA is removed and DI water vapors are introduced. The oxygen atoms of the water molecules react with the aluminum atoms and forms Al-O-Al bonds as shown by the following equation,

$$2H_2O + O - Al(CH_3)_2 + Al - O - H \rightarrow Al - O - Al(OH)_2 + 2CH_4$$

The DI water thus forms a monolayer of hydroxyl group, and thereby a monolayer of Al_2O_3 is formed, where the extra water molecules do not contribute to the additional growth. By reintroducing TMA and DI water vapors in repetitive sequence, one may grow alumina atomic layer by layer as shown in Fig. 9.20d, which justifies the name ALD.

While ALD is an excellent bottom up growth technique, it does have some bottlenecks. The growth rate is rather small due to inherent layer by layer growth, leading to low throughput. The ALD materials and supplies could be expensive, depending upon the film growth recipe. However, the advantages outpace the bottlenecks. To name a few, the yield is high with good repeatability and low defect density. The step coverage and conformality is also good to say the least as shown in Fig. 9.21, where the deep trenches in porous anodic alumina have been covered very well with 80 nm of ALD grown alumina film.

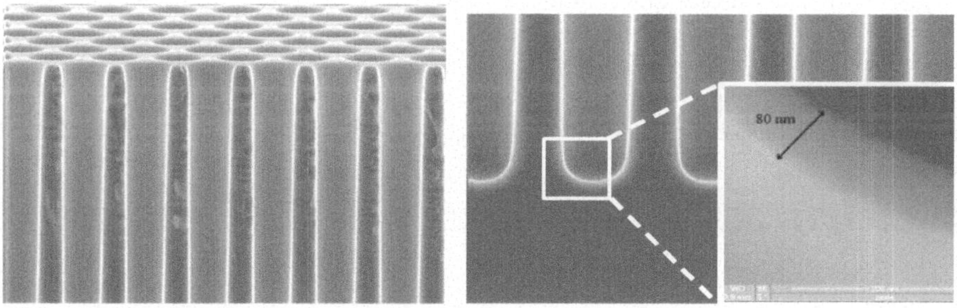

Fig. 9.21 ALD. Excellent conformality in high aspect ratio nanostructures due to bottom up material synthesis at the atomic scale. (Courtesy of Oxford Instruments)

9.11 Etching

In nanofabrication and microfabrication, selective material removal succeeds lithography. This may be achieved by liftoff or etching. In liftoff process, the material is removed from the region covered by the photoresist by dipping the sample in a solvent to dissolve the photoresist. In etching, the material is removed from the area not protected by the photoresist by using chemicals, reactive plasma, or high energy ions, called chemical etch, ion induced etch, or physical etch, respectively. The key etch parameters include etch rate, selectivity, undercut, and whether etchant leads to isotropic or anisotropic etch.

The chemical etch process involves dipping the patterned wafer in etchant bath of a certain composition under the right temperature conditions. The etch is isotropic leading to undercut as shown in Fig. 9.22. The etch rates are fast, 1 μm/min rates are not unheard of. The etchants may be engineered to be highly selective. However, the etching process may be nonuniform leading to rough films and edges. HF (hydroflouric) acid acts as an excellent etchant for silicon oxide. For etching silicon, mixture of HF and nitric acids may be used for nondirectional etch, whereas KOH (potassium hydroxide) based recipe etches at 35.26° angle along (111) plane. Peroxide etches GaAs well. For etching metals, various acids are used and the etch recipes are widely available.

Ion Induced Etch is highly anisotropic leading to minimum undercut and hence is ideal for etching small features. The etch rates are medium and the etch products are less volatile and coercive than chemical etch. The etch products tend to stick to the etch surface, forming a protective sidewall passivation layer, giving more vertical etch results. By applying a sample bias, ion bombardment can be facilitated, which helps to remove the etch products from the etch surface, and to allow the etch to proceed. Two most commonly used recipes are based on fluorine and chlorine gases. While fluorine based recipes are ideally suited for silicon related materials, chlorine based recipes are more applicable to GaAs related materials. Chlorine based chemistry is far more dangerous than fluorine one, and should be dealt with care. Fluorine based chemistry also has severe environmental impact.

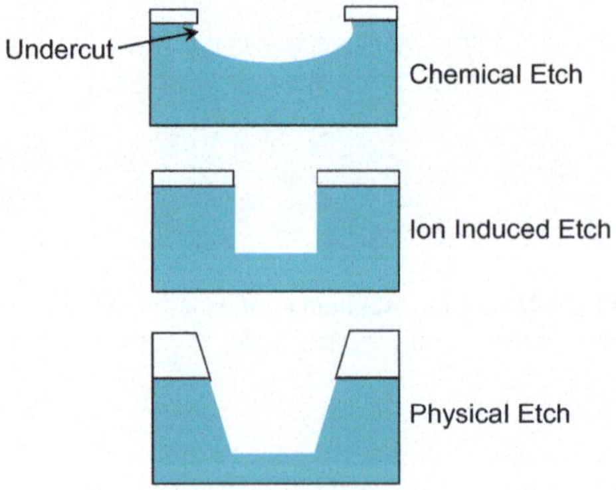

Fig. 9.22 Comparison of various etching techniques

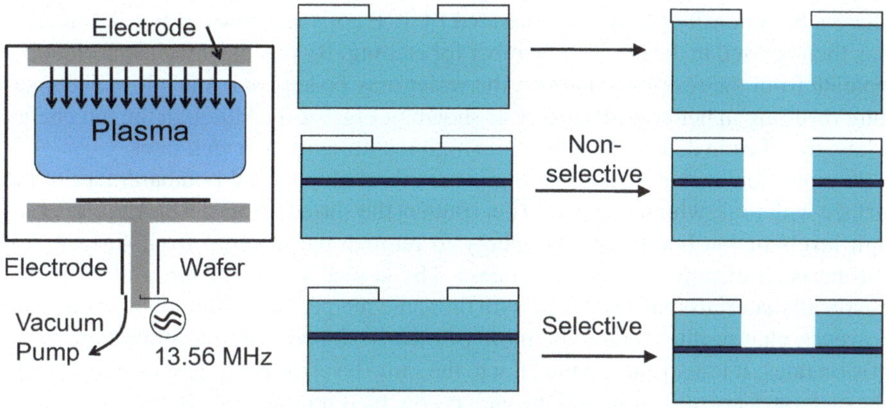

Fig. 9.23 Reactive Ion Etching. Selective and nonselective

A typical parallel plate reactive ion etch (RIE) setup is shown in Fig. 9.23. The sample is loaded under inert gases and the whole chamber is pumped down. The etch gases are introduced after establishing a vacuum. DC or RF power is applied to produce plasma, which helps to breakdown the bands of the material to be etched and removed. After the desired etch time, DC or RF power is stopped and the etch gases are purged. Samples are pulled out of the chamber after cooling down to room temperature. RIE is in general nonselective. However, another layer may be introduced in the film stack that not only etches with the gases used in the etch recipe but the etch automatically stops at this layer and hence stops the process.

While in a parallel plate RIE, the plasma is produced at the same place where reaction takes place, an improvement is ICP (inductively coupled plasma) etcher,

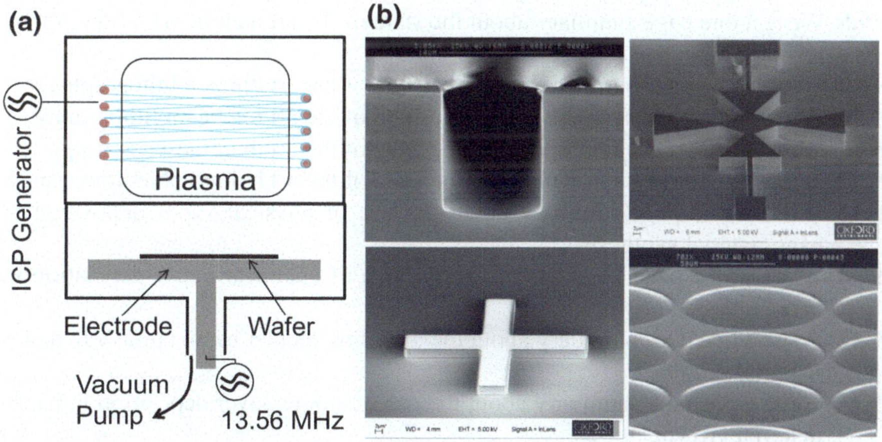

Fig. 9.24 Inductively coupled plasma (ICP) etching. **a** Equipment setup. **b** High aspect ratio features by virtue of ICP etching. (Courtesy of Oxford Instruments)

where a high density plasma is generated by ICP coils as shown in Fig. 9.24a. This may then be used in the process chamber for etching. By placing the plasma chamber separate from the reaction chamber, the wafer may be kept at a much lower temperature resulting in better etch profiles as shown in Fig. 9.24b. Although the ion energy is low, the plasma density may be quite high resulting in high etch rates.

In a physical etch process, etching occurs through physical bombardment of the surface with ions, which knock off (i.e. sputter) the surface atoms. This requires very high levels of ion bombardment energy to remove the material from surface. Ion Milling is an example of such a process. The key advantage of ion milling is that mostly all materials may be etched with the same recipe, i.e. it is highly nonselective. However, etch profiles tend to be more positively sloped due to faceting/high mask erosion rates. It is also anisotropic, not to the same level as that of RIE or ICP though. The etch rates are quite low, and the etch products, if produced at all, are nonvolatile.

Research Assignments

R9.1 Write a one page summary about the clean room classifications and technology.
R9.2 Write a one page summary about the new GHS system. How is it different from the old MSDS system?
R9.3 Write a one page summary about the state of the art photolithography technology.
R9.4 Write a one page summary about computational photolithography technology, in particular phase contrast computational lithography.
R9.5 Write a one page summary about the use of interferometry in lithography.
R9.6 Write a one page summary about the state of the art interferometric stages commercially available.
R9.7 Write a one page summary about the state of the art electron beam lithography technology.
R9.8 Write a one page summary about the state of the art nanoimprint lithography technology.
R9.9 Write a one page summary about the state of the art dip pen lithography.
R9.10 Write a one page summary about the state of the art ion beam lithography.
R9.11 Write a one page summary about the state of the art direct laser writing.
R9.12 Write a one page summary about the state of the art holographic lithography.
R9.13 Write a one page summary about the use of physical vapor deposition in nanomaterial synthesis.
R9.14 Write a one page summary about the use of chemical vapor deposition in nanomaterial growth.
R9.15 Write a one page summary about the use of molecular beam epitaxy in nanomaterial growth.
R9.16 Write a one page summary about the use of atomic layer deposition in nanomaterial growth.
R9.17 Write a one page summary about the use of novel etching techniques in fabricating nanostructures.

Chapter 10
Microscopy and Spectroscopy

In this chapter, we introduce various microscopy and spectroscopy techniques. Spectroscopy is the discipline of analyzing the response of a material to various stimuli.

The comparison and scope of various microscopy techniques is shown in Fig. 10.1. With an unaided adult eye, one may make observations at the length scale of about 200 μm, whereas by using the optical microscope, the spatial resolution may be pushed down to about 0.2 μm, which is limited by the diffraction limit.

Ever since the early days of quantum mechanics, it was realized that electrons have much smaller wavelength than that of the optical photons. An electron with 100 keV kinetic energy has 4 pm wavelength. At the atomic and the nanoscale, the use of electron microscopy thus becomes inevitable. In this context, resolutions of up to 0.1 nm = 1 Å are routinely achieved these days by using electron microscopes. The electron microscopes also have the advantage of greater depth of focus and hence are also used at much higher resolution of up to 100 μm to image biological entities.

10.1 Scanning Probe Microscopy

We start the discussion with scanning probe microscopy, where a probe is scanned over the sample. The interaction of the two provides the imaging information. An analogous instrument at microscale is based on the mechanical profilometer as shown in Fig. 10.2a, where a stylus scans over the surface of the sample by physically touching it. The deflection of the stylus is recorded by the trajectory of a light beam on a photographic plate. The resulting profiles of a polyester mesh is shown in Fig. 10.2b. The apparatus for SPM looks very similar to that of the mechanical profilometer, however there are key differences. In SPM techniques, very delicate forces or signals (currents for example) are monitored between the tip and the substrate on the atomic scale. By making use of this information, nano, atomic, or even subatomic scale imaging may be performed.

H. Raza, *Nanoelectronics Fundamentals*, NanoScience and Technology,
https://doi.org/10.1007/978-3-030-32573-2_10

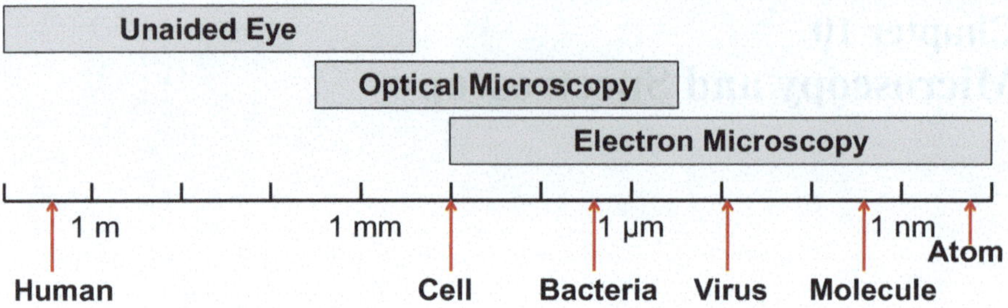

Fig. 10.1 Comparison and scope of various microscopy techniques

Fig. 10.2 Mechanical profilometer as a precursor for scanning probe microscope (Courtesy of KLA)

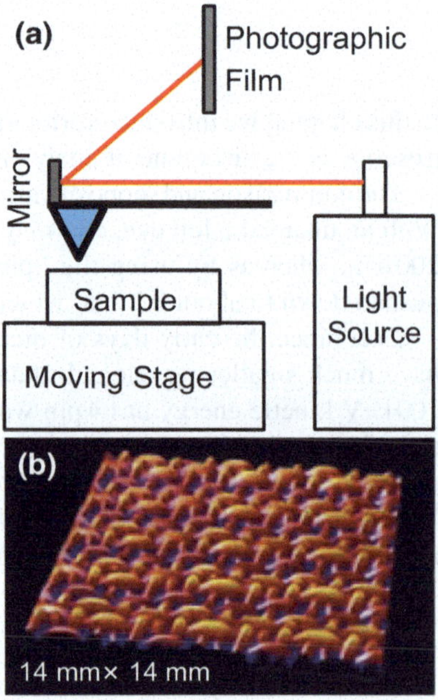

Scanning Tunneling Microscopy

Consider the experimental apparatus shown in Fig. 10.3a for a scanning tunneling microscope (STM), which is routinely used to obtain atomic resolution as shown in Fig. 10.3b. This is in fact one of the earliest images obtained by STM. The tip may consist of a cut platinum-iridium wire, electrochemically etched tungsten wire or tungsten wire sharpened with ion milling as shown in Fig. 10.3c. The tip is attached to a three-axis piezoelectric tube, which may scan in three dimensions with sub-nm precision. By applying a voltage to the piezo tubes, expansion or contraction may be caused in the X, Y and Z directions, which gives the XY scanning on the sample surface and Z motion to control the tip height over the sample surface. At the atomic scale, one may easily find a single atom at the end of the tip. By applying volt-

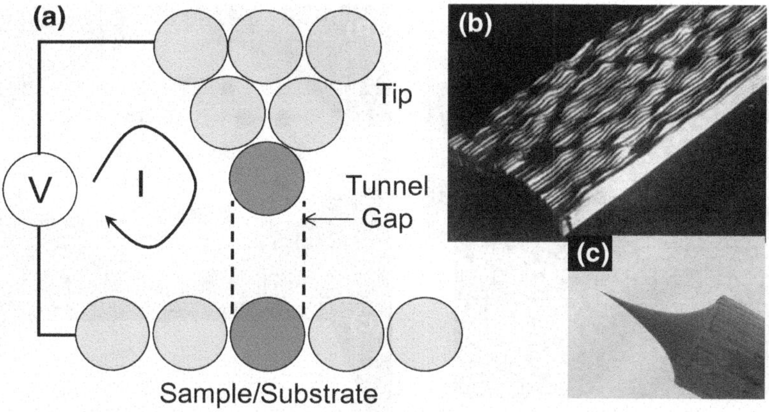

Fig. 10.3 STM. **a** Working principle. **b** Si surface visualized by using STM [1]. **c** STM tip

age across the tip and the substrate under UHV conditions, a quantum-mechanical tunneling current flows if the gap d is on the order of few Angstroms (Å).[1] This tunneling current is indeed a purely quantum mechanical phenomenon that exponentially depends on the gap between the tip and the sample, i.e. $I = I_o e^{-2\alpha d}$, where α is the decay constant given as, $\alpha = \sqrt{2m\Phi}/h$, and Φ is the barrier height between the sample and the STM tip.

Since the current depends exponentially on the barrier thickness, i.e. the gap between STM and substrate, about 90% current flows through the last atom on the tip into the nearest substrate/sample atom. If the tip atom is between the two atoms on the substrate as compared to exactly being in alignment with the substrate atom, the current is smaller due to the exponential dependence of the tunnel current on the barrier width. By using this difference in the current, one may deduce the tip atom location when it is directly above a substrate atom or in between the atoms, thus reproducing the substrate atomic structure. Furthermore, STM is essentially performed under ultra-high vacuum conditions, since minute perturbations in the form of gaseous atoms may interfere with the quantum mechanical tunneling, thereby resulting in image artifacts. There are two modes of STM operation, namely constant height mode and constant current mode. As the name suggests, in the constant height mode, the tip-sample separation is fixed and the tunneling current is monitored as a function of time during scanning. The varying current amplitude corresponds to the surface features. In the constant current mode, a current feedback mechanism is used to adjust the tip height and hence the tip-sample separation to keep the tunnel current constant. The tip height thus indirectly corresponds to the surface morphology, since the tip probes the local density of states (LDOS) of the surface and not necessarily the atomic locations.

[1] 1 Å = 0.1 nm.

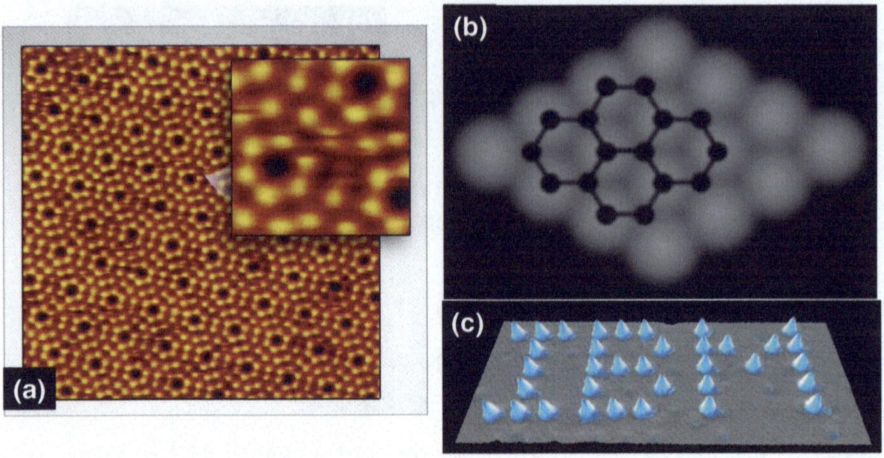

Fig. 10.4 STM images. **a** Si(111) − (7 × 7) surface (Courtesy of RHK Tech). **b** Graphite with overlaying atomic structure (Courtesy of Omicron). **c** Atomic control by using STM (Courtesy of IBM)

We show the STM viewgraph of a Si(111) − (7 × 7) reconstructed surface in Fig. 10.4a. In fact, after STM invention, the studies on this surface provided a compelling evidence of the instrument's capabilities. Earlier theoretical work had predicted various other reconstructions on this surface, but the experimental verification of the (7 × 7) reconstruction provided a very counter-intuitive conclusion.

Moreover, STM images should be interpreted in the right fashion. Since STM does not directly probe atoms, rather the atomic and molecular orbitals around the atoms, the orbital shapes and the resulting density of states becomes an important parameter. Consider the STM viewgraph of graphene in Fig. 10.4b, where the information about every other atom is missing in the STM image, as compared to the physical image overlaid, which is shown by the ball and stick model of the hexagonal arrangement of carbon atoms. Apart form this, one should also be considerate of the fact that the tip structure is amorphous or polycrystalline at best, which leads to various artifacts. Not to mention the effect of surface contaminants on both the sample surface as well as the tip surface.

STM may also be used for controlling atoms at the surface. Consider the arrangement of xenon atoms on the nickel surface in Fig. 10.4c. Such single atom lithography techniques have been extended to creating zero-dimensional quantum dot, and one-dimensional quantum wire structures on semiconductor substrates. While the throughput is low, these techniques provide an excellent mechanism to study quantum structures and nanomaterials at this scale. Once fabricated, the STM may also be used to study charge transport through them, a technique known as scanning tunneling spectroscopy (STS), where the tip is used as a probe and not as a contact in the traditional sense.

Atomic Force Microscopy

While STM is an excellent tool for studying conducting materials at the nanoscale, it may not be directly applied to the study of non-conducting materials. Filling this void was the primary motivation behind the invention of the atomic force microscope (AFM), which is based on the principle of the tip-sample interactions at the atomic scale as shown in Fig. 10.5a. A typical AFM tip is shown in Fig. 10.5b. A useful analogy of AFM is to that of a blindfolded person who may *feel* the features of a surface by touching it with the tip of his/her finger. In AFM, a laser beam tracks the tip motion by using a quadrant photodetector.

Depending on the tip-sample separation and tip excitation, there are various modes in which an AFM may operate. In the contact mode, the tip physically makes a contact with the sample as shown in Fig. 10.5c. The force between the tip and the sample is repulsive in this case. However, despite being the most sensitive technique, it may still damage the substrate. In the non-contact mode, the tip and the sample maintain a gap, while making sure that the sample is not damaged, but at the cost of its sensitivity. In the intermittent contact mode or tapping mode, the tip makes contact with the sample temporarily and for the rest of the time is isolated from the sample. The tapping mode is indeed a trade off between the contact and the non-contact mode. The 3D scanner of an AFM works the same way as that of STM. It consists of a piezo tube with three independent degrees of freedom. Thus by applying voltage to piezo tube, motion in X, Y and Z directions may be controlled independently with sub-nm precision. The output of the photodetector is then fed to a differential amplifier to

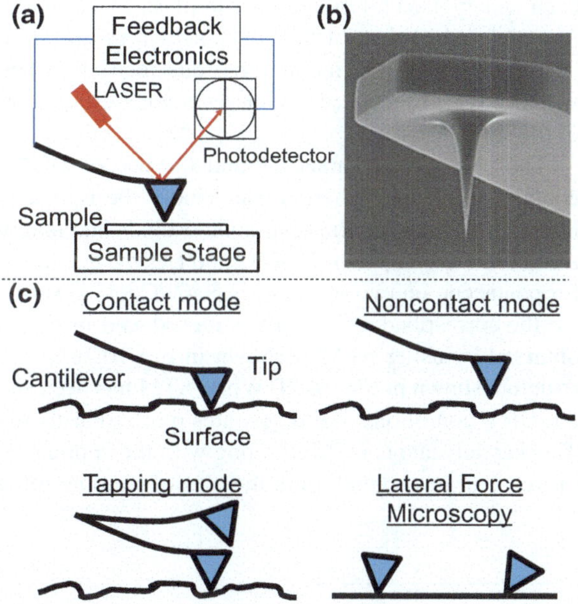

Fig. 10.5 AFM. **a** Working principle. **b** AFM tip. **c** Various operating modes of AFM

(a) (b)

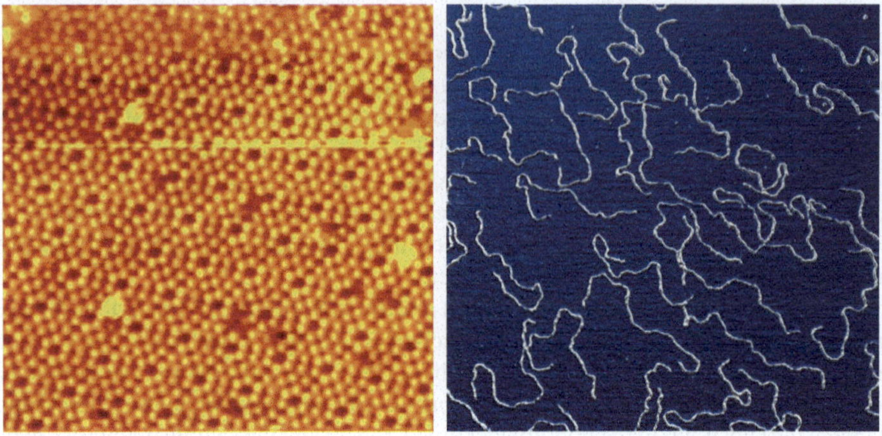

Fig. 10.6 UHV AFM. **a** UHV AFM viewgraph of Si(111) surface (courtesy of Unisoku). **b** AFM of ligands (Courtesy of Nanosurf)

get the error signal and control the PID (proportional-integral-derivative) feedback loop of the tip motion. The parameters of this feedback loop are quite important for proper functioning of the AFM.

An AFM image of the Si(111) $-$ (7×7) is shown in Fig. 10.6a. Under UHV conditions, atomic resolution is obtained for this surface, the viewgraph of which is comparable to the one obtained by using STM as shown in Fig. 10.4a. AFM has been widely used for biological samples as well, both under dry and wet conditions. A sample AFM scan of ligands is shown in Fig. 10.6b. AFM may also be used by keeping track of the lateral force due to friction between the tip and the sample surface.[2] This mode leads to interesting features about the stickiness and the friction of the sample surface.

The primary advantage behind the use of UHVAFM chamber is to ensure a contamination-free environment. Under the right conditions of stability, even the interactions between the tip atomic orbitals and that of the substrate may be visualized. The viewgraphs of HOMO and LUMO are shown in Fig. 10.7a, c, respectively for pentacene on two-monolayer-NaCl/Cu(111) substrate. The theoretical rendition for the corresponding orbitals is reproduced in Fig. 10.7b, d. The atomic structure obtained by using AFM is shown in Fig. 10.7e in comparison with the theoretical structure shown in Fig. 10.7f. While STM has been shown to be effective only under the UHV conditions, AFM operates quite robustly even under ambient conditions. This key advantage of AFM along with the applicability to the insulating substrates makes it an important tool in the area of nanometrology and nanocharacterization.

---

[2]This area of research is known as tribology.

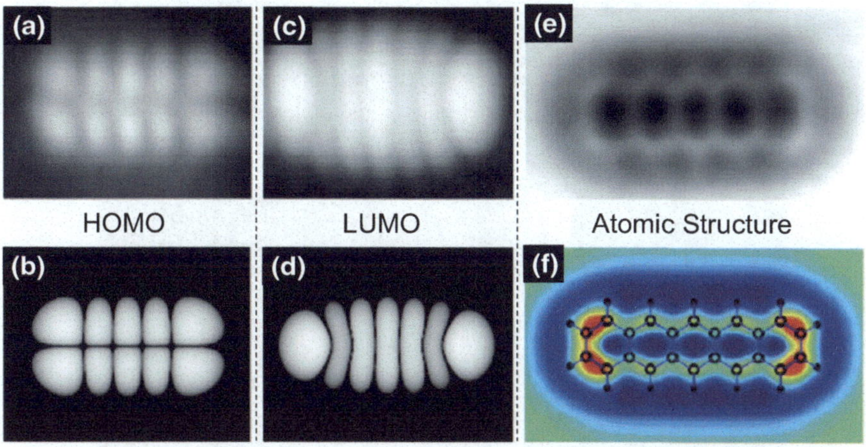

Fig. 10.7 UHV AFM of atomic and electronic structure of pentacene on two-monolayer-NaCl/Cu(111) substrate. **a** Experimental and **b** theoretical viewgraph of HOMO's electronic structure. **c** Experimental and **d** theoretical viewgraph of LUMO's electronic structure. **e** Experimental and **f** theoretical viewgraph of the atomic structure [2]

10.2 Electron Microscopy

With the notion established by quantum mechanics in the early part of the twentieth century that electrons behave as waves, it was evident even in early days that electron microscopes based on this novel idea can be made with tremendous possibilities. One compelling reason was that electrons have small rest mass and hence may be easily focused and manipulated by using electromagnetic lenses as shown in Fig. 10.8. As noted earlier in this chapter, 100 keV electrons have 4 pm wavelength, which sets the diffraction limit way above the atomic scale.

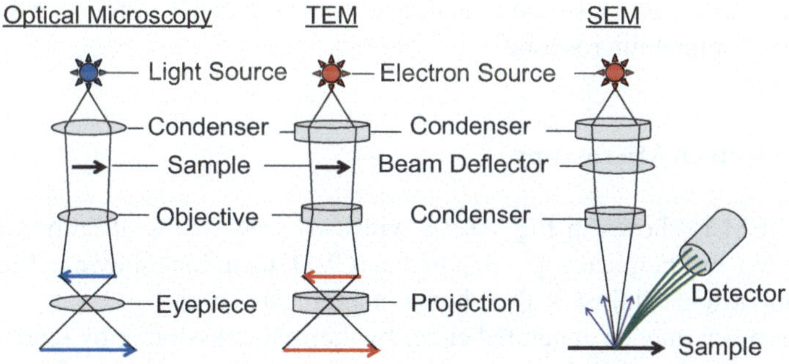

Fig. 10.8 Comparison of electron microscopy with optical microscopy

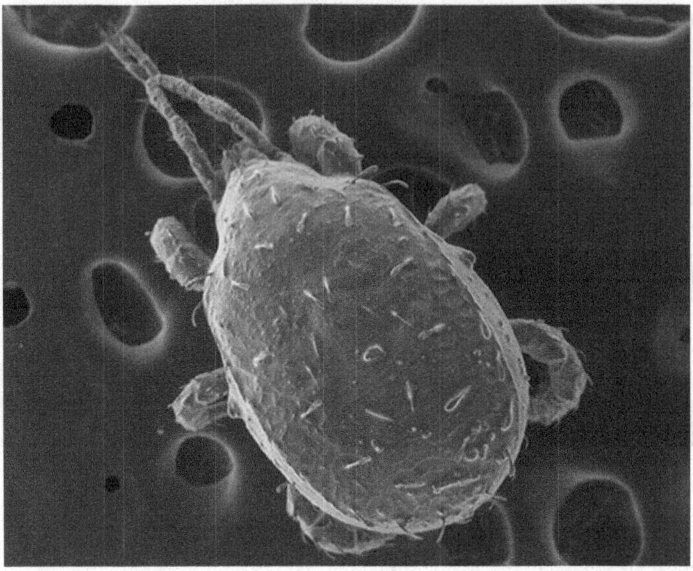

Fig. 10.9 One of the key advantages of electron microscope is the depth of focus (Courtesy of Hitachi)

In Fig. 10.8, a comparison of optical microscope with two kinds of electron microscopes namely, scanning electron microscope (SEM) and transmission electron microscope (TEM) is shown. There is almost a one-to-one correspondence between the optical microscopy and the TEM, given the optical lenses are replaced by electromagnetic ones. Although similar, SEM does have some unique features.

An electron microscope has excellent depth of focus, which is an important parameter while imaging materials of large dimensions, e.g. biological samples. Therefore, these microscopes find ample applications even at microscale for biological samples. Consider the electron microscope viewgraph of a mite shown in Fig. 10.9, where the whole mite is focused in a single scan due to very high depth of focus, a feature not achievable by optical microscope.

Scanning Electron Microscopy

A typical SEM is shown in Fig. 10.10a with two viewgraphs of carbon nanotube samples grown by chemical vapor deposition (CVD) technique shown in Fig. 10.10b, c. The nanoparticle catalyst is also shown in aggregate form.

Electron beam may be generated either by thermal emission or by field emission. In thermal emission, a tip is heated beyond a certain temperature that gives rise to the thermal emission of electrons over the tip-vacuum potential barrier. These thermal electrons are not coherent. In field emission, a voltage is applied across the tip and

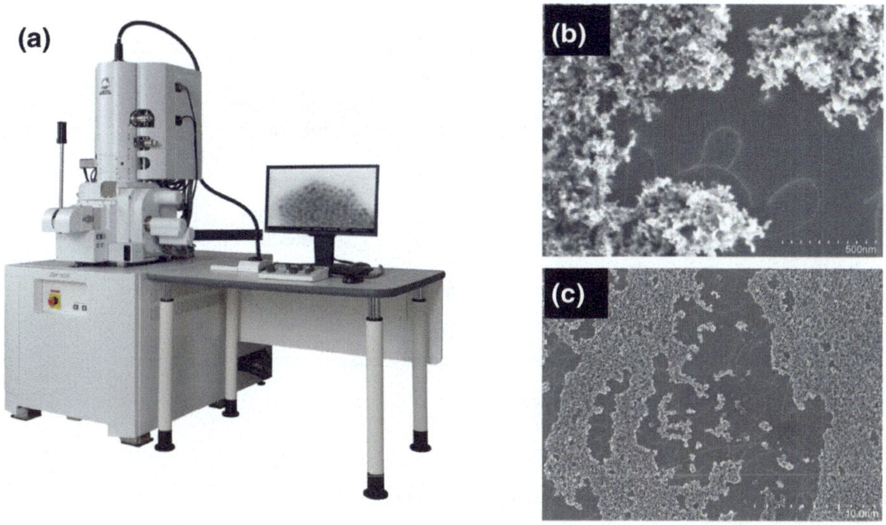

Fig. 10.10 Scanning electron microscope. **a** Commercially available SEM (Courtesy of JEOL). **b**, **c** Viewgraphs of carbon nanotubes with growth catalysts at two magnifications

anode that gives rise to very high electric field for a sharp tip. This field imparts enough energy to the tip electrons to become free and form a beam under the right conditions. A typical voltage of 5 kV yields electrons of about 2.5 Å wavelength, resulting in nanoscale resolution. Furthermore, the beam size is on the order of a few nm, which gives the required lateral resolution. The field emitted electrons are coherent and therefore are preferable over the ones obtained by the thermal emission.

Once electrons are emitted from the tip, the condenser lens, very similar to that of an optical microscope, focuses the beam as shown in Fig. 10.8. After the beam deflectors (scanning coils), another condenser lens focuses the beam further before the beam hits the sample. These lenses may be made by using fixed magnets, electromagnets or could simply be electrostatic lenses.

As a result of the beam-sample interaction shown in Fig. 10.11, X-rays are generated which contain chemical information about the substrate. This information may be used in EDS (energy dispersive X-ray spectroscopy), which is further discussed in Sect. 10.6. The most important part of the interaction is the emission of the secondary electrons, which carry topographical information. When this information is plotted as a function of raster beam scan, it gives the topographical information in the SEM viewgraphs. The backscattered electrons not only give topographical information, but also carry the chemical information about various kinds of atoms. The Auger electrons carry the surface sensitive compositional information. The rest of the electrons simply contribute to the sample current (not shown) that may be collected.

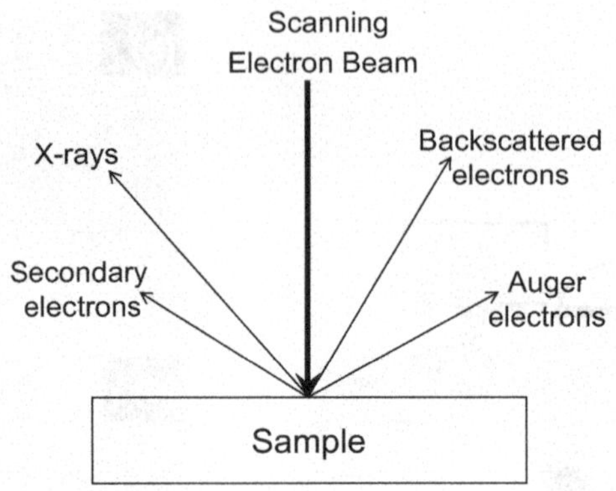

Fig. 10.11 Electron sample interaction

Transmission Electron Microscopy

In TEM, the sample is placed after the first condenser lens and the transmitted electrons carry the sample information, which is passed through the objective lens and finally through a projection lens onto the fluorescent screen as shown in Fig. 10.8. The magnified view of the sample at the atomic scale is thus displayed. A commercially available TEM is shown in Fig. 10.12a.

In TEM, the sample preparation is the key. The idea is to have the sample thickness down to 100 nm or less. While various membranes could qualify for a TEM scan, the case of heterostructures is particularly intriguing for cross-sectional TEM. The sample needs to be ground and polished down to about 100 nm, so that the electrons may transmit through the sample.

Once the sample is prepared, atomic scale viewgraphs may be obtained as shown in Fig. 10.12b. In the reported MOS structure, one may see silicon atoms with perfect atomic periodicity in the single crystalline silicon. Since silicon oxide is amorphous,

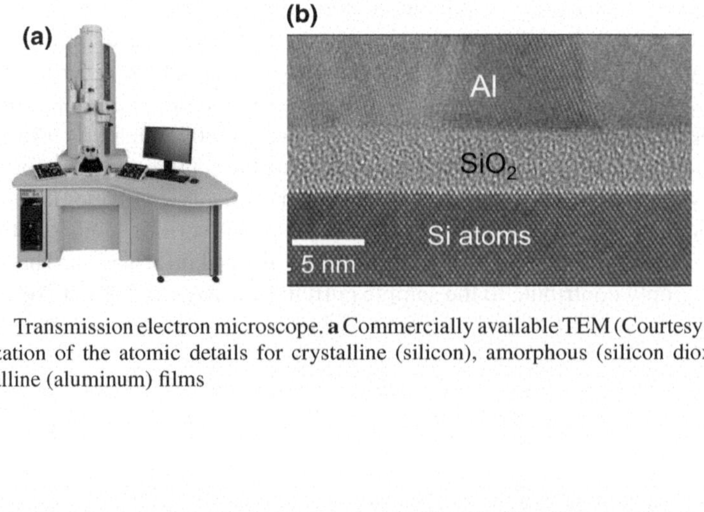

Fig. 10.12 Transmission electron microscope. **a** Commercially available TEM (Courtesy of JEOL). **b** Visualization of the atomic details for crystalline (silicon), amorphous (silicon dioxide), and polycrystalline (aluminum) films

no order is observed. Aluminum is however polycrystalline. Not only short range order is observed with some atomic granularity, but the grains are also visible, which are different across the grain boundaries.

In order to obtain this atomic precision and resolution in TEM, the electron wavelength has to be much smaller than that of an SEM. High acceleration voltage of up to 1 MeV is used, which gives wavelengths on the order of 1 pm $= 10^{-3}$ nm.

Scanning Transmission Electron Microscopy

A relatively recent development in electron microscopy is that of scanning transmission electron microscopy (STEM), where the advantages of both SEM and TEM are combined in one instrument. STEM consists of high acceleration voltages on the order of 100 kV, same as that of a TEM. However, while the electron beam in TEM has large size, the beam in STEM is only a few nm in size, similar to SEM. As the beam is scanned in a raster scan mode, the chemical information may also be obtained for a smaller specimen size. Additionally, EDS may be added to STEM, thereby making it even more powerful instrument. We discuss EDS in Sect. 10.6.

10.3 Optical Microscopy

Optical microscopy is the art of seeing small objects with magnification by using photons in the optical range of 400–700 nm wavelength. The two important lenses are the objective and the eyepiece that are vital for the image formation as shown in Fig. 10.8. Magnifications on the order of 150 may be routinely obtained.

There are primarily four types of optical microscopes, namely bright-field, dark-field, phase-contrast, and confocal. In a bright field microscope, the contrast is obtained by the absorption of white or near-white light. The reflected or transmitted light from an illuminated object is collected by the objective lens and then projected onto the human eye or a CCD (charge coupled device) sensor by using the eyepiece lens. CCD sensors are thoroughly discussed in Chap. 8. In dark-field microscope, the objective only collects the scattered light and hence forms the image by using diffracted light only. The dark field microscopy thus produces a bright image of the object against a dark background.

The phase contrast microscope is based on the phase change as the light passes through a sample. As the name suggests, the contrast is obtained precisely by this phase change. By using a phase ring, the light passing through two paths interfere, where through one path, a phase delay of $\lambda/4$ or $\pi/2$ is introduced. The resulting interference depends on the phase change through the sample. Phase contrast is ideally suited to image heterogeneous materials with varying refractive indices.

We next discuss the comparison of various optical microscopy for a PVA (polyvinyl alcohol) fiber sample. In the bright-field image, the viewgraph appears darker against a brighter background as shown in Fig. 10.13a, whereas the view-

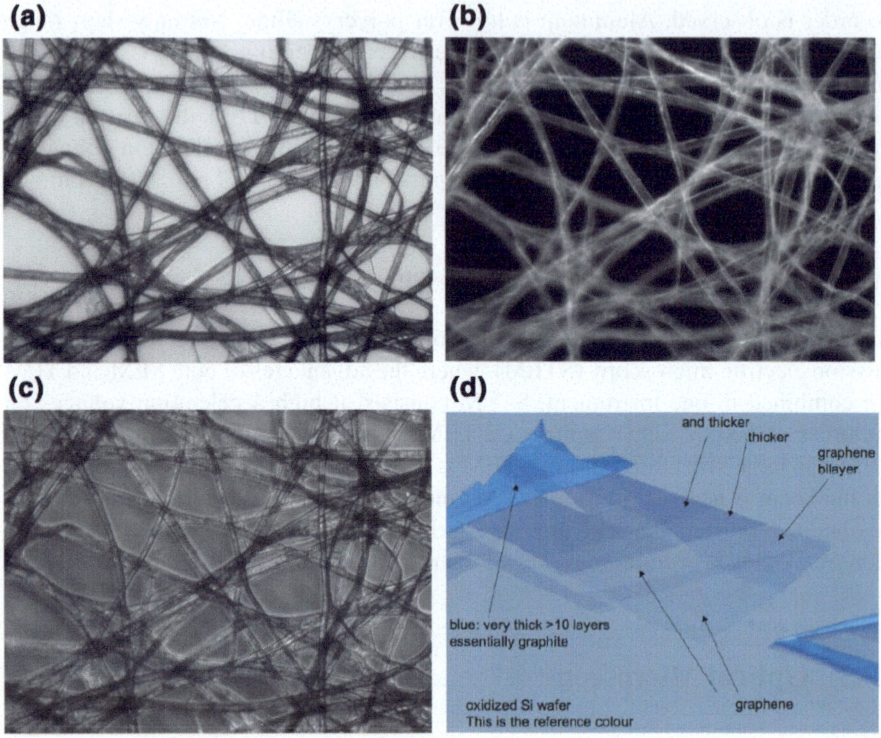

Fig. 10.13 Comparison of various optical microscopy techniques. **a** Bright field. **b** Dark field. **c** Phase contrast. **d** Differentiation of varying layers in a graphene sample by using optical microscopy (Courtesy of Zeiss and Graphene Industries)

graph is brighter against a darker background in the dark-field image as shown in Fig. 10.13b. Due to the nature of diffraction, objects with sub-diffraction resolution may be imaged by using dark-field microscope. It is quite common to image multi-wall carbon nanotubes with 50 nm diameter in the dark-field mode, which is simply not possible in the bright-field mode. In the phase contrast image, greater detail within an object may be seen due to interference caused by the variations in the refractive indices as shown in Fig. 10.13c. Apart from this, it is feasible to distinguish between various layers of graphene on a 300 nm SiO_2/Si substrate by using optical microscopy. This is due to the varying reflection and hence a difference in color for various graphene layers as shown in Fig. 10.13d.

Confocal microscopy has the following three types, laser scanning, spinning disc, and two photon. In this chapter, we discuss only the laser scanning confocal microscope to describe the principle, although all three types are equally important. The problem in wide field optical microscopy discussed so far, whether bright field, dark field or phase contrast, is that not only the in focus light is captured but the out of focus light is also collected. This makes the image out of focus and the features thus

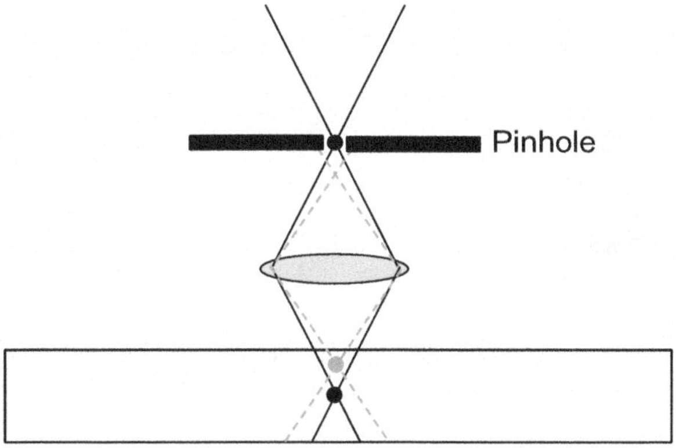

Fig. 10.14 Working principle of a confocal microscope

are not very sharp. In the confocal microscope, this problem is solved by using a pinhole to block the out of focus light as shown in Fig. 10.14. As a result, the confocal images are sharp with impressive details. However, due to the use of pinhole arrangement, the collected photon number density is rather small. This bottleneck is addressed by amplifying the signal by using photomultiplier tubes.

10.4 Photoemission Spectroscopy

We next discuss a class of spectroscopy techniques whose working principle is based on Einstein's photoelectric effect as shown in Fig. 10.15.

Incident photons transfer their energy to the electrons inside the sample. If the energy of the incident photons ($hf = \hbar\omega$) is higher than the surface barrier (ϕ_f), the electrons may overcome the surface barrier, and become free with a finite kinetic energy (E_k) which equals the difference between the photon energy and the surface barrier, i.e. $E_\mathrm{k} = \hbar\omega - \phi_f$. By analyzing the electron energy distributions, useful information may be deduced about the electron distribution in the sample.

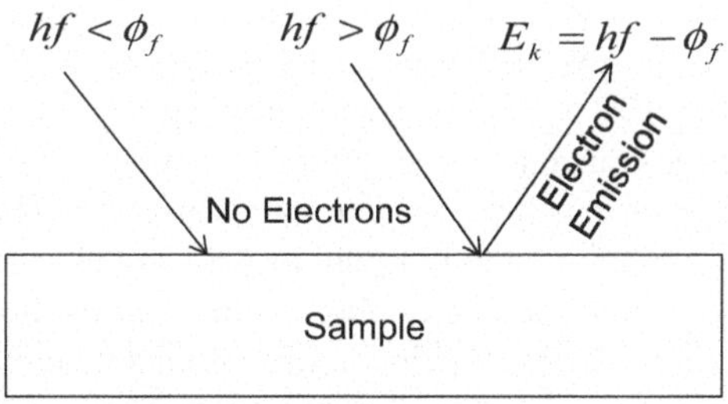

Fig. 10.15 Photoelectric effect forms the basis of photoemission spectroscopy

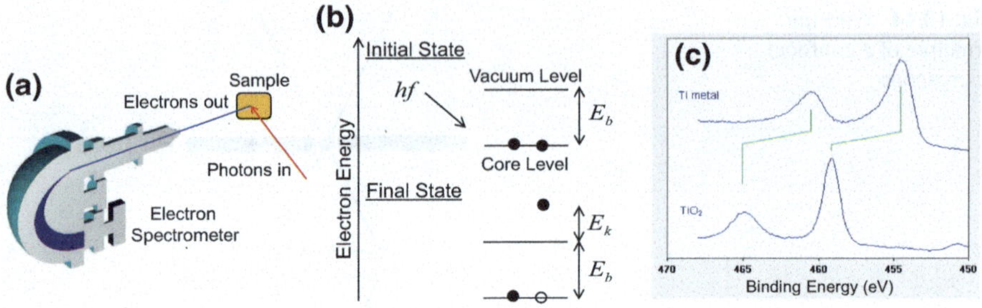

Fig. 10.16 XPS. **a** Equipment setup. **b** Working principle. **c** Chemical composition leads to various features in the binding energy spectrum (Courtesy of ThermoFisher)

X-ray Photoemission Spectroscopy

In the X-ray photoemission spectroscopy (XPS), X-ray photons are used to probe matter. The high energy photons impart enough energy to the core electrons to make them free. In the free state, the kinetic energy (E_k) of the electron is given by,

$$E_k = hf - E_b$$

where f is the photon frequency, h is the Planck's constant and E_b is the binding energy of the electron inside the sample.

The emitted electrons are then analyzed by using the electron spectrometer as shown in Fig. 10.16a, which consists of an electromagnet that bends the beam in a circle by applying a magnetic field density (B) with direction perpendicular to the circular planar path of the electrons. The radius (r) of the circle is proportional to E_k, given as follows $r = \sqrt{2mE_k}/qB$, and hence provides basis for the experimental observation of the spectrum as a function of the binding energy.

The mechanism behind the XPS spectrum is shown in Fig. 10.16b. In the ground state, the electrons are residing in the core level, where the energy difference between the vacuum level and the core level is called the binding energy. An electron needs energy equal to the binding energy or more to become free. These electrons absorb X-ray photons and become free with a kinetic energy, which may be analyzed by using a spectrometer. For the same frequency of the X-ray photons, different core levels give different peaks. A typical XPS spectrum for Ti and TiO_2 is shown in Fig. 10.16c, where the binding energy axis is in eV. Various materials have distinct peaks that may be quantified to distinguish between them or even distinguish between the various phases of the same material. The peak locations of the binding energies for titanium and titanium dioxide are also different.

While the X-ray photons provide useful information about the core levels of a material, which helps in the elemental analysis, it does not provide any information about the valence electrons or the valence band, which determine the electrical and chemical properties. This problem may be overcome by using the UV photons, instead of X-ray as we discuss next.

Ultra Violet Photoemission Spectroscopy (UPS)

The working principle of UV Photoemission is the same as that of XPS, however the energy scale is off by a few orders of magnitude. In a typical UPS setup, UV photons of few tens of an eV are used. Since energy values are rather small, the photons may only have access to the valence electrons of a material. The kinetic energy E_k of the emitted electrons is given as,

$$E_k = hf - \phi_f$$

where ϕ_f is the material work function. The UPS is essentially a three-step process. In the first step, the UV photons impart enough energy to the valence electrons that their energies become greater than the vacuum energy, a prerequisite to become free as shown in Fig. 10.17a. In the second step, these electrons transfer to the surface, in which process, secondary electrons are generated, which have a spread of energy leading to a tail in the UPS spectrum as shown in Fig. 10.17b. Finally in the third step, the electrons become free, and are in the vacuum. As shown in Fig. 10.17c, these electrons may be analyzed by using a spectrometer very similar to the one used for XPS. However, the design is much simpler as compared to that of XPS, since the energy range is only a few eV.

For both XPS and UPS, the quality of the photon source is quite important. Usually it is desirable that the photon source be monochromatic (i.e. have single frequency) with enough number density. For this reason, the synchrotron facilities are favored for these studies due to access to monochromatic light sources of appreciable intensities.

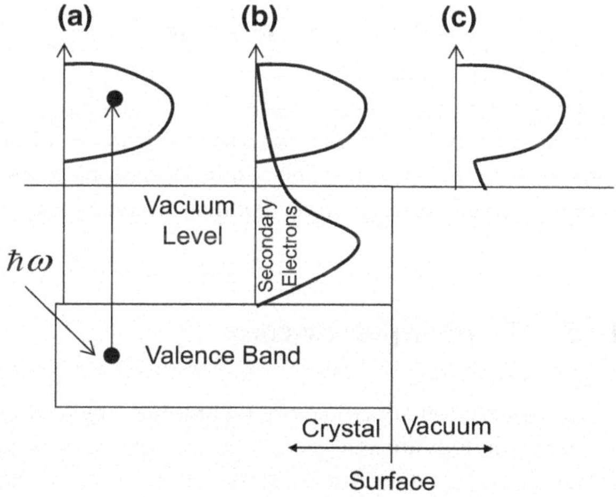

Fig. 10.17 Working principle of UPS. **a** Photon absorption. **b** Secondary electrons. **c** Free electrons to be analyzed

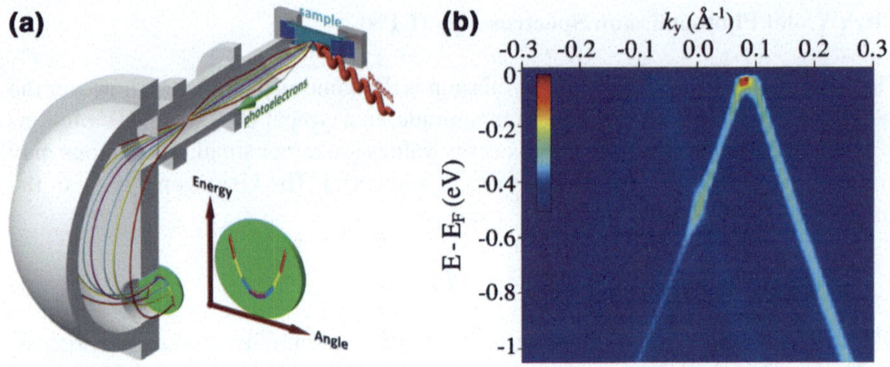

Fig. 10.18 ARPES. **a** Equipment setup. Photoelectron spectra is collected at varying angles [3]. **b** Experimental band structure of graphene [4]

Angle Resolved Photo Emission Spectroscopy

While UPS determines the information about the electron dynamics, one may add the reciprocal space information by using a unique technique called ARPES (angle resolved photo emission spectroscopy). The experimental setup is shown in Fig. 10.18a. For the photons incident at an angle (θ), the momentum conservation rules for the parallel and perpendicular components of the initial state of electron ($\hbar k_i$, inside the material) and the final state ($\hbar k_f$, in vacuum) dictate that,

$$k_{||i} = k_{||f} = \sqrt{2mE_\mathrm{k}}\sin(\theta)$$
$$k_{\perp i} = \frac{1}{\hbar}\sqrt{2m(E_\mathrm{k}\cos^2\theta + V_o)}$$

Thus, for varying angles, the wavevector k changes. By keeping track of this book keeping, one may study the electron spectrum as a function of the wavevector k for a given photon energy E_k. The resulting photoelectron spectrum at various angles is then used to generate the $E(k)$ band structure as shown in Fig. 10.18b for graphene.

10.5 Photon Spectroscopy

In various photon spectroscopy techniques, the scattered, reflected, transmitted, or absorbed photons are analyzed to extract sample information. A molecule has distinct vibration and rotational modes. We show some of these vibrational and rotational modes in Fig. 10.19a for water. Such modes may include in plane symmetric and asymmetric stretching movements, bending movements like scissoring and rocking as well as out of plane bending movements like twisting and wagging. In Fig. 10.19b,

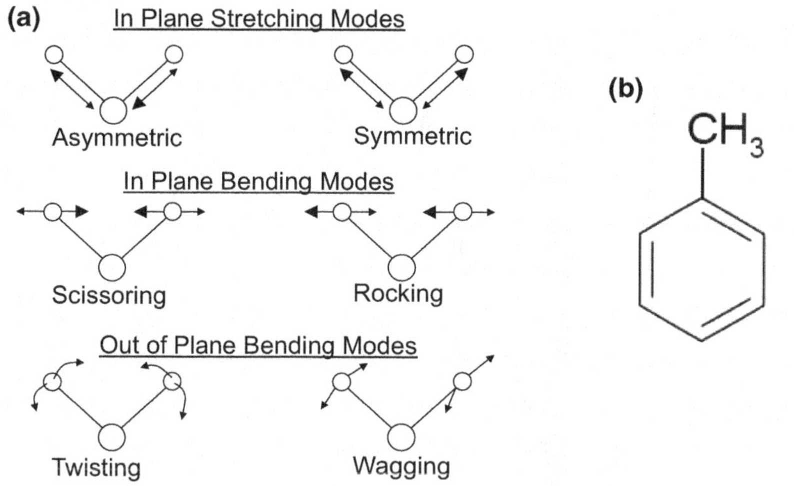

Fig. 10.19 Vibrational and rotational modes. **a** Various modes for water molecule. **b** Toluene molecule

the methane radical in the toluene molecule is shown, which has a rotational mode around the benzene molecule.

Each of these modes have a characteristic vibrational energy and polarization (dipole moment), which gets reflected in how the incident photons interact with the material, since photons are electromagnetic in nature and hence possess distinct oscillatory electric field.

UV-Vis Spectroscopy

Various molecules have fairly good absorption in ultra violet (UV), visible (Vis) and infrared (IR) bands. Our body extensively uses such molecules to take advantage of this behavior. In the visible range, the eye molecules absorb 400–700 nm wavelength photons. Molecules in our skin absorb UV photons to make Vitamin D. Another set of molecules absorb IR radiation for warmth.

Consider water molecule for which the absorption spectrum is shown in Fig. 10.20. In particular, water has a high absorption coefficient in UV (shorter wavelength) and IR (longer wavelength) range, whereas in the visible range, the absorption coefficient is rather small. This is not surprising since water is transparent in the visible range. One should also note that the top horizontal axis is the wavenumber, which is simply inverse of the bottom horizontal axis of the wavelength, without the 2π factor.[3] While we discuss the IR spectroscopy next, it is clear that these molecules have

[3]In photon spectroscopy, the wavenumber is defined as $\nu = 1/\lambda$, one should note that the factor of 2π is missing in this definition. $2\pi/\lambda$ gives the number of radians per unit length, and $1/\lambda$ gives the number of wavelengths per unit length.

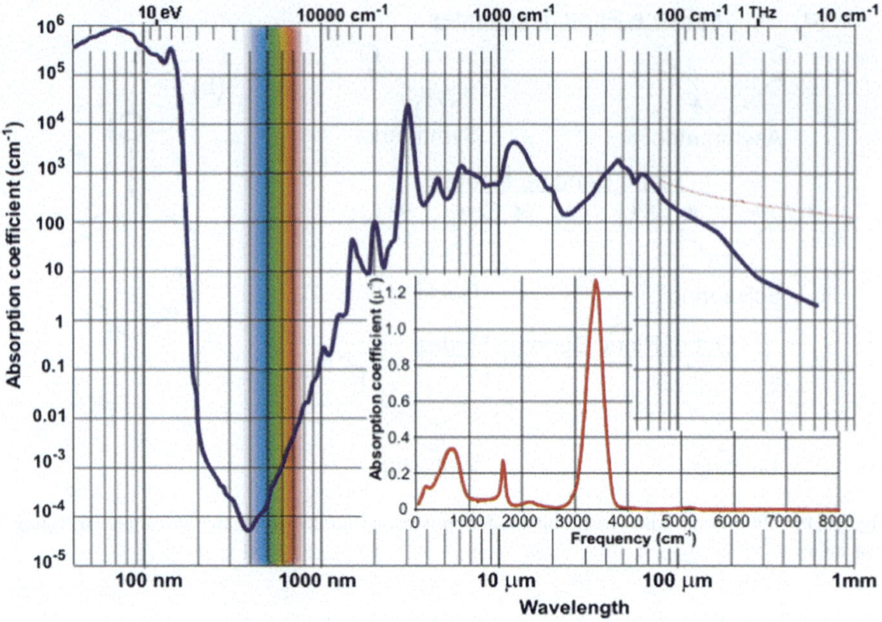

Fig. 10.20 UV-Vis absorption spectrum of water (Courtesy of OMLC)

characteristic absorption behavior in the ultraviolet and visible spectrum. In the UV-Vis spectroscopy, we precisely quantify this information to detect the presence of various species.

Infrared and Fourier Transform Infrared Spectroscopy

Next we discuss IR spectroscopy, which is the study of how various molecules interact with the infrared radiation. We discuss FTIR (Fourier transform IR) spectroscopy which has become quite popular due to the widespread availability of computers.

While various molecules have vibrational and rotational modes that fall in the IR wavelength range, not all of the modes are IR active. To qualify as an IR active mode, the vibrational pattern should have a well-defined dipole moment and hence polarization, which may interact with the incident photon's polarization. The symmetric vibrational model of CO_2 molecule does not have a well-defined polarization as shown in Fig. 10.21a, where the center electropositive carbon atom is visualized to be fixed in position, the two electronegative oxygen atoms move and hence change the bond length. The polarization or dipole moment is defined as the product of charge and separation (in this case, it is the bond length). The net polarization for this mode is zero. Hence, it is not an IR active mode. Consider the asymmetric and bending modes of CO_2, as shown in Fig. 10.21b, c, respectively. Clearly the two

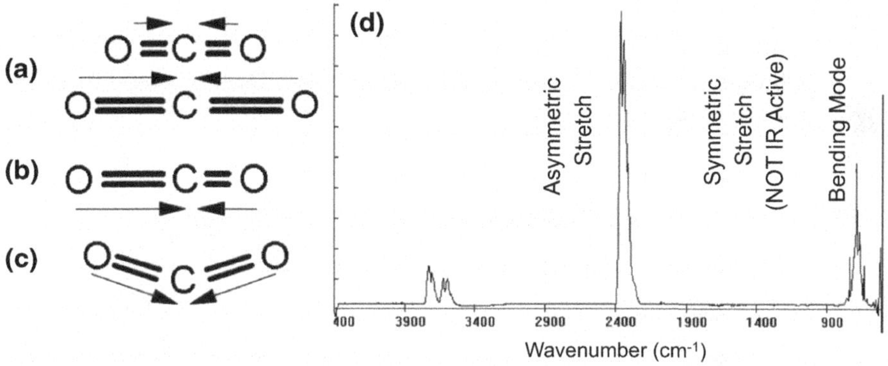

Fig. 10.21 Infrared active modes. **a** Symmetric stretch mode. Arrows show the direction of the dipole moment. **b** Asymmetric stretch mode, and **b** Bending mode. **d** IR spectrum (Courtesy of ThermoFisher)

modes have a well-defined dipole moment, which may interact with the polarization of an IR photon. These two modes are thus IR active. If a wide band IR signal of 400–4000 cm^{-1} wavenumbers is incident on a CO_2 specimen, the asymmetric stretch and bending modes absorb the radiation corresponding to the mode energy, which is observed in the transmission spectrum as dips (inverse peak). These dips may be used as a characteristic feature of CO_2 presence as shown in Fig. 10.21d.

While IR spectroscopy is a very useful tool for chemical identification and in fact extracting bonding information, including alignments and bond angles, it is indeed time consuming due to scanning over a large wavenumber range (usually 400–4000 cm^{-1} as discussed above). With the advent and subsequent widespread use of computers, another IR spectroscopy technique based on the use of Fourier transform has become quite popular due to its simplicity, yet rigor and precision. As shown in Fig. 10.22a, the output of an IR source is fed to an interferometer that generates an interferogram, which is incident on the sample. The transmitted wave is detected and sent to the computer along with the original interferogram. The computer takes the inverse Fourier transform with respect to the reference interferogram to generate the transmission spectrum. One such spectrum is shown in Fig. 10.22b for polystyrene, which has characteristic absorption peaks due to bond stretching and rocking amongst various other modes.

Raman Spectroscopy

When a photon interacts with matter, it may be absorbed or scattered through elastic and/or inelastic scattering. UV-Vis spectroscopy and IR spectroscopy are examples of inelastic scattering due to the absorption of photons by various modes of a molecule. Rayleigh scattering, shown in the energy level diagram of Fig. 10.23 is a purely elastic (without energy exchange) process that Rayleigh used to explain *why the sky appears*

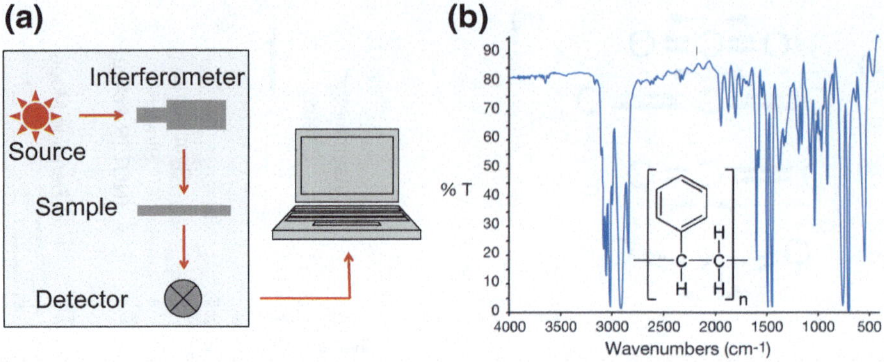

Fig. 10.22 FTIR spectroscopy. **a** Working principle of FTIR. **b** Infrared spectrum of polystyrene (Courtesy of ThermoFisher)

blue. Atmospheric molecules scatter blue light more efficiently than the light of other colors and hence photons of energy corresponding to blue color reach us.

In yet another process called Raman scattering, the incident photon interacts with the matter and induces a dipole moment in the molecule due to the polarization of the incident photons. While doing so, a small energy exchange is inevitable that becomes the feature of what is called Raman Spectroscopy. As the energy exchange excites a vibrational mode, not all modes are Raman sensitive.

One should note that the notion of energy exchange is such that in Rayleigh scattering, if the incident photon is of green color, the scattering photon would also be of green color. However, in Raman scattering, the scattered photon could be red (if energy is absorbed by the material) or blue (if energy is released by the material) for the green color incident photon. As shown in Fig. 10.23, the Rayleigh scattering is a perfectly elastic process. The Raman scattering leads to either anti-Stokes peaks that represent energy released by the molecules, and Stokes peaks that depict energy absorbed by the molecules as shown in Fig. 10.23. Since Stokes peaks have larger amplitude than that of anti-Stokes peaks, the Stokes peaks are usually kept track of in

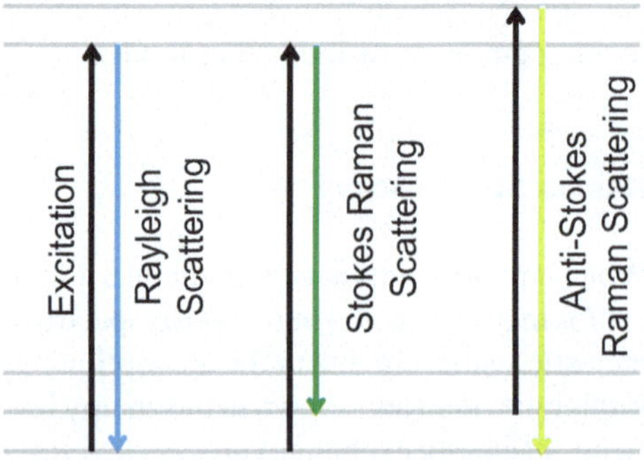

Fig. 10.23 Working principle of Raman spectroscopy. Rayleigh, Stokes Raman, and Anti-Stokes Raman peaks

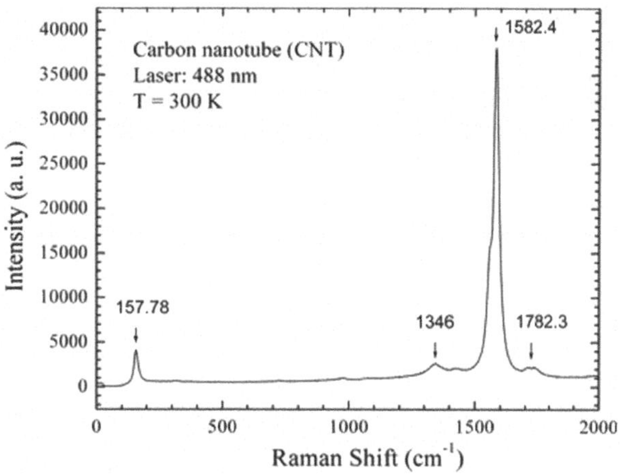

Fig. 10.24 Raman spectrum of carbon nanotubes (Courtesy of ThermoFisher)

the Raman spectrum. Apart from this, instead of the absolute wavenumber, the shift in the wavenumber due to the Raman scattering process is of greater importance, which is called the Raman shift. The Raman spectrum thus consists of anti-Stokes peaks plotted as a function of the Raman shift in wavenumbers.

Raman spectroscopy has been widely used to quantitatively characterize carbon based nanomaterials like Bucky balls, carbon nanotubes, graphene, etc. Not only the number of walls in carbon nanotubes and the number of layers in a graphene sample may be determined by using Raman spectroscopy, but the defect densities may also be measured experimentally.

A typical Raman spectrum for a single wall carbon nanotube at 300 K with a 488 nm laser excitation is shown in Fig. 10.24. By using the peak location of 1582 cm^{-1}, one may determine the diameter of the carbon nanotube as $d(nm) = 248/\upsilon(\text{cm}^{-1})$. Furthermore, the peak at 158 cm^{-1} shows the defect density in the carbon nanotube.

10.6 Electron Spectroscopy

In electron spectroscopy, electrons are used as a probe to quantify specimen information. In electron energy loss spectroscopy (EELS), an electron beam of a known energy is incident on the specimen under ultrahigh vacuum (UHV) conditions. The vibrational modes in a molecule and phonon modes in the condensed matter phase interact with the electron beam. As a result, some of the electrons in the beam loose energy to the vibrational or phonon degree of freedom. The energy spectrum of the transmitted electron beam is obtained by using an electron spectrometer. As shown in Fig. 10.25, the electron energy loss peaks carry information about the energy lost to the vibrational modes or the phonons. One may use these features to distinguish between amorphous carbon, graphite, and diamond. High resolution electron energy

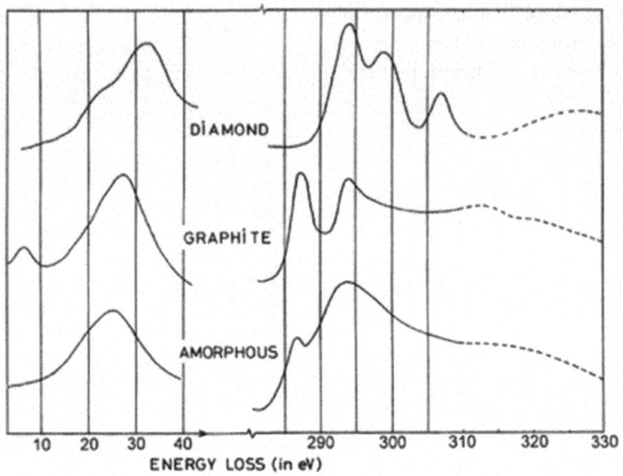

Fig. 10.25 EELS. The electron energy loss is used to distinguish between amorphous carbon, graphite, and diamond [5]

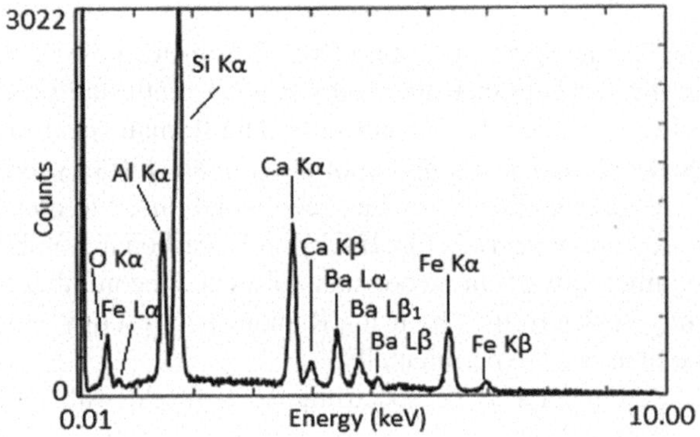

Fig. 10.26 EDS spectrum of an alloy [6]

loss spectroscopy (HREELS) is a version of electron spectroscopy with higher electron energy resolution. The primary feature of HREELS is a better electron spectrometer with higher resolution in meV.

In scanning electron microscope (SEM), if an X-ray detector is used to collect the generated X-rays during electron beam specimen interaction, it is called EDS (energy dispersive spectroscopy). The resulting spectrum in shown in Fig. 10.26 that has characteristic peaks corresponding to the core levels of various materials in a specimen. The sample consists of an aluminum, silicon, iron, calcium and Ba alloy with oxygen content. Although EDS is a very powerful technique for material analysis, the disadvantage is that it may damage the sample due to high electron energies being used.

In a way, EDS is an inverse process of XPS. While in XPS, X-rays are incident on a sample and electrons come out which are analyzed for extracting useful information, in EDS, high-energy electrons are incident on the sample, which result in X-ray emission which are subsequently used for elemental analysis. Along the lines of inverse XPS, one may imagine inverse UPS and inverse ARPES, techniques which are routinely used in laboratories for niche applications.

Research Assignment

R10.1 Write a one-page summary about the state of the art scanning tunneling microscopy.
R10.2 Write a one-page summary about the state of the art atomic force microscopy.
R10.3 Write a one-page summary about the state of the art scanning electron microscopy.
R10.4 Write a one-page summary about the state of the art transmission electron microscopy.
R10.5 Write a one-page summary about the state of the art scanning transmission electron microscopy.
R10.6 Write a one-page summary about the state of the art optical microscopy.
R10.7 Write a one-page summary about the working principle of phase contrast optical microscope.
R10.8 Write a one-page summary about the working principle of the three kinds of confocal microscopy techniques.
R10.9 Write a one-page summary about the state of the art X-ray photoemission spectroscopy.
R10.10 Write a one-page summary about the state of the art UV photoemission spectroscopy.
R10.11 Write a one-page summary about the state of the art angle resolved photoemission spectroscopy.
R10.12 Write a one-page summary about the state of the art UV-Vis spectroscopy.
R10.13 Write a one-page summary about the state of the art FTIR spectroscopy.
R10.14 Write a one-page summary about the state of the art Raman spectroscopy.
R10.15 Write a one-page summary about the state of the art EELS and/or HREELS.
R10.16 Write a one-page summary about the state of the art EDS.
R10.17 Write a one-page summary about the state of the art LEED and RHEED—a technique we have not discussed in this chapter.

References

1. Binnig et al., Phys. Rev. Lett. **50**, 120 (1983)
2. F. Moresco, A. Gourdon, Proc. Natl. Acad. Sci. **102**, 8809 (2005)
3. H. Raza, *Graphene Nanoelectronics* (Springer, 2012)
4. M. Sprinkle et al., Phys. Rev. Lett. **103**, 226803 (2009)
5. Li et al., *Electron Energy Loss Spectroscopy (EELS)*, in Encyclopedia of Tribology, eds. by Q. J. Wang, YW. Chung. (Springer, 2013)
6. J. Goldstein et al., X-Ray Spectral Measurement: EDS and WDS, in *Scanning Electron Microscopy and X-ray Microanalysis*. (Springer, 2017)

Index

A
Absolute zero, 83, 86
Absorption, 127
Accumulation, 150
Algorithms, 6, 122
Alignment, 225, 230
Amorphous, 12
Amplifier, 156
Analog, 185, 208
Angle resolved photoemission spectroscopy, 8, 258
Angular frequency, 30
Angular momentum, 30
Annealing, 220, 223
Antiferromagnetic, 166
Antiparallel, 166
Application Specific IC (ASIC), 209
Architectural design, 207
Arrangement, 6
Atmospheric pressure, 235
Atomic force microscopy, 8, 247
Atomic layer deposition, 8, 189, 238
Atomic structure, 6
Auger electrons, 252
Avalanche breakdown, 145
Avalanche process, 145

B
Backscattered electrons, 252
Band
  conduction, 72, 139–141, 153, 205
  diagrams, 11, 29, 148
  linear, 98
  nonparabolic, 98
  parabolic, 95
  semimetal, 67
  structure, 11, 29, 58, 163
  valence, 72, 139–141, 205
Bandgap, 92, 147, 191, 205
Bandwidth, 60, 63
Barrier height, 148, 156
Basis set, 49
  Gaussian, 57
  nonorthogonal, 104
  orthogonal, 36
  orthonormal, 36
  p_z-orbitals, 56, 67
  real space, 42
  slater type orbital, 56
  sp^3d orbitals, 56
  sp^3d^5 orbitals, 59
  sp^3 orbitals, 56, 58
Basis vector, 13
Binary, 183
Bipolar, 139
Bistable, 183, 187
Bit lines, 187
Bloch equation, 176
Block diagram, 187
Body Centered Cubic (BCC), 17, 74
Body factor, 153, 156
Bohr, 30
Bohr magneton, 175
Boltzmann's constant, 87, 210
Boolean, 185, 199
Born approximation
  selfconsistent, 134
Bose factor, 128
Bosons, 128
Bottom up, 5, 215
Boundary conditions, 40
Bravais lattice, 13, 22
Breakdown, 146

H. Raza, *Nanoelectronics Fundamentals*, NanoScience and Technology,
https://doi.org/10.1007/978-3-030-32573-2

Bright-field, 253
Brillouin zone, 22, 58, 70, 73
Buckyball, 4
Buried channel, 205, 206
Butticker probe, 131

C
Capacitance, 150
Cartesian, 11, 34
Causality, 100, 101
Center for disease control, 217
Characteristics, 155
  CV, 192
  IV, 192
  voltage transfer, 184
Charge based devices, 139
  three terminal, 139
  two terminal, 139
Charge coupled devices, 8, 205
Charge loss, 207
Charge transfer efficiency, 207
Charging energy, 134
Chemical etch, 240
Chemical safety, 217
Chemical vapor deposition, 231, 234, 250
Circuit design, 207
Classical, 29
Clean room, 215
Clock tree synthesis, 207
Coercive field, 161
Color chart, 223
Combinational circuits, 198
Common mode noise, 185
Complementary MOS (CMOS), 183, 185, 197
Compound semiconductors, 237
Computer Aided Design (CAD), 209
Conduction energy window, 82
Configuration space, 136
Confocal, 253
Contact resistance, 145
Continuum mechanics, 31
Control dielectric, 190, 192
Cost, 225
Coulomb blockade, 136
Crystal, 12, 17
Crystalline, 12
Current
  conventional, 80
  electron, 80
  probability, 80
Current in plane, 166
Current perpendicular to plane, 166
Cylindrical, 34, 42
Czochralski method, 218

D
Damping, 176
Dark-field, 253
Deal's triangle, 223
De Broglie-Bohm, 33
Defect, 146
Degenerate, 39, 41
Deionized (DI) water, 219, 224
Delta function
  Dirac's, 90
  Kronecker's, 23, 51
Density
  electron, 177
  spin, 177
Density functional theory, 57
Density matrix, 133
Density of states, 91, 102
Dephasing, 79
  elastic, 79
  inelastic, 79
Depletion, 150
Depletion capacitance, 153
Depth of focus, 229
Design rules, 208
Device, 3
  fabrication, 215
  scaling, 156
  three terminal, 7, 139, 147
  two terminal, 3, 102, 139
D-flip flop, 204
Diamond crystal, 18
Dielectric, 140
Dielectric constant, 150, 191
Diffraction limit, 229, 243
Digital, 208
Dimensional, 5
  one, 4, 59, 115
  three, 4, 73, 97, 120
  two, 4, 39, 67, 95, 120
  zero, 93, 112
Diode
  pn junction, 140
  Zener, 140, 146
Dirac-Feynman, 33
Dirac points, 70
Dispersion, 164, 170
Displacement vector, 58
D-latch, 203

Domain
behavioral, 205
algorithms, 198
Boolean logic, 198
register transfer, 198
energy, 11
frequency, 11
physical, 198
physical/geometric, 205
structural, 205
ALU, 198
CPU, memory, 198
time, 11
Dopant
acceptors, 92
donor, 92
Doping, 146, 219
degenerately, 146
n-type, 92
p-type, 93
Double edge triggered flip flop, 204
Down-spin, 32, 161
Drain, 147
Drain Induced Barrier Lowering (DIBL), 156, 157
Duplexer, 202
Dynamic power, 209

E
Effective mass
conduction band, 94
valence band, 95
Eigen, 90
energy, 90
functions, 37, 53, 177
values, 37, 46
vectors, 46
Electric field intensity, 133, 141
Electric flux density, 133, 190
Electron beam, 232
Electron Energy Loss Spectroscopy (EELS), 264
Electron-hole pair, 145, 205
Electron-hole symmetry, 92
Electronic charge, 84
Electronic structure, 6, 29
Electron microscopy, 249
scanning, 8, 236, 250
scanning transmission, 8, 253
transmission, 8, 236, 252
Electron spectroscopy, 263
Electrons, 92, 139, 205
Electrostatic potential, 161
Electrostatics, 38, 148
Ellipsometry, 223
Emission, 127
Energy diagram, 29
Energy Dispersive Spectroscopy (EDS), 264
Entropy, 210
Boltzmann-Gibbs, 210
Shannon, 210
Environmental protection, 217
Equilibrium, 79, 89, 92, 140
Equivalent direction, 20
Equivalent plane, 19
Erase operation, 192
Error, 185
Etching, 8, 240
Evaporation, 231
Exchange bias, 166
Expectation value, 35
Extended Hückel theory, 73

F
Fabrication, 208
Face Centered Cubic (FCC), 17
Feature size, 226
Feedback loop, 248
Fermi-Dirac statistics, 89
Fermions, 89, 163
Fermi's functions, 87, 122, 142
Ferromagnetic, 5, 7, 74, 161
Field Effect Transistor (FET), 3, 7, 140, 147
Field emission, 250
Field Programmable Gate Array (FPGA), 209
Finite difference, 42, 43, 59, 108
Finite element, 43, 47
Fixed layer, 166
Flash memory, 4, 183, 190
Flat band, 150
capacitance, 150
voltage, 150
Float Zone method, 218
Floor planning, 207
Fock space, 136
Forbidden, 31
Fourier transform, 11, 121
Fourier transform infrared spectroscopy, 260
Free electrons, 257
Free particle, 38
Function
broadening, 102
correlation

electron, 103
hole, 103
inflow, 103
outflow, 103
Functional, 57, 207

G
Gajski Kuhn chart, 197
Gate, 148
Gauss's law, 133
Generation, 145
Giant magnetoresistance, 7, 161, 165
Gilbert damping parameter, 179
Global Distribution System (GDS), 208
Globally Harmonized System (GHS), 217
GMR ratio, 167
Graphene, 4, 16, 25, 67, 70, 254, 258
Green's function, 99, 170
advanced, 100
nonequilibrium, 102
retarded, 100
Group velocity, 39, 82, 95, 164

H
Half metallic, 174
Hamiltonian, 33, 34, 49, 104, 163
Hard bake, 225
Hard disks, 4, 183
Hard layer, 166
Hardware Description Languages (HDL), 209
Hartree Fock theory, 57
Hazard, 217
Heat dissipation, 146
Heisenberg, 31, 33
Hermitian Hamiltonian, 53
Heterostructures, 237
HIGH, 185
High Efficiency Particulate Air (HEPA) filters, 216
Highest Occupied Molecular Orbital (HOMO), 56, 248
High frequency, 152
High-K dielectric, 189
Hilbert transform, 103, 127
Holes, 92, 139, 205
Hopping parameter, 56, 59, 109
Hückel model, 55
Hydrogen spectrum, 30

I
Identity matrix, 36
Image reversal, 225
Impulse response, 100
Inductively Coupled Plasma (ICP) etcher, 241, 242
Inelastic interband scattering self energy, 146
InElastic Tunneling Spectroscopy (IETS), 128
Information, 183
Infrared spectroscopy, 260
Integer quantum Hall effect, 31
Integrated circuit, 215
Interconnects, 215
Interferometric stage, 228
Inverse subthreshold slope, 154, 156
Inversion, 150, 192
Inverter, 185
Ion induced Etch, 240
Ion milling, 242
Ion sensitive FET, 4

K
Kondo effect, 136

L
Landauer-Butticker, 131
Landauer's approach, 86, 104, 125
Landau-Lifshitz-Gilbert (LLG), 178
Laplace's equation, 133
Laplacian, 34
Larmor frequency, 175
LASER, 225
Lattice, 12
Lattice unit vectors, 13
Leakage current, 209
gate, 139
source-drain, 139
LEDs, 4
Lenses
electromagnetic, 251
electrostatic, 251
optical, 250
Lift off, 225, 233
Linear dispersion, 70
Linear momentum, 31
Linear screening, 133
Liquid crystal display, 216
Liquid crystals, 12
Lithography
contact, 225

direct write, 229
electron beam, 229
immersion, 228
nanoimprint, 230
projection, 226
proximity, 225
Local density of states, 91
Logic, 185, 198, 207
Logic 1, 185
Logic operation, 198
Longitudinal, 120
Louis de Broglie, 31
LOW, 185
Low bias conductance, 79
Lowest Unoccupied Molecular Orbital (LUMO), 56, 248
Low frequency, 151
Low-*K* dielectric, 191
Low pressure, 235

M
Magnet
hard, 167
soft, 167
Magnetic field intensity, 161
Magnetic flux density, 162
Magnetic moments, 161, 174
orbital motion, 174
spin, 174, 177
Magnetic polarization, 163
Magnetic susceptibility, 162
Magnetic tunnel junction, 161, 168
Magnetization, 161, 163
Many electron systems, 57
Mask aligner, 225
Materials, 3
one dimensional, 59, 94
three dimensional, 97
two dimensional, 67, 95
zero dimensional, 93
Material safety data sheet, 217
Matlab, 58
Matrix approach, 33
Matter waves, 31
Mean free path, 132
Memory
nonvolatile, 139
volatile, 139, 183
Metal, 190, 225
Metal induced gap states, 119
Metal nanocrystal, 190
Metal-oxide-semiconductor, 140
Microfabrication, 215
Microscopy, 6, 243
Midgap, 92
Miller index, 19
Mold, 231
Molecular beam epitaxy, 8, 237
Molecular self assembly, 238
Momentum relaxation, 79
MOS capacitors, 149, 205

N
NAND gate, 199
Nanocrystalline, 12
Nanocrystals, 4
Nanoelectronics, 3, 197
Nanofabrication, 6, 8, 215
Nanoribbon, 64
armchair, 15, 70
graphene, 4, 15, 70
zigzag, 15, 70
Nanoscale, 3
Nanotube, 64
armchair, 15
carbon, 4, 15, 20, 56, 70, 235
zigzag, 15
Natural response, 100
n-channel MOS (nMOS), 148, 152, 184
Nearest neighbor, 45
Negative differential conductance, 158
Negative differential resistance, 140, 158
Negative edge triggered flip flop, 204
Newtonian mechanics, 29
Noise, 185
immunity, 185
margins, 185
suppression, 185
Non Hermitian, 102
Nonequilibrium, 7, 142
Nonequilibrium Green's function, 80, 133
Nonorthogonal, 17
Nonvolatile, 183, 190
NOR gate, 199, 200
Normalization, 34
NOT gate, 198
Numerical aperture, 226, 229

O
Occupational safety and health, 217
OFF states, 139
Ohm's law, 121
One dimensional conductor, 115

Onsite energy, 56, 109
ON states, 139
Open boundary conditions, 79
Operators, 33
Optical microscopy, 253
Optoelectronic, 216
Orbitals, 49
Orbital space, 49
Orbital symmetry, 52, 169
Orthogonal, 17, 41
Output, 155
Output resistance, 155
Overlap, 36, 49, 105
Overlay, 230
Oxidation, 220
  dry, 221
  pyrogenic, 221
  wet, 221

P
Packaged, 208
Parabolic dispersion, 60, 70
Parallel, 166
Paramagnetic, 7, 163
Particle in a box, 40
Partitioning, 207
Pauli matrices, 176
Pauli's exclusion principle, 57
p-channel MOS (pMOS), 147, 184
P configuration, 171
Peak current, 158
Permittivity, 133
Personal Protection Equipment (PPE), 217
Phase, 39
Phase-contrast, 253
Phase factor, 58
Phase velocity, 39
Photodiode, 205
Photo emission spectroscopy, 255
Photolithography, 224
Photomultiplier tubes, 255
Photon, 30
Photon spectroscopy, 258
Photoresist, 224
  negative, 224
  positive, 224
Physical design, 207
Physical Etch, 240
Physical Vapor Deposition (PVD), 8, 231
Physical verification, 208
Piezo tubes, 244
Pilot wave theory, 33
Pinned layer, 166
Placement, 207
Planck's constant, 11, 30, 84
Plasma, 137
Plasma enhanced CVD, 235
Plasmonics, 137
pn junction diodes, 7, 205
Polycrystalline, 12
p-orbital symmetry, 164
Positive edge triggered flip flop, 204
Potential, 141
Potential energy, 36, 140, 170
  chemical, 83, 87, 92, 142, 157, 164
  Hartree, 133
  Poisson's equation, 133
  selfconsistent loop, 133
Power dissipation, 209
Precession, 176
Primitive basis vectors, 14
Primitive unit cell, 12, 14
Probabilistic nature, 35
Probability density, 34
Profilometer, 244
Program process, 192
Program pulse, 193
Propagator, 100
Proportional-Integral-Derivative (PID), 248

Q
Quantum efficiency, 207
Quantum mechanical reflections, 34
Quantum mechanical tunneling, 34, 46
Quantum mechanics, 29
Quantum number, 31, 41
  azimuthal, 49
  magnetic, 49
  principle, 49
  spin, 49
Quantum of conductance, 84
Quasi crystals, 12

R
RAM, 7
  dynamic (DRAM), 183, 188
  static (SRAM), 183, 185
Raman spectroscopy, 261
Rapid thermal annealing, 223
Rapid thermal processing, 223
Rayleigh's criteria, 226
Reactive ion Etching, 241
Read Only Memory (ROM), 183

Real space, 11, 22, 58
Real space translation vector, 22
Reciprocal space, 22, 58, 94
Reciprocal space basis vectors, 22
Recombination, 145
Recombination generation, 151
(2 × 1) reconstruction, 21
(7 × 7) reconstruction, 21
Rectifier, 140
Recursive surface Green's function, 107
Reflection, 34, 85
Reflective High resolution Electron Energy Diffraction (RHEED), 237
Refractive index, 226
Refreshing, 189
Register transfer, 197
Relaxation matrix, 176
Remanence, 161
Reset operation, 192
Resistivity, 219
Resolution, 225
Resonant tunneling diodes, 7, 156
Retention, 193
Reticle, 226
RHEED, 237
Robotic automation, 215
Rutherford, 30

S
Saturation, 156
Scanning probe microscopy, 8, 243
Scanning tunneling microscopy, 8, 21, 236, 244
Scattering, 124, 145
  elastic, 129
  function, 128
  incoherent, 79
  inelastic, 79, 126
  interband, 126
  intraband, 126
  momentum relaxation, 126
  phase breaking, 126
  Raman, 262
  Rayleigh, 262
  spin flip, 165
Scattering function, 128
Schottky gate transistor, 3
Schrödinger equation, 33, 80
  time dependent, 34
  time independent, 34
Secondary electrons, 252, 257
Secular equation, 37
Selfconsistent solution, 135
Self energy, 105, 170
Self limiting, 239
Semiconducting, 190
Semiconducting channel, 139
Semiconductor, 156
  dopant, 92
  n-type, 93, 140, 147
  p-type, 93, 140, 147, 148
  wide bandgap, 216
Semi-infinite, 107
Sensitivity, 6
Sequential circuits, 203
Series resistance, 145
Shannon-von Neumann-Landauer (SNL) limit, 210
Shift register, 207
Signal routing, 207
Sign off, 208
Silicon nanowires, 4
Simple cubic crystal, 17
Single electron picture, 57
Si/SiO_2 interface, 205
Slater determinant, 57
Soft bake, 224
Soft layer, 166
Source, 147
Source drain tunneling leakage, 157
Spectral function, 102
Spectroscopy, 6, 243
Speed of light, 70
Spherical, 11, 34, 42
Spin, 7
  based devices, 161
  current, 177
  degeneracy, 84, 104
  down, 32, 161
  majority, 165, 170
  minority, 165
  operator, 176
  up, 32, 161
Spin based devices, 161
Spin current, 177
Spin degeneracy, 84, 104
Spin operators, 176
Spinors, 177
Spin transfer torque, 7, 161, 174, 178
Sputtering, 231, 234
Standing wave, 41
States
  Eigen, 91
  occupied, 81
  unoccupied, 82

Static power, 209
Step coverage, 233
Steppers, 226
STO-nG, 57
Storage gate, 188
Storage node, 190
Superparamagnetic limit, 5
Superposition, 40
Surface charge density, 191
Surface Green's function, 107
Surface reconstruction, 20
Switching, 179
Symmetry directions, 26, 74
Symmetry points, 26
Synchronous, 204
System on chip, 8
System specifications, 207

T
Taylor series, 72
Tensor, 128
Thermal emission, 250
Thermal NIL, 230
Threshold voltage, 151, 184, 191, 193
Throughput, 225
Tight binding, 54, 109, 163
  empirical, 56
  nonorthogonal, 55
  orthogonal, 55
  semi empirical, 56
Time evolution operator, 100
Timing closure, 207
Timing diagram, 203
TMR ratio, 169
Tomonaga–Lüttinger liquid, 137
Top down, 215
Torque, 175
Transconductance, 154, 155
Transfer characteristics, 153
Translation vectors, 13, 22
Transmission, 84, 89, 104, 169
Transmission gate, 201
Transport
  ballistic, 121, 132
  coherent, 79, 111, 122
  diffusive, 132
  incoherent, 79, 124
  mean field, 133, 136
  quantum, 6, 7, 79
  selfconsistent, 133
Transverse, 120
Traveling wave, 41
Trench capacitor, 189
Tribology, 248
Triode region, 155
Truth table, 198
Tunnel dielectric, 190
Tunneling, 147
Tunneling probability, 169
Tunnel magnetoresistance, 7, 168
Tunnel transistors, 7, 147

U
Ultra high purity, 220
Ultrahigh Vacuum (UHV), 237, 245, 248
Ultra violet, 224
Ultra violet photo emission spectroscopy, 257
Ultra violet-visible spectroscopy, 259
Uncertainty principle, 31, 130, 210
Undercut, 225, 240
Unipolar, 139, 152
Unitary transformation, 104
Unit cell, 12
UV catastrophe, 29
UV NIL, 230
UV photoemission spectroscopy, 8

V
Valley current, 158
Van Hove singularity, 94, 111
Very Large Scale Integration (VLSI), 207
Virtual contact, 125
Voltage gain, 185

W
Wafer, 215
  double side polished, 218
  single side polished, 218
Wavefunction, 34, 85, 100
  antibonding, 53, 54
  atomic, 51
  bonding, 53, 54
  molecular, 56
  reflected, 178
  symmetry, 52
  transmitted, 178
Wavelength, 31
Wave mechanics, 29, 40
Wavevector, 11, 38, 85
Wigner Seitz method, 12, 22
Wigner solid, 137
Word, 187

X
X-ray photoemission spectroscopy, 8, 256

Y
Y chart, 197, 207
Yield, 225

Z
Zener breakdown, 147
Zener diode, 7
Zero bias conductance, 79

The manufacturer's authorised representative in the EU is Springer Nature Customer Service Centre GmbH, Europaplatz 3, 69115 Heidelberg, Germany. If you have any concerns regarding our products, please contact ProductSafety@springernature.com

Printed and bound by CPI Group (UK) Ltd, Croydon, CR0 4YY
05/10/2026
02228123-0010